Ethics and the Life Sciences

Edited by
Frederick Adams

Special Supplement
Journal of Philosophical Research

Philosophy Documentation Center
Charlottesville, Virginia
2007

This volume has been published as a service to the profession, and is available to subscribers of the *Journal of Philosophical Research* and to all members of the American Philosophical Association at a reduced price. For more information, please contact the Philosophy Documentation Center at the following address:

Philosophy Documentation Center
P.O. Box 7147
Charlottesville, Virginia 22906-7147

Web: www.pdcnet.org
E-mail: order@pdcnet.org
800-444-2419 (U.S. & Canada); 434-220-3300

ISBN-10 1-889680-53-2
ISBN-13 978-1889680-53-8

Related Publications:

Journal of Philosophical Research, ISSN 1053-8364

Philosophy in America at the Turn of the Century, ISBN 1-889680-33-8

Ethical Issues for the Twenty-First Century, ISBN 1-889680-37-0

© 2007 by the Philosophy Documentation Center

All rights reserved. Copyright under Berne Copyright Convention, Universal Copyright Convention, and PanAmerican Copyright Convention. No part of this book may be reproduced, stored in a retrieval system, or transmitted in any form, or by any means, electronic, mechanical, photocopying, recording, or otherwise, without prior permission of the publisher.

Ethics and the Life Sciences

TABLE OF CONTENTS

FREDERICK ADAMS, University of Delaware
    Preface... 1

Animals

CATHERINE M. KLEIN
    Creation and Use of Transgenic Animals in Pharmaceutical and Biomedical Research: Animal Welfare and Ethical Concerns 7

ROGER WERTHEIMER, Agnes Scott College
    The Relevance of Speciesism to Life Sciences Practices.................... 27

DAVID DETMER, Purdue University Calumet
    Vegetarianism, Traditional Morality, and Moral Conservatism 39

NATHAN NOBIS, Morehouse College
    A Rational Defense of Animal Experimentation................................. 49

ROBERT STREIFFER, University of Wisconsin, Madison
    At the Edge of Humanity: Human Stem Cells, Chimeras, and Moral Status 63

Environment

STEVE VANDERHEIDEN, University of Minnesota Duluth
    Climate Change and the Challenge of Moral Responsibility 85

GERALD J. KAUFFMAN, University of Delaware
    Perspectives on Ethics and Water Policy in Delaware........................ 93

Food

MATTHEW LISTER, University of Pennsylvania
    Well-ordered Science: The Case of GM Crops..................................... 127

JENNIFER WELCHMAN, University of Alberta
   Frankenfood, or, Fear and Loathing at the Grocery Store 141
J. ROBERT LOFTIS, St. Lawrence University
   The Other Value in the Debate over
   Genetically Modified Organisms 151
WHITON S. PAINE AND MARY LOU GALANTINO,
Richard Stockton College of New Jersey
and University of Pennsylvania
   Biomarketing Ethics, Functional Foods, Health, and Minors 163
DAVID KAPLAN, University of North Texas
   What's Wrong with Functional Foods? 177
DANE SCOTT, University of Montana
   The Magic Bullet Criticism of Agricultural Biotechnology 189
CHRISTINA PIÑA, Swarthmore College
   The Dietary Limitations Imposed by Mexico's Social Structure 199

Human Health

FRITZ ALLHOFF, Western Michigan University
   Germ-line Genetic Enhancements and Rawlsian Primary Goods 217
FRITZ ALLHOFF, Western Michigan University
   Telomers and the Ethics of Human Cloning 231
GREGOR DAMSCHEN AND DIETER SCHÖNECKER,
Universtät Halle-Wittenberg and Universität Siegen
   Saving Seven Embryos or Saving One Child?
   Michael Sandel on the Moral Status of Human Embryos 239
KATHERIN A. ROGERS, University of Delaware
   A Clone by any Other Name:
   The Delaware Cloning Bill as a Model of Misdirection 247
DAVID K. CHAN, University of Wisconsin-Stevens Point
   Wrongful Life, Wrongful Disability, and the
   Argument against Cloning 257
JAKOB ELSTER, University of Oslo
   Wrongful Life, Suicide, and Euthanasia 273
LISA BELLANTONI, Albright College
   Are We Good Enough? The Paradox of Genetic Enhancement 283

BRAD F. MELLON, Bethel Seminary of the East
   Learning to Cope with Ambiguity:
   Reflections on the Terri Schiavo Case 291

GARY FULLER, Central Michigan University
   PVS and the Terri Schiavo Case: A Reply to Brad Mellon 299

ALEXANDER R. COHEN, University of Virginia
   Truly Human Reproduction 305

PETER DANIELSON, RANA AHMAD, ZOSIA BORNIK, HADI DOWLATABADI, AND EDWIN LEVY,
University of British Columbia
   Deep, Cheap, and Improvable:
   Dynamic Democratic Norms and the Ethics of Biotechnology 315

KRISTIN LEFEBVRE, Widener University
   An Ethical Evaluation of the Supreme Court Decision
   Regarding ERISA Interpretation 327

DALE MURRAY AND HEATHER CERTAIN,
University of Wisconsin-Baraboo and University of Wisconsin-Richland; William S. Middleton Memorial Veterans Hospital, GRECC and University of Wisconsin Center for Women's Health Research
   Pharmaceutical "Gift-Giving," Medical Education,
   and Conflict of Interest 335

NEIL A. MANSON, University of Mississippi
   Why Shouldn't Insurance Companies
   Know Your Genetic Information? 345

ETHICS AND THE LIFE SCIENCES

INTRODUCTION

The current volume is a collection of invited papers resulting from a conference held at the University of Delaware in the Fall of 2004. This was the second of three conferences jointly planned and co-sponsored by the University of Delaware and the American Philosophical Association. The first was in October of 2001 and the last of the series is planned for Fall Semester of 2007. The agreement that we conceived included three conferences to be held at the university between 2001 and 2007, after which we would decide whether these conferences were in the mutual interest of both institutions and should be considered. A volume from the first conference *Ethical Issues for the 21st Century* was published in 2005, and this volume focusing on Ethical Issues in the Life Sciences is the second volume of papers to be published.

Since the time of the first conference, the University of Delaware has instituted the Delaware Interdisciplinary Ethics Program (of which I am co-director). This program involves representatives from each college of the seven colleges of the University of Delaware and the American Philosophical Association's Executive Director. It also includes representatives from Delaware State University, Delaware Technological University, and Christiana Care Hospital. This year we will expand representation to include Wesley College, Widener University Law School (in Delaware), and A. I. DuPont Children's Hospital. So the Delaware Interdisciplinary Ethics Program is expanding and very much looks forward to continuation of this valuable joint partnership with the American Philosophical Association. Inspiration for creating this program came from the 1955 graduating class of the University of Delaware. In their wisdom they endowed an ethics fund and put it in the hands of the philosophy department. I inherited their generosity and good guidance when I came to the university as chair in 1997.

As we planned the conference from which the current volume derived, we decided to focus on the area of the Life Sciences. Biotechnology is on our minds here in the pharmaceutical corridor of the north-eastern portion of the United States. We know that this new technology will allow modification of animals, food, the environment, and of human beings. We also know that ethical considerations will apply to every kind of application of this new biotechnology. So we decided to focus the topic of the 2004 conference on these areas. I'm delighted to report that the turnout from around the world was excellent and the papers described below represent the best of the papers presented at the conference.

The Ethical Treatment of Animals: Ethical issues involving the moral treatment of animals include *Catherine Klein*'s paper on the transgenic manipulation of animals. She considers the various technologies for creation and use of transgenic animals. She then considers the ethical issues that result from the new technology. She notes the need to weigh the potential benefits to human health against the potential harms to the animals in question. Klein revisits old ethical questions about the uses of animals and raises new ones that arise uniquely out of new technology. *Roger Wertheimer* considers the charge that speciesism is similar to racism. He questions whether showing a preference for one's own species is necessarily indefensible and whether it really is inconsistent with much of what animal protectionists believe. He accepts that there may be non-human moral persons (such as extra-terrestrials). He describes speciesism as a special concern for fellow humans, without disdain or indifference to the plight of non-humans. *David Detmer* argues for vegetarianism as a morally conservative (not a radical or politically liberal) view. He shows how the view is implied by other principles that are clearly conservative moral principles, and ones that surprisingly nearly everyone already accepts. Detmer's arguments directly conflict with considerations raised by Wertheimer. *Nathan Nobis* considers current justifications for causing pain and suffering and death in animals for the knowledge we gain to benefit humans. He finds the current justifications lacking and presents a challenge to find better ones, if we are to continue our current practices. *Robert Streiffer* considers the moral status of the beings we would create if we allow transplantation of human stem cells into nonhuman animals for therapeutic reasons. What will we have created? What will their moral status be? Can human moral status be conferred because of transfer of significant numbers of human genes? What policy decisions should be made, on the basis of the answers to these questions?

Ethical Issues Related to the Environment: *Steve Vanderheiden* asks us to suppose that humans really are causally contributing to global warming. How is moral responsibility for harmful effects that we could have prevented to be distributed: individually, collectively? In this paper, Vanderheiden begins to sort out these pressing moral issues—tracking the mathematically small contributions that individuals make to large global problems. Can we be held morally responsible for the causal contributions we make individually? *Gerald Kauffman* looks at ethical considerations as a basis for water policy decisions. His is a case study of the rise of environmental ethics and its impact on Delaware Law and water policy. He considers the current laws that decide water use in Delaware and examines whether decisions made on the basis of these laws takes ethics into consideration, attempting to balance economic interests of the individual and environmental needs of the community. Kauffman examines the degree to which these conflicting interests are balanced in recent water policy in Delaware.

Ethical Issues Related to Food: *Matthew Lister* looks at the science of GM crops. The science is serving the interest of whom? Lister uses the case of GM foods to raise the broader question of the role of a well-ordered science in serving the public good. Who sets the agenda for the development and deployment of GM crops? Are the views of those who will use and be affected by GM crops

INTRODUCTION 3

being represented? Is the public being kept in the dark, suffering the tyranny of the ignorant? *Jennifer Welchman* considers genetically modified foods. Genetically modified foods have been called 'frankenfoods,' for emotional effect. How much, if any, of our emotional response is telling us something about these foods? Welchman considers popular conceptions and misconceptions of monsters, nature, and genetically modified foods, and their consequences for humans, society, and our view of our relationship with our environment. *J. Robert Loftis* considers whether economic liberty is decisive in deliberations over the moral permissibility of genetically modified organisms. He looks at this issue in regards to genetically modifying herbicide-resistant soy and concludes that, at most, regulatory oversight is justified and that considerations of economic liberty justify the use of many such GMOs. *Whiton S. Paine* and *Mary Lou Galantino* consider the food industries practices of marketing functional food products to children, products that go beyond basic nutrition. These products will be marketed as having benefits ranging from promoting health and preventing disease to aesthetic, behavioral, or bodily enhancement. Whiton and Glantiono consider how the food industry should be regulated or regulate itself to prevent ethical abuses in marketing products to minors. *David Kaplan* points out distinct problems with functional foods that need to be considered: viz., their advertised benefits are mostly exaggerated or non-existent, the fact that they are of medicinal value blurs the distinction between food and drugs, without proper regulatory oversight, and their popularity and advertising is fueled by the food industry not by the medical profession. These important concerns need to be brought to the attention of consumers and policy makers, for they may not now be in the hands of those who have the best interest of the consumers in mind.

Dane Scott considers whether GMOs are "magic bullets." Genetically modified organisms (GMOs) are on the rise due to modern biotechnology. Some opponents of GMOs argue that they are being used as magic bullets (techno-fixes) for agricultural or environmental ills. Magic bullets are supposed to target one and only one problem or benefit, but of course they do not always succeed in limiting their effects. And then we need new magic bullets to repair the bad consequences of the last. Scott examines this criticism of GMOs and finds that in some cases the criticisms may be historically correct, but there is no necessary flaw in GMOs in principle, when they are developed and used properly. *Christina Piña* studies the societal and political impact in Mexico upon the food and dietary choices of its citizens. She finds faults in the economic, social and political structure of the society that goes beyond the eating habits or choices of the individual—faults that contribute to poor eating habits and poor health.

Ethical Issues for Human Health: *Fritz Allhoff* tackles the arguments against genetically enhancing humans. He also examines arguments in its favor and concludes that the stronger arguments are there. In his second paper, Allhoff discusses the problem of telomere shortening during the process of cell replication during cloning. He discusses whether, short of a future technological fix, this shortening and consequent life-shortening undermines the moral permissibility of cloning (especially human cloning). *Gregor Damschen* and *Dieter Schönecker* set up something of a "trolley problem" where you have to choose between saving a human child and saving seven human embryos during a fire in an fertility clinic. Suppose you would choose to save the child. What does that show? Sandel believes it shows that human embryos do not enjoy the same moral status as the human child. Damschen and Schönecker challenge the assumption that it does show that and they offer other scenarios of their own, where one might choose to save the embryos, to attempt to demonstrate this. *Kate Rogers* considers a bill before the Delaware State Legislature to allow therapeutic cloning. Rogers argues that the true nature of the bill is being hidden behind new terminology and that, when unmasked, the burden of proof to justify use of this type of cloning has not been met. *David Chan* asks what it takes for a person to have come into existence in such a fashion that one is entitled to claim wrong doing on the part of those responsible for one's birth. Is it always a good thing to bring a new human life into existence? If not, how do we decide? Can one sue one's parents for "wrongful life?" Chan explores these issues within the context of the potential for new problems due to new cloning technology. *Jakob Elster* explores whether there is any connection between one's claim to have been brought to life wrongfully (such that one would have been better off to die or not to have lived) and suicide or euthanasia. Elster denies that there is any such direct connection. He also denies that wrongful life entails that one would be better off dead. *Lisa Bellantoni* asks whether, if humans have the technology to enhance ourselves genetically, should we use it? And is it wrong to change the human genome? Bellantoni considers arguments for and against genetic enhancement and finds that the arguments against fail to be persuasive. *Brad Mellon* considers ethical issues surrounding one's being in a persistent vegetative state. He claims it is ambiguous, physically and morally, and that we have to learn to cope with the ambiguity. *Gary Fuller* replies to the considerations Brad Mellon raises (this volume), claiming that if there are ambiguities in these cases, they are not irresolvable. Fuller argues that there is actually much rational agreement in the Schiavo case, contrary to Mellon's arguments, and that we may indeed know the right thing to do in these very difficult situations. *Alexander R. Cohen* could have titled his paper "The Case for Designer Babies." Cohen argues that new technology to help humans reproduce should be seen as "truly human" and not worse or lesser than reproducing the old fashioned way—by having sex. In fact, Cohen argues that reproducing the new technological way can even be better for mankind. Having sex to reproduce leaves far too much to chance, says Cohen. Cohen considers and rejects several objections. *Peter Danielson, Rana Ahmad, Zosia Bornik, Hadi Dowlatabadi*, and *Edwin Levy* raise important questions about the input the public has to advances in biotechnology. What input should the public have in a democracy? To have input, the public must be informed of the science and then have an avenue of influence on decision-making in business, industry, and government development of biotechnology. This paper presents a web-based mechanism of input that is, as the authors say, 'deep, cheap, and improvable.' *Kristin*

INTRODUCTION

Lefebvre considers how the Employee Retirement Income and Security Act affects patient autonomy, physician malpractice, informed consent and human health. Looking at specific court decisions, Lefebvre investigates the ethical precedents being set by Supreme Court decisions regarding ERISA legislation. *Dale Murray* and *Heather Certain* turn their attention to how the acceptance of gifts by doctors from pharmaceutical companies may affect a physician's practice and decision-making. Is this necessarily bad? Murray and Certain argue that it is. They argue that there is an implicit expectation on the part of the pharmaceutical companies that physicians will prescribe their products. They argue that physicians are not entitled to these 'gifts,' and that by accepting them physicians compromise the integrity of the patient-physician relationship because decisions are made without regard to patients' health. And finally, *Neil A. Manson* challenges the received view on why people should be able to hide their genetic information from insurance providers: genetic discrimination. When Manson finds the most important justifications for legislation to protect against genetic discrimination fail, he then considers a free market solution to the problem: genetic insurance.

I would very much like to thank the Executive Directors of the American Philosophical Association with whom I have worked and who have vigorously supported this cooperation. They include Eric Hoffman, Elizabeth Radcliffe, Michael Kelly, and the current Director, David Schrader. I must also thank the members of the APA office staff for all of their support, and especially Katina Saunders, Janet Sample, and Linda Smallbrook. I must also thank my office staff and members of my department who helped with logistics, publication, and refereeing of papers. Nearly all helped, but in particular I could not have put this volume together without help from: Darlene Reynolds, Maria Verderamo, Kate Rogers, Richard Hanley, David Haslett, David Silver, Mark Greene, and Robin Andreasen. At PDC, I need to thank George Leaman, Marty Klaif, and Greg Swope.

Frederick Adams
Professor of Cognitive Science and Philosophy
Chair of Philosophy
University of Delaware

ETHICS AND THE LIFE SCIENCES

CREATION AND USE OF TRANSGENIC ANIMALS IN PHARMACEUTICAL AND BIOMEDICAL RESEARCH: ANIMAL WELFARE AND ETHICAL CONCERNS

CATHERINE M. KLEIN

ABSTRACT: The creation of transgenic animals has application in the following areas of pharmaceutical and biomedical research: the production of biopharmaceuticals for human use; the production of organs for xenotransplantation; and the generation of animal models for human genetic diseases. Nuclear transfer technology offers a more precise and efficient way of performing genetic modification and creating transgenic animals than the more traditional method of pronuclear microinjection. This paper will review nuclear transfer as a means of producing transgenic animals; introduce advantages nuclear transfer technology offers in the field of animal transgenesis; and highlight some of the animal welfare issues and ethical concerns raised by the generation and use of transgenic animals in the aforementioned fields of study. Finally, the influence of objectifying language and terminology used to describe transgenic animals will be considered, and the impact of phrases such as "living bioreactor" and "spare part supplier" examined.

I. INTRODUCTION

Recent developments in animal transgenesis and cloning have application in several areas of pharmaceutical and biomedical research including: the production of biopharmaceuticals for human use; the production of organs for xenotransplantation; and the generation of animal models for human genetic diseases. In each field of research, nuclear transfer offers a more precise and efficient way of performing genetic modification and creating transgenic animals, thereby presenting certain advantages over the more traditional method of pronuclear microinjection. Efficient application of transgenic technologies will reduce the negative impact of such methods upon animals used in research; however, the effect of the transgene itself continues to jeopardize animal welfare. Regardless of the method of genetic modification utilized, serious animal welfare and ethical concerns remain. Examples set forth within the aforementioned fields of research will illustrate the importance of the ethical debate and provide a foundation for understanding concerns arising from the genetic modification of animals for human benefit.

The influence of the terminology and language used to describe transgenic animals will also be examined. For example, terms such as "living bioreactor" and "spare part supplier" connote the non-living and may negatively impact one's perception of the inherent value of these living creatures. Use of objectifying terminology and desensitizing words invoke fear that transgenic animals will be viewed as mere commodities, and will perhaps limit society's ethical questioning of these technologies. As science paces forward, a reexamination of fundamental ethical questions in light of emerging technologies will be vital to discern the impact upon animal welfare and to guide society into the future.

II. METHODS OF GENE TRANSFER: PRONUCLEAR MICROINJECTION VS. NUCLEAR TRANSFER

Several methods are used to produce transgenic animals including pronuclear microinjection, embryonic stem-cell mediated gene transfer, viral vectors, sperm-mediated transgenesis and somatic cell nuclear transfer (NT). Gene transfer by pronuclear microinjection has been the principal method used to produce transgenic farm animals; therefore, this discussion will focus on a comparison of microinjection with the more recent and promising method of NT.

A. Pronuclear Microinjection

Traditionally, transgenic animals have been created by pronuclear microinjection of one-cell embryos. This process uses a fine needle to inject DNA into recently fertilized eggs, which are then cultured and implanted into surrogate mothers. This technique, however, has proved to be rather inefficient. Successful integration of the transgene into the host genome is a hit-or-miss event—approximately 1 to 5 percent of resulting offspring carry the transgene and only a proportion of the transgenic progeny express the added gene in a desired manner and at a high level.[1] Furthermore, integration of the transgene into the host DNA is a random process, which may occur anywhere within the genome. The expression of the transgene is influenced by sequences surrounding its insertion

ANIMAL WELFARE AND ETHICAL CONCERNS

site; therefore random insertion may produce position effects that contribute to the unpredictable and variable expression of the transgene between transgenic founders.[2] Multiple lines of animals must therefore be tested for proper gene expression when this technique is employed. Random incorporation may also induce mutations that disrupt the function of host DNA coding sequences. Since insertional mutations are oftentimes recessive, their deleterious effects cannot be detected until the animals are bred to transgenic relatives.[3]

Pronuclear injection also leads to the generation of mosaics, which impedes growth of the transgenic herd. If the transgene is incorporated into the host genome before the zygote undergoes its first division, copies of the added gene should appear in all cells of the developing animal, including its eggs or sperm. However, if the transgene is not integrated into the host chromosome until after the zygote has divided, the added DNA will appear in some, but not all, cells of the developing animal. The resulting mosaics will produce two different kinds of germ cells—some contain the added transgene, while others do not.[4] Therefore, even if an individual animal expresses the transgene, it may not transmit the transgene to its offspring.

Various reproductive manipulations (e.g., superovulation, artificial insemination, embryo collection, and embryo transfer) are used to produce transgenic offspring and breeding animals may be repeatedly exposed to these procedures. As noted above, only a small percentage of embryos created by pronuclear injection carry the transgene of interest. In order to reduce the number of non-transgenic pregnancies developing to term, recipient cows, for example, may be subject to transvaginal amniocentesis to verify whether transgenes have integrated into the genome.[5] Non-transgenic fetuses are aborted and the surrogate reused as a recipient. While this approach limits the number of animals used as recipients, it also raises welfare concerns, as individual animals may be repeatedly subject to "procedures likely to cause pain and distress."[6] In contrast, as discussed below, only those cells exhibiting the desired genetic modification are selected to create embryos with NT. The use of NT to create transgenic animals, therefore, could eliminate the problem of repeated elective abortion and reuse of recipient animals.[7]

B. Nuclear Transfer

NT promises to facilitate genetic transformations and increase the efficiency of transgenesis to 100 percent. In pronuclear microinjection, the transgene is injected into the pronucleus of a single-celled fertilized egg; however, in NT, genetic material is transferred to cell lines in culture. After addition of the transgene, cells can be cultured further and analyzed to be certain they contain the added genetic material. The genetic material from the nucleus of the cultured donor cell can then be transferred to an enucleated recipient egg. Embryos are implanted in surrogate mothers and all animals born will be transgenic. NT also eliminates the problem of founder mosaicism. Genetic modification of the donor nucleus together with NT should introduce the genetic change into every cell of the resulting offspring, including its eggs or sperm.

An additional benefit of NT is the apparent reduction in the number of animals and surgical procedures necessary to generate founders,[8] as only those cells with the desired genetic change are selected as donor cells. Producing transgenic animals by NT uses less than half the experimental animals than does pronuclear injection. Work performed by the Roslin Institute and PPL Therapeutics between 1989 and 1996, for example, required an average of 51.4 ewes per transgenic lamb produced by pronuclear microinjection.[9] Only 20.8 ewes were required per transgenic lamb produced by NT using donor fetal fibroblasts as reported by these groups in a later study.[10] As the researchers stated, "The most important difference is that no recipients are wasted gestating nontransgenic lambs in the nuclear transfer technique."[11] Furthermore, when zygotes are used in pronuclear injection, their sex is not known. There is a 50 percent chance the resulting offspring will be male, and for certain applications (i.e., the production of biopharmaceuticals in the milk), transgenic females are desired. Selection of a female transgenic cell line for NT eliminates the possibility of male offspring.

Importantly, NT permits production of transgenic founder animals in the first generation.[12] Normally, one must wait for a transgenic animal produced by pronuclear injection to mature and reproduce. However, once a transgenic cell line has been identified that expresses a human protein at the desired level, NT permits production of a number of transgenic founders in a single step.[13] Once the founder herd or flock is established, the genetically modified animals could then breed naturally to establish a transgenic line. This factor is particularly relevant for those companies engaged in the production of pharmaceuticals in the milk of transgenic animals. By reducing the time needed to produce a founder herd or flock of lactating females, the time to large-scale protein production, clinical trials, and commercial production will also decrease. As noted by one commercial biotech company, "Where it would normally require 44 months to reach production flock status in sheep, (78 months in cows), nuclear transfer technology can reach production flock status in 18 months for sheep, (33 months for cows)."[14]

While pronuclear microinjection permits only the addition of genetic material to the zygote, cells in culture used for NT can be manipulated to not only add new genetic material, but to delete or substitute specific genes. This will prove to be of significance in the field of xenotransplantation. A very strong immune response is stimulated when pig organs are transplanted into human recipients, leading to hyperactue rejection of the transplanted organ. Pig tissues display a carbohydrate epitope that reacts with human antibodies, stimulating this immune response.[15] A targeted deletion of the gene encoding the enzyme that produces this epitope should diminish hyperacute rejection.[16]

Furthermore, gene targeting[17] helps avoid those problems associated with the random incorporation of DNA observed with pronuclear injection. The position of the gene within the genome affects expression of the transgene; therefore, pre-selection of transgenic integration sites and precise placement of transferred genetic material into the host genome permits more predictable and controlled gene expression. In addition, the use of a clonal population of transgenic cells as nuclear donors will guarantee the same transgene insertion site for each clone, thereby decreasing animal to animal variation in transgene expression levels.[18] Finally, cultured nuclear donor cells can be frozen and used when desired to generate identical cloned transgenic offspring over a prolonged period of time.[19]

III. ETHICAL ISSUES AND ANIMAL WELFARE CONCERNS

Several general questions are raised in an examination of ethical issues in the field of animal transgensis and cloning: Does animal cloning in pharmaceutical and biomedical research raise new ethical questions or are previously introduced issues being re-examined in the context of a new, developing technology? If our concerns are similar to those elicited by other uses of animals in pharmaceutical research, does cloning intensify or heighten these concerns? If we find there are objections that are unique to cloning for pharmaceutical and biomedical research, do those concerns stem from the nature of cloning itself, or do they arise from the consequences—potential harms and benefits—of this research? Finally, will routine cloning and mass production of genetically identical copies have a negative impact on the value of animal life? In other words, will cloning further advance the commodification of living creatures? Many ways in which our society currently uses animals promote commodification. Therefore, we may again want to ask whether cloning exacerbates this concern more so than other uses of animals.

A. Pharming

Biopharming refers to the production of pharmaceuticals from genetically modified plants or animals. Although the focus of this discussion will be on the production of human proteins in the milk of transgenic animals, there is also the potential for the production of pharmaceuticals in the urine, blood, or eggs.

A number of therapeutically valuable human proteins can be produced in the milk of transgenic sheep, goats, cattle, and even rabbits and pigs. Examples include human factor IX, used to treat hemophilia B[20]; alpha-1-antitrypsin to help counteract lung damage in patients suffering from emphysema and cystic fibrosis[21]; and antithrombin, a plasma protein with anticoagulant and anti-inflammatory properties.[22] Secretion of human proteins in the milk of transgenic animals has resulted in increased volume output and lower cost per unit as compared to traditional cell culture systems.[23] It has been estimated that several hundred transgenic pigs could provide enough factor IX to treat all the world's hemophiliacs;[24] and, theoretically, a herd of 600–700 transgenic cows could produce quantities of human serum albumin that would satisfy worldwide demand.[25] A second advantage over the use of cell culture systems is that the mammary gland is capable of producing complex proteins that require posttranslational modifications for full bioactivity. Finally, proteins produced in the milk of transgenic animals are free of potentially infectious agents that may be associated with human blood products.

The creation of transgenic animals that produce human therapeutic proteins in their milk appears to offer significant human benefit with arguably minor intervention in the animal—particularly once a herd or flock of founder animals has been established. Performance of a cost/benefit analysis is more difficult, however, in fields of animal biotechnology, "because the costs and benefits will be experienced by two different groups with different interests—human beings and animals."[26] There are also disparate subdivisions in our society that perceive and value the risks and benefits to humans and animals differently. Given these diverging value systems, the weighing of risks and benefits can vary and the ultimate outcome will depend upon which group performs the analysis.

Most people would likely agree that animals do have interests not to be caused pain and suffering. The question remains whether, and to what extent, these interests may be sacrificed for human interests.[27] Furthermore, in assessing potential risks to animals, all intermediate steps and all animals used in the creation of a founder animal must be considered in addition to the final genetically modified sheep, goat, pig, or cow. Reproductive procedures including administration of drugs to donor animals to induce superovulation, retrieval of donor eggs, and implantation of genetically modified embryos into surrogate mothers, as well as the accompanying stress of handling and post-operative pain must all enter the final analysis. This has been termed "procedural distress," and contrasts to other forms of animal experimentation in that several generations of animals may be subjected to pain, distress, and suffering during production of the final model.[28]

Some contend that NT is simply an extension of selective breeding that has been practiced throughout history. This argument, however, seems to imply that the status quo is an ethically acceptable standard, and it is from this baseline that new developments in animal biotechnology should be judged.[29] Many would argue that there are ethical objections to conventional breeding practices that have serious animal welfare implications. For example, turkeys are bred with such large breasts that they cannot naturally breed, and double muscling in cattle has led to problems during calving. Although man has consistently altered the genetic makeup of animals through selective breeding, and species change naturally through evolution, notable differences do exist between these "natural" events and direct genetic modification. First, transgenesis and cloning permit the transfer of genes between widely different species, while it is exceedingly difficult to cross species boundaries in selective breeding. The production of sheep and goats containing human genes that code for the production of human proteins in their milk, for example, could never be accomplished through selective breeding. Second, often unpredictable and extreme genetic changes may occur rapidly, in a single generation, providing little time to observe potentially deleterious effects upon the animals.[30] Selective breeding, in contrast, is a more gradual process that allows changes in animals to be observed and monitored over several generations. Finally, as with any developing technology, scientists and researchers do not have the benefit of previous experience and scientific knowledge, and unforeseen outcomes may generate substantial animal

ANIMAL WELFARE AND ETHICAL CONCERNS

suffering. This presents a difficult challenge to animal use committees and ethics committees evaluating proposed research protocols.

In general, the efficiency of reproductive cloning in animals remains low and cloned animals produced by NT have displayed a variety of anatomical and physiological abnormalities, high birth weight, and high pre- and postnatal mortality.[31] Other cloned animals do, however, appear quite normal.[32] Pregnancy complications can cause fetal loss and also result in increased morbidity and mortality in surrogate mothers.[33] Increased size of the fetus causes distress to both mother and fetus during parturition and often necessitates a C-section delivery. Researchers should take steps to document the health, physiology, and behaviors of their cloned animals, and studies should continue throughout the animals' life span.[34] Since the effects of genetic manipulation may not be apparent at all stages of life, animals must be studied at different stages, including the oldest age likely to be reached during usage.[35] Even those cloned animals that appear normal in the early stages of life should be monitored as they age and reproduce[36] as unanticipated side effects may not appear for several generations. Finally, it is important to monitor transgenic animals producing human proteins in their milk to determine if excessive production of an unnatural protein may cause any chronic health problems.[37]

One argument that calls into question the genetic modification of animals focuses not on the technology employed, but on the effect the transgene may have on the physical or physiological state of the animal. This appears to be a valid objection, particularly if the animal's metabolism is changed in a way that is not in the animal's interest or the protein is expressed in an organ or at a level that results in harm to the animal.[38] Bernard Rollin has set forth the "principle of conservation of welfare" as the standard of welfare for agricultural biotechnology, which states,

> Genetic engineering should not be used in ways that increase or perpetuate animal suffering.... Any animals that are genetically engineered for human use ... should be no worse off, in terms of suffering, after the new traits are introduced into the genome than the parent stock was prior to the insertion of the new genetic material.[39]

Applying the principle of conservation of welfare to transgenesis in biomedical research, it is important to note that the methods utilized in transgenesis do not necessarily have an adverse effect on animals' welfare; however, successful integration and expression of the transgene may negatively impact animal welfare.[40] To illustrate this point, certain human proteins produced in the milk of transgenic animals may cause harm to the animal if the transgene is expressed ectopically or the recombinant protein leaks from the mammary gland to the blood.[41] For example, a transgenic cow containing the gene for human erythropoietin has been created but never allowed to produce milk because studies have shown that human erythropoietin can have fatal effects in mice when it circulates in the blood.[42]

This example demonstrates potentially fatal consequences for a transgenic animal. However, as valuable human proteins proceed through clinical trials and are exploited commercially, will less severe effects on animal welfare be tolerated?

In pigs, for example, there has been evidence of abnormal mammary development due to expression of the transgene, which may have caused painful lactation.[43] Will financial incentives overtake concerns for the proper care and welfare of transgenic animals and will animal suffering of this magnitude be viewed acceptable? Will we be inclined to push transgenic animals even harder by increasing the frequency of milking or length of lactation? High rates of milk production in dairy cows have been related to an increased incidence of mastitis. Will this risk to the transgenic animal be recognized in a welfare analysis?

B. Xenotransplantation

Xenotransplantation is the transfer of cells, tissues, or whole organs from one species to another. The shortage of human organs and tissues available for transplantation is the most notable rationale given for xenotransplantation. It is estimated, for example, that there are currently over 17,000 patients awaiting liver transplants in the U.S.[44] In response to the shortage of human organs available for transplantation, researchers have expressed interest in using organs from animals to treat human patients. Pigs are favored as a potential source of transplantable organs because their organs are physiologically similar to those of humans; porcine organs are of an appropriate size; and pigs reproduce quickly and give birth to a large number of offspring.[45] An immediate immunological barrier to xenotransplantation, however, is the hyperacute rejection reaction provoked by porcine organs transferred to human recipients.[46] Pigs, therefore, must be genetically altered before their organs can be transferred to humans. NT is regarded as a means of introducing genetic modifications into an appropriate strain of pigs, in an attempt to combat rejection mechanisms. Researchers hope that cloning will enable them to knock out the pig gene that triggers rejection by the human body, as well as insert human genes more accurately to "humanize" pig organs to counter other human defense mechanisms.[47]

Additional animal welfare concerns related to xenotransplantation include the manner in which pigs are housed, handled, and treated prior to slaughter and organ retrieval. To minimize the risk for transmission of pathogens to human recipients, specific pathogen free (SPF) pigs are used as organ sources. In order to obtain SPF pigs, the pregnant sow is anesthetized shortly before she is to give birth and the entire uterus containing the piglet embryos is removed in a sterile "bubble."[48] (Alternatively, piglets may be born by cesarean section.) Piglets are then reared in isolation for fourteen days and the sow is typically slaughtered. Pigs are intelligent, social, and highly inquisitive animals and it has been demonstrated that piglets subjected to extremely early weaning, as is the case with SPF pigs, develop abnormal behaviors.[49]

Pigs also develop abnormal behaviors in confinement if not given the opportunity to root or build nests.[50] This suggests additional welfare issues since pigs intended for use as organ sources might be housed in extremely barren environments that are easily sanitizable.[51] The UK's Home Office Code of Practice for organ-source

ANIMAL WELFARE AND ETHICAL CONCERNS

pigs recommends that pigs be housed in stable social groups, provided adequate space to move around freely, and provided environmental enrichment, such as straw or other materials for manipulation, to satisfy pigs behavioral needs in terms of rooting and investigative behavior.[52] While it is recognized that the requirement to maintain SPF status may compromise the animal's behavioral needs to some extent, justification is needed if such a compromise becomes essential for a xenotransplantation protocol. The National Research Council has noted, "There are no comparable standards for pigs intended for xenotransplantation in the U.S., and the lack of standardization of housing and care among U.S. facilities for these pigs is a source of concern."[53]

As mentioned previously, pigs are highly inquisitive and intelligent animals. An important question to ask is: Will transgenic organ source pigs be kept isolated and confined in sterile environments, "with no opportunity to fulfill their behavioral and psychological needs?"[54] Furthermore, should the psychological suffering (i.e., frustration, anxiety, loneliness, boredom, fear) of transgenic animals, such as pigs, be acknowledged in a discussion of welfare issues? In addressing this concern, we may draw on the concept of an animal's telos, a term adapted from Aristotle's philosophy and defined by Bernard Rollin as,

> [T]he set of needs and interests which are genetically based, and environmentally expressed, and which collectively constitute or define the 'form of life' or way of living exhibited by that animal, and whose fulfillment or thwarting matter to the animal. The fulfillment of telos matters in a positive way, and leads to well-being or happiness; the thwarting matters in a negative way and leads to suffering.[55]

In this context, to prevent animal suffering and enhance happiness means that attention must be paid to more than the physical—the "behavioural, functional and cognitive drives"[56] of an animal of a given species are additional factors to be recognized in evaluating its welfare. Because pigs are naturally social animals, to isolate such an animal "does not cause it physical pain, but can cause psychological suffering because its telos is being ignored or violated."[57] Similarly, Gary Comstock, director of the Research Ethics Program at North Carolina, has also addressed the concept of "respecting" animals. He contends, "[t]he key to respecting animals, ..., is respecting their right to satisfy their primary desires."[58]

Another phenomenon to examine in the area of xenotransplantation is gradualism, "in which progressive increments are gradually made in an area of technology, each step being justified on the basis that it represents only a small change from the last."[59] However, when the overall change is evaluated after a period of time has passed, it may appear that an unacceptable change has taken place when compared with the original starting point. Consider, for example, the use of porcine heart valves in human patients, which has become common practice. Simply because this particular use of pig heart valves is generally accepted, does not mean we should condone all other uses of these animals. A rather significant leap is made from the acceptance of pig valves to the idea that whole animal organs may be transplanted to humans; however it may be presented as simply another small step. The human benefits of xenotransplantation are still largely potential, but even if xenotransplantation becomes medically feasible, it does not answer the question whether it is ethically acceptable. We must refer each step back to more fundamental values and focus not only on what the next step represents in its similarity to the last step. What is required is that we "step back and look at the complete sequence of steps and ask if the final end is in fact acceptable."[60]

Finally, an argument set forth in favor of raising pigs to supply organs is the so-called "ham sandwich" argument. Throughout history, animals have been raised as a source of labor, food, and clothing. It is estimated, for example, that 94.5 million pigs were born, raised, and slaughtered in the U.S. in the year 2002.[61] Since we slaughter pigs for food in order to live, how can we object to raising them and killing them for organs to save human life? One response to this question has been presented as a "naturalness" argument, which has been set forth as follows: All would agree that everyone must eat in order to live. Although it is debatable whether humans must eat animals, it could be said that it is "natural" to do so. This argument does not necessarily extend to the transplantation of animal organs to humans. Organ replacement has become possible only through human skill and scientific and medical advances. "It is not natural to use an animal as a spare part [supplier]. It is human artifice."[62] Although not arriving at the conclusion that it is wrong to use an animal in this way, the authors propose that "in ethical terms it is not the same as eating an animal."[63]

C. Models for Human Genetic Diseases

The third area in which NT will impact the generation of transgenic animals is the creation of animal models for human genetic diseases. This area poses perhaps the most serious animal welfare concerns since gene targeting and cloning may allow production of models for many debilitating human genetic diseases. Transgenic animal models will likely have no option but to suffer, no matter what the end. Is the pain and suffering of the animal justified by the potential benefit to human beings, or does this intervention fall into the category of "harms of a certain degree which ought under no circumstances to be inflicted on an animal?"[64]

Although some contend there is no ethical difference between chemically or surgically inducing a disease condition in a laboratory animal and modifying its genetic structure so as to cause it to develop a particular human disease, it appears as though certain differences do exist. Transgenic animals created to model human genetic diseases will be genetically programmed to suffer the effects of disease from birth. In contrast, those animals with disease conditions created in the laboratory will suffer effects of the disease only from the time it is actually induced.[65] Second, the advantages put forth to support genetic inducement of disease are the reliability and repeatability of the effect.[66] These factors themselves almost guarantee that the animal will suffer adverse effects of the disease condition they are created to model. If an experiment does not proceed for some reason, i.e., lack of continued funding, genetically altered animals will develop the disease condition and likely suffer symptoms even though the study has not proceeded and no useful data has been generated; whereas, experiments using animals whose disease condition has been chemically or surgically induced can be curtailed and additional animal suffering can be avoided. Finally, it has been suggested that moral questions arising from the development of animal models by transgenesis do not differ in kind from the questions arising from the development of models by chemical or surgical inducement, however they do differ in degree. "Transgenics provides the potential for generating vast numbers of animals modeling genetic disease and other diseases with devastating symptoms."[67]

Many researchers support the use of NT to create genetically uniform animal models of human diseases. They contend that these animals are more accurate disease models and therefore should generate more precise and reliable data. When scientists test a certain chemical agent or medical procedure, they will know that differences in test results are due to the procedure or drug, and not to genetic differences between the research animals.[68] As a result, the number of animals needed for research should decrease. Others however disagree.

First, it has been suggested that the total number of animals used in research may actually increase because there are thousands of genetic diseases that may potentially be created in animals. Second, the argument that the number of animals used in experimentation would decrease was the initial justification for the use of genetically engineered model mice in the early 1990s. Evidence shows, however, that there has been a significant increase in the number of mice used in research since the development of transgenic technologies. It appears that all over the world, research centers and animal facilities are filled to capacity with mutant mice and some laboratories are forced to turn down applications for storage simply because of lack of space and funding.[69] Rats and mice are exempt from the Animal Welfare Act and no government agency in the United States requires the reporting of mice numbers used in research. However, one author estimates that the number of mice and rats used in research increased from approximately eleven million to nineteen million in 1993 to eighty million in 2001.[70] Today, genetically transformed laboratory mice can be ordered on-line or via toll-free numbers as though they are mere items listed in a catalogue.[71] Referring back to the concept of gradualism first explored in the discussion of xenotransplantation, "having developed a culture which sees the use of disease model mice as a norm, the progressive extension of this could exceed ethical bounds by imperceptible steps."[72]

Transgenic technologies have contributed greatly to the production of mouse models of human diseases; however, mice sometimes fail to provide a complete model of the human phenotype. The term "phenotype gap" has been used to refer to the gulf between mouse mutant strains available for study and the full range of phenotypes necessary to exploit the mouse as an animal model.[73] Differences in human and mouse life span as well as differences in anatomy and physiology have contributed to the failure to produce a mouse model that resembles certain human diseases.[74] Certain livestock species are considered better models because they appear to be more similar to humans with respect to size, anatomy, physiology, and life span.[75] For example, mouse models of cystic fibrosis fail to exhibit the same lung pathology seen in humans, and researchers have turned to the sheep as a potential animal model.[76] Other scientists have turned toward creating a porcine or bovine model of the genetic disease ataxia-telangiectasia, as mice do not display the neurodegenerative phenotype seen in humans.[77] Using larger animals as models may raise greater welfare concerns than the use of smaller animals, such as rodents.

Additional limitations of mouse models for various neurogenetic disorders[78] and neurodegenerative diseases[79] have been identified. First, a mouse ortholog to a human gene of interest may not exist.[80] Second, as mentioned above, mouse models do not exhibit the same phenotype observed in humans.[81] Mouse models may exhibit only some symptoms of the disease observed in humans, or they may exhibit no symptoms at all. Furthermore, there are limited cognitive and behavioral tests available for rodents, which may not be applicable to the study of neurodegenerative diseases.[82] As a result, some researchers contend that non-human primates (NHPs) are necessary to study these neurological disorders because mouse models are simply not suitable.[83] In contrast to mouse brains, NHP brains are more complex and display greater similarities to the human brain. Rhesus macaque models are favored because these NHPs display "perceptual, cognitive and behavioral plasticity not observed in mice."[84] The qualities that make NHPs more desirable as models for these diseases are the same qualities that give rise to greater welfare concerns. NHPs are more sentient beings and have higher cognitive capacities and engage in more complex social interactions than small rodent research models.

Although the production of genetically modified cloned NHPs still poses significant challenges, a team at the University of Pittsburgh has made significant steps toward successful therapeutic cloning of nonhuman primate embryos with the hope of producing embryonic stem cells.[85] These researchers are also working towards cloning nonhuman primates as a way to generate genetically uniform animals for experimentation.[86] If researchers are successful in overcoming the obstacles encountered in NHP cloning, concerns arise that the number of primates used in research will increase as we are able to generate models of more and more human genetic diseases. The use of NHPs may rise significantly if they become exploited on a long-term and widespread basis. Not only will these animals suffer debilitating disease symptoms, but the difficulty in satisfying the social and behavioral requirements of NHPs in the laboratory setting will add to their potential for psychological suffering.

The severity of symptoms of many genetic disorders must also enter a welfare analysis. Lesch-Nyhan Syndrome, for example, is a rare genetic disorder caused by a deficiency of the enzyme HPRT.[87] Symptoms of the disease include joint pain, kidney problems, muscle weakness, and uncontrolled spastic muscle movement.[88] The most striking aspect of the disease is the development of self-mutilating behaviors, such as lip and finger biting, which begin in humans during the second year of life.[89] While there is certainly the potential for significant human benefit arising from the study of this disease, there is a concern over the welfare of animals used as models. Again, researchers may turn to NHP models of this syndrome since mice containing the genetic mutation which leads to Lesch-Nyhan Syndrome do not demonstrate the phenotype typical of this neurogenetic disorder in humans.

Patients diagnosed with Lesch-Nyhan Syndrome do not exhibit symptoms from birth, but develop them later. Death usually occurs in the first or second decade of life due to kidney failure.[90] Therefore, in order for researchers to study the full course of the disease, animals will need to be kept alive for as long as possible. It does not seem possible to study these diseases in acute, terminal, or short-term experiments,[91] and the potential exists for a considerable amount of animal pain, distress, and suffering as these animals will show symptoms displayed by humans with the same syndrome over a long period of time. Although studying diseases such as LN syndrome may benefit humans, it should not be forgotten that the increased use of TG technologies to model human genetic diseases in research animals has its costs.

IV. COMMODIFICATION AND OBJECTIFICATION

Even the absence of welfare problems in transgenic animals, however, does not necessarily imply the absence of a moral problem. In addition to respect for nature, and respect for the natural way, other arguments stem from the mass production of identical genetic copies. Some authors argue that cloning may further dilute the "essence" in copies of the same creature.[92] Similar concerns have been expressed by others, who suggest that NT will encourage animals to be treated increasingly as commodities and will negatively impact the value of animal life. In our market economy, the perceived value of consumer goods decreases as they are produced on a large scale and the number of identical copies increases. As one author has stated in reference to routine cloning for animal production: "To clone routinely would apply a factory model of mass production too far into the realm of living creatures. We need to remind ourselves we are not dealing with identical widgets on a production line, but living creatures, useful to us, but still creatures."[93] The same author notes, however, that small scale special cloning, i.e., use of NT to produce five to ten founder transgenic animals that would then breed naturally, lacks something of the "instrumentality" of other cloning applications.[94] The primary aim of such work is not to clone as such, but to more efficiently perform a genetic modification that could not occur naturally.

A final consideration centers on the choice of terminology used to characterize transgenic animals. The use of phrases such as "living bioreactor" to refer to transgenic animals whose mammary function is used for the production of human proteins, and "spare parts supplier" to describe transgenic pigs whose organs may be used for xenotransplantation, depicts an extremely instrumental view of these animals. The use of this terminology may contribute to even greater objectification of living animals. As stated by one working group,

> In calling an animal a bioreactor, or spare part supplier, it is described primarily for what it is functionally—as a means to an end—not what it is as an animal.... By extrapolating from concepts of the factory, a statement is made that [animals] are more closely related to the non-living world than the living.[95]

Such an instrumental view, which depicts transgenic animals more as production machines than living creatures, conveys a degree of disrespect for these animals. Continued use of these terms may cause them to become more commonplace, affirm questionable attitudes and truncate the ethical debate. Will word choice be responsible for a slow indoctrination and assimilation of such viewpoints? Widespread acceptance of such language may contribute to a sense of normality, obfuscate ethical questioning, and lead to fear of having the insensitivity of the terminology hived on to public acceptance.

V. CONCLUSION

Advances in NT are proceeding at a rapid pace. Accompanying these great strides is the potential for significant animal suffering. Society's views of non-human animals continue to evolve and animal welfare issues concern a growing percentage of our population. Consequently, the impact of emerging technologies on animal welfare will likely influence public acceptance of new scientific breakthroughs. NT offers significant advantages over pronuclear injection for the generation of transgenic animals. Nonetheless, concerns remain regarding both the animals' physical and psychological well-being. Cloning raises issues regarding the increased objectification and commodification of living creatures. Although not unique to the applications of cloning technology discussed above, these matters demand further examination. Importantly, society's view on what is ethically acceptable can change over time. Therefore, it is essential that we revisit old questions, raise new ones, and reassess often as cloning research progresses and animals created by NT are monitored over longer periods of time. The ethical dialogue must remain dynamic as science progresses and the public must be engaged as well as educated. As experience with this technology increases and information is gathered, it should be possible to better anticipate risks and benefits to both humans and non-human animals.

NOTES

1. A. Dove, "Milking the Genome for Profit," *Nature Biotechnology*, 18 (2000) pp. 1045–1049.

ANIMAL WELFARE AND ETHICAL CONCERNS

2. A. Colman, "Somatic Cell Nuclear Transfer in Mammals: Progress and Applications," *Cloning*, 1 (1999/2000) pp. 185–200.

3. National Research Council. *Animal Biotechnology: Science-based Concerns*, pp. 43, 97 (The Nat'l Academies Press 2002).

4. See I. Wilmut, K. Campbell, and C. Tudge. *The Second Creation: Dolly and the Age of Biological Control* (Farrar, Straus and Giroux 2000) p. 34 (hereinafter *The Second Creation*).

5. M. F. Brink, et al. "Developing Efficient Strategies for the Generation of Transgenic Cattle which Produce Biopharmaceuticals in Milk," *Theriogenology*, 53 (2000) pp. 139–148.

6. National Research Council, *supra* note 3, at p. 94.

7. Ibid.

8. See E. Behboodi, et al. "Transgenic Cloned Goats and the Production of Therapeutic Proteins," in *Principles of Cloning* (Elsevier Science, 2002) pp. 459–474.

9. *The Second Creation*, *supra* note 4, at pp. 237–238.

10. Schnieke, A., et al. "Human Factor IX Transgenic Sheep Produced by Transfer of Nuclei from Transfected Fetal Fibroblasts," 278 *Science*, pp. 2130–2133 (1997).

11. Ibid.

12. S. L. Stice, et al. "Cloning: New Breakthroughs Leading to Commercial Opportunities," *Theriogenology*, 49 (1998) pp. 129–138.

13. Dove, *supra* note 1, at p. 1046.

14. PPL Therapeutics, "What We Do: Nuclear Transfer," available at http://www.ppl-therapeutics.com/what/what_2_content.html.

15. Dove, *supra* note 1, at p. 1047.

16. Ibid.

17. National Research Council, *supra* note 3, at p. 97. Gene targeting is defined as the controlled integration of transgenes into specific, predetermined locations within the genome.

18. Craig A. Hodges and Steven L. Stice, "Generation of Bovine Transgenics Using Somatic Cell Nuclear Transfer," *Reproductive Biology and Endocrinology*, 1 (2003), available at http://www.rbej.com/content/1/1/81.

19. L. M. Houdebine, *Animal Transgenesis and Cloning*, p. 73 (John Wiley & Sons Ltd. 2003).

20. I. Wilmut, "Cloning for Medicine," *Scientific American*, (Dec. 1998) pp. 58–63.

21. PPL Therapeutics, "Products: Alpha-1-Antitrypsin," available at http://www.ppl-therapeutics.com/products/products_1.html.

22. GTC Biotherapeutics, "ATryn®—Recombinant Human Antithrombin," available at http://www.transgenics.com/products/atryn.html.

23. Dove, *supra* note 1, at p. 1046. ("A transgenic goat, for instance, produces protein at a unit cost of $10–25/g compared with $100–1000/g for cell culture.")

24. "Biotech in the Barnyard: Implications of Genetically Engineered Animals." Proceedings from a workshop sponsored by the Pew Initiative on Food and Biotechnology, available at http://www.pewagbiotech.org/events/0924/proceedings1.pdf.

25. Christopher Bowe, "Biotech Companies Plan to Milk Herds of Cloned Cows for Human Drug Needs," *Financial Times*, Oct. 13, 1999, World News: U.S. and Canada, at p. 6.

26. R. Straughan, "Ethics, Morality and Animal Biotechnology," available at http://www.bbsrc.ac.uk/tools/download/ethics_animal_biotech/ethics_animal_biotech.pdf. See also, "Working Group of the Society, Religion and Technology Project, Church of Scotland," *Engineering Genesis: The Ethics of Genetic Engineering in Non-Human Specie*, ed. Donald and Ann Bruce, (Earthscan 1998) p. 103 [hereinafter *Engineering Genesis*]. It has been proposed that the term "risk-benefit" analysis is preferable to "cost-benefit" analysis, since the latter suggests monetary evaluation and economic worth, and "Ethical assessment can never be quantitative in this sense."

27. "Genetic Engineering: Animal Welfare and Ethics," a discussion paper from the *Boyd Group* (Sept. 1999), available at http://www.boyd-group.demon.co.uk/genmod.htm [hereinafter *Boyd Group*].

28. A. Van't Hoog, et al., "Dolly's Deceiving Perfection: Biotechnology, Animal Welfare, and Ethics," *J of Applied Animal Welfare Science*, 3 (2000) pp. 63–69.

29. *Engineering Genesis*, supra note 26, at p. 91.

30. Ibid. ("Direct modification of the genome can produce more novel, surprising and wide-ranging phenotypic effects, with greater potential to compromise welfare, in one step."). Pronuclear injection is less precise than NT, as the transferred genetic material is randomly integrated into the recipient animal's genome. Therefore, the unpredictability of adverse effects on animal welfare is arguably a more relevant concern when microinjection is used to create transgenic animals.

31. *Scientific and Medical Aspects of Human Reproductive Cloning* (National Academy Press, 2002) p. 40. See also, F. B. Garry, "Postnatal Characteristics of Calves Produced by Nuclear Transfer Cloning," *Theriogenology*, 45 (1996) pp. 141–152. Anatomical and physiological abnormalities include lung and cardiovascular problems, joint and limb abnormalities, immune system dysfunction, and liver and kidney problems.

32. See, Robert P. Lanza, et al., "Cloned Cattle Can Be Healthy and Normal," *Science*, 294 (2001) pp. 1893–1894.

33. *Scientific and Medical Aspects of Human Reproductive Cloning*, supra note 31, at p. 40. Pregnancy complications include abnormal placentation, pregnancy toxemia, and hydroallantois.

34. I. Wilmut, "Are there any Normal Cloned Mammals?" *Nature Medicine*, 8 (2002) pp. 215–216.

35. D. Broom, "The Effects of Biotechnology on Animal Welfare," in *Animal Biotechnology and Ethics*, ed. Alan Holland and Andrew Johnson (Chapman & Hall 1998) p. 73.

36. J. Cibelli, et al. "The Health Profile of Cloned Animals," *Nature Biotechnology*, 20 (2002) pp. 13–14.

37. A. George, "Animal Biotechnology in Medicine," in *Animal Biotechnology and Ethics*, ed. Alan Holland and Andrew Johnson (Chapman & Hall 1998) 32.

38. *Engineering Genesis*, supra note 26, at pp. 135–136.

39. B. E. Rollin, *The Frankenstein Syndrome: Ethical and Social Issues in the Genetic Engineering of Animals* (Cambridge University Press 1995) p. 179 [hereinafter *The Frankenstein Syndrome*].

40. See Paul B. Thompson, "Research Ethics for Animal Biotechnology," in *Ethics for Life Scientists*, ed. M. Korthals and R. J. Bogers (Springer 2004) pp. 112-113.

41. L. M. Houdebine, "Transgenic Animal Bioreactors," *Transgenic Research*, 9 (2000) pp. 305-320.

42. *Engineering Genesis*, *supra* note 26, at p. 118. To eliminate or reduce potentially deleterious side effects, a genetic change can be limited to certain tissues of interest. See M. B. Dennis, "Welfare Issues of Genetically Modified Animals," *ILAR J*, 43 (2002) pp. 100-109. Milk-specific gene promoters are used to direct expression of the human gene only in the mammary gland of the transgenic animal.

43. National Research Council, *supra* note 3, at p. 102.

44. "U.S. Transplantation Data. United Network for Organ Sharing Data," available at http://www.unos.org/data/default.asp?displayType=usData.

45. *The Second Creation*, *supra* note 4, at 255.

46. See J. A Platt, "Primer on Xenotransplantation," in *Xenotransplantation 8*, ed. Jeffrey L. Platt (ASM Press 2001).

47. D. M. Bruce, "Ethics Keeping Pace with Technology," in *Beyond Cloning: Religion and the Remaking of Humanity*, ed. Ronald Cole-Turner (Trinity Press Int'l 2001) pp. 34-49. As noted above, *supra* p. 6, pig tissues display a carbohydrate epitope that reacts with human antibodies, stimulating hyperacute rejection of the transplanted organ. Researchers hope to lessen this immune response by deleting the gene encoding the enzyme that produces this epitope. One researcher has noted, however, "the [carbohydrate] structure may provide some essential biological function in pigs and thus destroying the alpha-1,3 GT enzyme could be deleterious to the animals." Dove, *supra* note 1, at 1047, quoting Irina Polajaeva (former head of the Cell Biology Group and project manager for the Porcine Nuclear Transfer Program at PPL Therapeutics Inc.). This reemphasizes the need for long-term monitoring of modified animals.

48. J. D'Silva, "Campaigning against Transgenic Technology," in *Animal Biotechnology and Ethics* 98, ed. Alan Holland and Andrew Johnson (Chapman & Hall 1998).

49. National Research Council, *supra* note 3, at p. 103, citing D. M. Weary, et al., "Responses of Piglets to Early Separation from the Sow," *Applied Animal Behavior Science*, 63 (1999) pp. 289-300. See also, Alexander Tucker, et al., "The Production of Transgenic Pigs for Potential Use in Clinical Xenotransplantation: Microbial Evaluation, in *Xenotransplantation*, 9 (2002) pp. 191-202. These authors described custom built nursery isolator tanks designed to provide a germ free environment while attempting to minimize any impact on piglet behavior or welfare. The isolator tanks were designed to hold five piglets each and provided a heated enclosed bed area, feeding area, a storage area, and piglet segregation area. The tanks were constructed of stainless steel with PVC canopies. Unopened bags of sterile water were positioned in the bed area to provide comfort and a nosing substrate, thereby minimizing behavioral stereotypes such as umbilicus sucking and belly nosing.

50. National Research Council, *supra* note 3, at p. 103.

51. Ibid. See also, Nuffield Council on Bioethics, "Animal-to-Human Transplants: The Ethics of Xenotransplantation," (March 1996), at p. 64, available at http://www.nuffieldbioethics.org/fileLibrary/pdf/xenotransplantation.pdf. (5.20 "Even if isolation is not required, in order to keep animals free from infection, the environment will have to be kept relatively sterile and therefore be easy to clean. So it is likely to consist of monotonous textures and to be free of items which might enrich the life of the animal, but which might also harbour infectious organisms. Human contact, which can be advantageous for animals in captivity, may have to be minimized since human beings harbour some diseases (such as influenza) that can be passed on to pigs.")

52. Her Majesty's Government, "Home Office Code of Practice for the Housing and Care of Pigs Intended for Use as Xenotransplant Source Animals" (2000), available at http://www.homeoffice.gov.uk/docs/xenopig.pdf.

53. National Research Council, *supra* note 3, at p. 103.

54. *The Frankenstein Syndrome*, *supra* note 39, at p. 195.

55. B. Rollin, "On Telos and Genetic Engineering," in *Animal Biotechnology and Ethics* 162 (1998), ed. Alan Holland and Andrew Johnson (Chapman & Hall).

56. See Thompson, *supra* note 40, at p. 114.

57. Straughan, *supra* note 26.

58. *Biotech in the Barnyard*, *supra* note 24.

59. See *Engineering Genesis*, *supra* note 26, at p. 80.

60. Ibid., p. 137.

61. National Pork Producers Council, "Issue paper #4," June 5, 2003, available at http://www.nppc.org/issue_brief/2003/brief_060503.html.

62. *Engineering Genesis*, *supra* note 26, at p. 137.

63. Ibid.

64. M. Banner, "Report of the Committee to Consider the Ethical Implications of Emerging Technologies in the Breeding of Farm Animals" (Banner Report). Ministry of Agriculture, Fisheries and food, HMSO, London (1995).

65. See Boyd Group, *supra* note 27.

66. *Engineering Genesis*, *supra* note 26, at p. 139.

67. *The Frankenstein Syndrome*, *supra* note 39, at p. 204.

68. *The Second Creation*, *supra* note 4, at p. 245.

69. J. Knight and A. Abbott, "Full House, *Nature*, 417 (2002) pp. 785–786.

70. L. Carbone, *What Animals Want: Expertise and Advocacy in Laboratory Animal Welfare Policy* (Oxford University Press 2004) p. 27. See also, Dennis, *supra* note 42, at pp. 102–103. The author noted an increase in mouse populations at the University of Washington of more than 23 percent per year during the eight year period from 1993–2001. In addition to the increase in numbers of animals on studies, the author attributes this growth in numbers of mice used in research to the following: First, a large number of animals are necessary to create each genetically modified line which is then bred to produce animals to study. Second, in order to maintain a genetically modified line, it is necessary to continue breeding animals

that are not studied. These animals have compromised health but do not produce any data directly.

71. See The Jackson Laboratory, Jax® Mice Online Order Request Form, available at https://secureweb.jax.org/jaxmice/order/orderform.pdf.

72. *Engineering Genesis*, *supra* note 26, at p. 70.

73. Steve D. M. Brown and J. Peters, "Combining Mutagenesis and Genomics in the Mouse—Closing the Phenotype Gap," *Trends in Genetics*, 12 (1996) pp. 433–435.

74. Kelly S. Swanson, et al., "Genomics and Clinical Medicine: Rational for Creating and Effectively Evaluating Animal Models," *Experimental Biology and Medicine*, 229 (2004) pp. 866–875.

75. Alison J. Thomson and Jim McWhir, "Biomedical and Agricultural Applications of Animal Transgenesis," *Molecular Biotechnology*, 27 (2004) pp. 231–244.

76. Ann Harris, "Towards an Ovine Model of Cystic Fibrosi"s, *Human Molecular Genetics*, 6 (1997) pp. 2191–2193.

77. Hodges, *supra* note p. 18. See also, Swanson, *supra* note 74, at pp. 870–871.

78. Robert B. Norgren, "Creation of Non-human Primate Neurogenetic Disease Models by Gene Targeting and Nuclear Transfer," *Reproductive Biology and Endocrinology*, 2 (2004) pp. 40–48, available at http://www.rbej.com/content/2/1/40. Examples of neurogenetic disorders include Kallman's syndrome, Lesch-Nyhan's disease, and Ataxia-Telangiectasia.

79. Anthony W. S. Chan, "Transgenic Nonhuman Primates for Neurodegenerative Diseases," *Reproductive Biology and Endocrinology*, 2 (2004) pp. 39–46, available at http://www.innovitaresearch.org/news/04082501.html. Examples of neurodegenerative diseases include Parkinson disease, Alzheimer's disease, and Huntington disease.

80. Norgren, *supra* note 78. See, B. Alberts, et al., "Molecular Biology of the Cell" 4th ed. (Garland Science, New York 2002) p. 22. An ortholog has been defined as, "Genes in two separate species that derive from the same ancestral gene in the last common ancestor of those two species."

81. Norgren, *supra* note 78.

82. Chan, *supra* note 79.

83. See Norgren, *supra* note 78. See also, Chan, supra note 79.

84. Norgren, *supra* note 78.

85. "Efforts to Clone Primates Move Forward: Results Represent Significant Development toward Therapeutic Cloning of Stem Cells," *UPMC News Bureau*, available at http://news-bureau.upmc.com/PDF/SchattenPrimateCloneStudy2004.pdf.

86. Ibid.

87. National Institute of Neurological Disorders and Stroke, NINDS Lesch-Nyhan Syndrome Information Page, available at http://www.ninds.nih.gov/disorders/lesch_nyhan/lesch_nyhan_pr.htm.

88. National Organization for Rare Disorders, Lesch-Nyhan Syndrome, available at http://www.rarediseases.org/search/rdbdetail_abstract.html?disname=Lesch%20Nyhan%20Syndrome.

89. National Institute of Neurological Disorders and Stroke, *supra* note 87.
90. Ibid.
91. *The Frankenstein Syndrome*, *supra* note 39, at p. 204.
92. Van't Hoog, *supra* note 28, at p. 65.
93. D. Bruce, "Polly, Dolly, Megan, and Morag: A View From Edinburgh on Cloning and Genetic Engineering," *J of the Society for Philosophy and Technology*, 3 (Winter 1997), available at http://www.scholar.lib.vt.edu/ejournals/SPT/v3n2/BRUCE.html.
94. Ibid.
95. *Engineering Genesis*, *supra* note 26, at p. 136.

ETHICS AND THE LIFE SCIENCES

THE RELEVANCE OF SPECIESISM TO LIFE SCIENCES PRACTICES

ROGER WERTHEIMER
AGNES SCOTT COLLEGE

ABSTRACT: Animal protectionists condemn speciesism for motivating the practices protectionists condemn. This misconceives both speciesism and the morality condoning those practices. Actually, animal protectionists can be and generally are speciesists. The specifically speciesist aspects of people's beliefs are in principle compatible with all but the most radical protectionist proposals. Humanity's speciesism is an inclusivist ideal encompassing all human beings, not an exclusionary ethos opposing moral concern for nonhumans. Anti-speciesist rhetoric is akin to anti-racist rhetoric that condemned racists for regarding people as moral inferiors because of their skin color. Actually, racists never thought that skin color is itself a reason for discounting someone's interests, just as humans have never thought that only a human can be a proper object of moral concern. Some speciesists have great concern for animal suffering; some don't. Animal protectionists have yet to show that a lack of concern is due to some false assumptions.

Much human activity harms nonhuman animals. We routinely target them for foodstuff and apparel, as experimental subjects and objects of sport. They may suffer even more as collateral damage from our onslaught on the world we both occupy. The life sciences contribute to all the harm by their means of acquiring knowledge of the mechanisms of life and still more through the applications and other consequences of that knowledge.

Protests against human activities harming animals have become progressively more popular and influential. This essay is studiously noncommittal on specific proposed reforms of common practices. Its subject is the role and accuracy of one recurrent premise of the critiques of popular practices. Criticisms of any practice may come from very disparate reasoning, with specific directives derived from competing, incompatible conceptions of morality that deny the legitimacy of each other's inferences. Despite their conceptual incompatibilities, ethical theories may come to close convergence—or wide divergence—on concrete reforms of our practices regarding animals, as they can be brought to converge or diverge on practical specifics on most other matters.

Along the way from competing great abstract generalities of principle to concrete particulars of action, animal protectionists unite with a shared conception of their common enemy. They premise that the popular moral indifference about animal well being is a consequence or expression of a prevalent moral prejudice these critics call "speciesism": a favoritism of, by, and for members of a species. This premise of animal protectionist rhetoric is erroneous and irrelevant. Animal protectionists misconceive the structure and content of humanity's morality. The specifically speciesist aspects of common attitudes may be compatible with all but the most radical animal protectionist ideals. The explanation of the practices they want reformed lies elsewhere, so they needn't consider our concern for our conspecifics objectionable and indefensible. They'd do better to understand and confront what they're really up against.

Much that animal protectionists say about speciesism is unexceptionable. Certainly, our culture seems dominated by an ideology of our all being bonded by our birth into a species whose members are obliged to care for one another. Looked at in the large, our species seems naturally disposed to evolve an ideal of the species as a single family, with all members brothers or sisters who are naturally predisposed (upon normal development) to care for each other in ways that work to perpetuate the species' flourishing. This seems to be our human nature partly because it makes sense that things be that way. It seems that nature would favor the development of a species that markedly increased its chances of perpetuation by its members' acquiring a native propensity to care for their kind. Nature should also favor the development of a species that acquired the capacity and propensity to protect and promote the species by prescribing and praising the protection and promotion of, for, and by its members. However we came to have this nature, it seems a neat bit of intelligent design. However accidental its origins, it seems like it should work to perpetuate itself, unlike many other persisting species traits that may once have been functional adaptations but are no longer advantageous. It may not be the optimal design (whatever that might mean), but it seems good enough for now for us to be satisfied with this aspect of our nature.[1]

This bit of arm-chair potted natural history makes no pretense of being any more than a plausible story. Even if it is sufficiently accurate, it doesn't entail that the speciesist principle is objectively true. No matter. The aim of this essay is not to demonstrate the objective truth of our speciesism but only to make good enough sense of what people are believing and explain why it is not something animal protectionists have reason to condemn.

The critics' neologism "speciesism" has helped to fix attention on this conspicuous, basic, pervasive species-centrism of humanity's evolving morality that has matured in modern times. Many friends of speciesism have betrayed it by continuing to deny the critics' accusation that it clashes with our high culture's theories that profess to make systematic sense of humanity's morality. Previously, ethical theorists had claimed close concordance with humanity's morality while actually they had not given or really taken any account of speciesism, but instead fobbed off one or another ersatz principle and derived mankind's moral equality from some psychological properties of normal adult humans, like sentience, self-consciousness or rationality. Plainly no such property is universal in our species and perhaps none is unique to it.[2]

Our speciesist concerns have often been rationalized as responses to some alleged extra-empirical attribute of all humans, like inherent worth, dignity, or natural moral rights. But unless those attributions are themselves predicated on possession of some psychological capacities, they are comparable to their religious translations in terms of sanctity, sacredness, ensoulment, God's image, His children, or the like. Such talk serves to *express* the idea of the moral brotherhood of humankind. It does nothing to justify the idea or explain how to make good, rational sense of it.

We're naturally drawn to justify favoring humans by looking for some property essentially and inalienably possessed by each human being, for that's the basic structure of justifications elsewhere. Such reasoning does speciesism no favor, for it fails to appreciate that speciesism is the primary, fundamental principle of humanity's morality, and that it couldn't be fundamental if it were derivable from our possession of some other property, mental, spiritual or whatever, even if there were such a defining property. How to justify a truly fundamental principle, in morals, math, logic, or elsewhere has been an unsolved puzzle for a few millennia. Among the few certainties is that such truths cannot be proven like a theorem, corollary, or subordinate, derivative truth.

Ethical theories about persons (psychologically specified) cannot match humanity's morality because of the conceptual and metaphysical independence of biological categories from psychological categories. Our psychological abilities and processes are explained by our biology, but they aren't defined or identified by them, so any human psychological ability or process could be possible or normal for creatures of some other actual or conceivable species. Conversely, our psychological processes may affect and partly explain some organic events, but they don't define or identify biological categories. Any *homo sapiens* could have been what some of us are, a congenitally insensate human being, and anyone with a mind could entirely lose it, as some do. No human life has any experience necessarily.

Ethical theories and principles formulated in psychological, non-biological, terms depart from humanity's morality in their sense, and sanction significantly different practices and attitudes. The clash is most evident at the beginning of a human life, and again at its end, and also in the kinship discriminations structuring human life. The rhetoric of intra-speciesist debates on matters like abortion has exhibited a stable consensus over paradigm instances of human being and considerable continuing dissension about "marginal" cases. What—if anything—are the essentials to qualify as a human being remains controversial.[3] What is clear is that the qualifications ethical theorists propose for moral equality do not capture or closely correlate with the popular speciesist notion of humanity as essentially a biological kind.[4]

Theorists now appreciate that 'person' and 'human being' are distinct terms despite being freely interchangeable in most everyday discourse. Our religions tell us of spiritual persons; our legal systems recognize institutional persons; and our fictions, awake and asleep, are filled with nonhuman persons of every imaginable variety, biological and otherwise. A creature can be a person just because of its psychological capacities. So, ethical theorists define the term by psychological properties. Their paradigm of a person is a morally accountable agent with the sophisticated cognitive and motivational capacities for the self-conscious, rational self-control essential for morally accountable agency. The population of persons is expandable by reducing the requirements. The conception can be broadened to encompass any creature with a subjective life able to experience some benefits or harms.

Philosophies present competing conceptions of personhood, but each conception strives to be unified. Meanwhile, in the language of our common culture and as formalized in our law courts, the term 'person' is disjunctive. An individual is a person, whatever its biology, just because of its psychology, and also, instead or in addition, just by being of our specific biological kind, whatever its own mental capacities. Our legal system allows withholding treatment of anacephalic infants—not by pretending that they are not human beings nor by deeming them non-persons—but only because of the futility of treatment. So too, for withholding or withdrawing treatment of those in a permanent vegetative state: the termination of their mental activity does not terminate their membership in our biological species or their legal status as persons. (What some folks seem unwilling or unable to understand is that we may be obliged to terminate a human life out of respect for its being the life of a person in a permanent vegetative state.) And so also, many people recognize a zygote or embryo as a person without pretending it has some kind of mental life. Instead they suppose that a human life begins at conception (or implantation or the like), and *ipso facto* so does the life of a person. Whatever ethical theorists may say, no one in our judicial or legislative halls would seriously propose legally recognizing human embryos as human beings while denying that they are persons.

Humanity's morality, like the legal systems that evolved with it, is unequivocally disjunctive. It recognizes moral personhood on grounds of both minded agency and biological kinship. Human beings have never supposed that only human beings can be our moral equals. No society on record has lacked our current culture's propensity to imagine non-human persons—alter-specific characters like the crowd at a Star Wars cantina—whom we'd all recognize as our moral equals. (Such recognition is consistent with our triumphantly exterminating the congenitally homicidal, and shunning any persons, however harmless, incarnated in revolting globs of hermaphroditic pus.) Human beings everywhere, whatever their regard for the animals around them, seem to display a deep need to imagine—in fictions and fantasies, awake or asleep—our having moral relations with nonhuman agents. Our science fictions standardly assume, rightly I think, that our legal systems would readily recognize rationally self-controlled extra-terrestrials as persons.

Such conceptions come easily because an agent competent to make contracts and commitments, and thus be held accountable by us, could properly hold us accountable. Such competence for moral agency is a matter of psychological capacities other conceivable creatures might have. Moral agency and the moral standing it can demand cannot be monopolized by a biological species.

Humanity's speciesism is an insistently inclusivist ideal, not an exclusionary ethos. It is a concern for fellow human beings, not an innate or principled indifference or antipathy for nonhuman things.

When Richard Ryder coined the term 'speciesism' and declared that "[s]peciesism means hurting others because they are members of another species"[5] he was free to play with words—if he was only playing games. He wasn't. (And if he were, we needn't play along.) He meant his term to be both pejorative and explanatory. He meant to be referring to the beliefs and attitudes actually motivating the common practices he condemned—and he meant it to refer to beliefs and attitudes that are vicious and indefensible. Ryder was wrong, and his error was not verbal. Rhetorically his verbal invention has been a boffo success.

Human beings (e.g., intro ethics students) can rather easily get befuddled into thinking that they've been doing something very wrong, and that there's something wrong with themselves. Taking advantage of that vulnerability might be justified by some consequent greater good. Still, getting people to wrongly accuse themselves is a nasty business, best to be avoided.

It might sometimes be justified. Animal protectionists condemn speciesism as akin to racism. Animal protectionist rhetoric models itself on anti-racist rhetoric. We've all heard (and perhaps ourselves uttered) the principle that discounting people's interests just because of their skin color is unjust and unreasonable. That principle is as true and obvious and undeniable and thus rhetorically effective as it is irrelevant. It is irrelevant since racists have not been motivated by a moral absurdity about skin color. They have thought that (a) your belonging to another natural kind is a reason for not considering your interests the same, and (b) skin pigmentation is a prominent (albeit imperfect) indicia of human racial kind. Abolitionist and anti-racist rhetoric avoided attacking those two assumptions, for they are not obviously crazy like the ideas they foisted on racists. Anti-racist rhetoric has stayed far way from any suggestion that every biological kinship relation is in itself morally irrelevant. Ready acceptance of that premise is not to be expected from the great mass of mankind for whom the relationships of father, mother, brother, sister, son, and daughter are the paradigm and strongest moral bonds.

Anti-racist rhetoric was sophistical but served a worthy cause. It helped motivate some people to cease their oppressive discrimination by inducing the misconception that they had been operating on an embarrassingly stupid, thoroughly senseless moral principle. They had actually gone wrong in their thinking, not by believing an absurd principle about skin color, but in thinking that the difference in kind was of a kind that really did and should matter. That idea is not so readily dispelled. It has taken a few millennia for educated people to rightly regard it as objective fact that race has little biological reality (we can and repeatedly do successfully interbreed) and even if it had more (e.g., if lineage lines were less entangled), we have no good reason to let it matter in our public world. Demonstrably, throughout our shared public world, whites and blacks and Asians can be equal partners in cooperative activities—political, commercial, religious, etc.—and their racial history is no reason not to. We all have compelling reasons for enforcing a race-blind society. It is more mutually beneficial, harmonious, productive, efficient, just, and natural. Something is wrong with someone who can't get that. He's not just mistaken about some fact, and it may not be his fault, but there is something wrong with him as a human being.[6]

To have a role in political debate "speciesism" has to refer to some historically significant ideas or attitudes, so it can't be a belief that only a human being could be a proper object of our moral regard, or any idea like that. *Contra* Ryder, "speciesism" cannot refer to some principle or propensity to aim at harming things because of their not being human. On that definition speciesism is an aberrancy more anomalous than misanthropy. We meat-eaters generally have nothing against other animals. We hurt or subsidize the hurting of some animals because (e.g.) we relish the taste of their flesh. We needn't relish their gruesome slaughter. Actually, we mainly prefer not to think about that. Anyway, their being nonhuman is not what sets us going. It means only that we don't constrain our penchants as we would if their flesh were human. We may hate ants for their picnic intrusions and crows for attacking crops, but hurting cows because they aren't human is pathological, like helping Herefords because they are hooved. Accusing the Macdonald's crowd of being speciesists in Ryder's sense is demagogic bullshit, as nonsensical as it is, and nasty.

Subsequent animal protectionists softened Ryder's language and accused our culture, not of malevolent sadism, but only some callous principled indifference. Actually, in the scholarly literature where speech is held to higher standards of precision, "speciesism" gets used and explicitly characterized in all sorts of inequivalent terms.[7] Most commonly, critics include the concern for conspecifics I've described, but they add something more, some pernicious attitude toward animals. What's problematic then is which of our attitudes are truly pernicious, which are truly directed at animals and which are truly speciesist.

Nature seems to favor our promoting concern for humans and treating the nonhuman world as a means to our individual and collective ends. Natural selection often favors a species preying on another species for food. It does not generally favor species sadism: there's rarely advantage for a species or its members to go out of the way "hurting others because they are not members of [their] species."

Rather, nature would seem to favor our generally regarding the nonhuman things of our world with indifference except where and how it affects our interests. That's one tendency.

Nature also seems to have favored our being triggered to respond to conspecific distress by cues quite like those with comparable functions in other species. Like it or not, our normal healthy human sensibilities are prone to respond empathically to our sensing suffering and fully recognizing real pain in nonhumans as well as humans. We seem predisposed to condemn unmitigated cruelty and to suppose that something is wrong with human beings who sadistically delight in causing or observing another creature's suffering. Some capacity for cross-species sympathy appears in some other species. Perhaps our rationality and its essential capacities for abstraction, imagination and distantiation enable us—and make us liable—to experience a far more extensive, trans-species empathy. Evidently, among humans the propensity for such feelings varies widely.

Most often (more often than with humans) when we see a suffering animal and see that it is suffering, we don't see its suffering. We don't sense its pain or have any sense of its sensations. When we do sense it, we're liable to react with resonance and be moved to ameliorate the suffering. Some of us are stimulated to make this a matter of policy calling for more than merely a random rescue of a stray kitty drowning in a swimming pool; they take up the cause of animal protectionism. Many of us prefer to avoid those sensations of vicarious distress. We'd rather not be bothered. Still, to be honest with ourselves and keep our integrity intact, we may prefer not to fool ourselves or allow ourselves blissful ignorance of the significance of our lives in the world, including the grisly facts of our food production. But we prefer not to dwell on all that. We're not interested in devoting greater attention and time and effort to being more sensitive to that sector of reality. Our sympathies may move us if we let them. But we see no need for that, no compelling reason to reschedule our priorities and restructure the economy of our consciousness.

Suffering can call out our compassion, not our respect. Enduring suffering with courage and dignity is worthy of respect, but the suffering itself is not something estimable or admirable. It confers no authority to command us to take more interest in the sufferer's interests. Of course a victim's injuries and death are evils for the victim. They needn't be bad for anything we care about. We may not have any interest in the victim's well being, and see no reason to take more interest. The victim needn't be human for us to have reason to take an interest, but whether and how the bare fact of its having interests is reason enough for all of us to take an interest has never been convincingly explained. (That is partly because the explanations offered compete and cancel each other out.)

Indiscriminate sympathy is not our only natural inclination regarding animals. Evidently, humans are prone to respond with affection and concern for "loyal" or "friendly" animals, and feel tugs to return their friendship and loyalty. When they provide us some service, whether pulling a wagon, rescuing us from danger or providing comfort by their mere presence, a rush of gratitude may feel well-deserved and obligatory. The enforced dependency of pets understandably generates a sense of responsibility. And so on. Such responses to individual animals engender quasi-moral concerns, bonds, a sense of obligation. These relationships and responses are, however, as they are in inter-human affairs, individualized and personal, not generalizable reasons for concern for every animal of its kind.

Some human responses seem natural and appropriate regarding particular kinds of animals. We may be struck with awe at the sheer immensity of whales as we may with redwoods and mountains, and think them worth preserving for their awesomeness. With other animals it's their beauty, or intelligence or some other marvel whose value we recognize and whose wanton destruction seems a terrible waste.

None of these attitudes are inconsistent with a principled concern for human beings. Speciesism is not a principled opposition to caring for nonhuman things. There's nothing inhuman or unnatural in our being open to all kinds of concerns about nonhuman things, sentient and insensate, living and nonliving. Whether and to what extent we should concern ourselves with nonhuman interests (other than as a means to serving human interests) is not determined by our speciesism—at least not in any immediate or direct way.

Humanity's speciesism seems compatible with all but the most radical policy animal protectionists might advocate—and most every policy they oppose: it is logically consistent with all of them. Rabid animal liberationists can be, and doubtless some are, equally rabid fetal liberationists who condemn vivisection, factory farming, and the like for the pain the animals endure and condemn abortion because they consider embryos human beings. If humanity's speciesism is as natural and generally beneficial as appears, vegetarians are likely to be speciesists at near the rate the rest of us are. After all, their rationale for solicitude toward animals almost always applies to humans as well, yet rarely does the quality and quantity of anyone's moral concern for animals (other than pets and the like) match their concern for humans with lesser mental capacities. Their speciesism explains their distinctive concern for their conspecifics with greater elegance and less strain than any philosophical invention, utilitarian, contractarian, or otherwise. As things are, animal protectionists drive themselves into deep denial about their own speciesist traits, and insist on implausibly interpreting their attitudes as motivated by some suppositious calculation that leaves them unembarrassed. Such self-distortion serves no one.

Most people reject the animal and fetal liberationist agendas, not *per se* because beasts and pre-babies aren't human (enough), but rather because they aren't like morally responsible ET's either, and their well-being has no other substantial enough basis for a claim on our concern, or so people suppose. We sense no ties that bind, no bonds of allegiance. Many of us are moved, some ways, to some degree, by squarely confronting the reality of the suffering we cause or contribute to by one means or another. Yet while our feelings may move us to make some minor adjustments, does cool reason compel us to coerce other people to endure (what they regard as) hardships for the sake of animals, or to allow others to coerce them to that end?

RELEVANCE OF SPECIESISM TO LIFE SCIENCES PRACTICES 35

Perhaps some coercive measures may become clearly justified by our own individual or collective well being (e.g., due to the global resource inefficiencies of meat diets.) As things are, animal protectionists cannot yet rely solely on that kind of derivative, incidental concern to make a compelling case for most of their reform agenda. More disinterested motivations are needed.

It may be that protectionist reforms would be less onerous for humans than is often imagined. We might all benefit from the nutritional advantages of a more vegetarian diet. Perhaps progress in the life sciences and product development need not be significantly impeded by further restricting animal experimentation. Perhaps the massive occupational dislocation due to animal protectionist reforms would be manageable. Perhaps the loss of certain pleasures is bound to be decreasingly onerous: much of taste is merely habit. Perhaps in a world ruled by animal protectionists, many people might grumble about various deprivations, but after suitable accommodations, few people could truthfully complain of severe hardship.

All that is arguable and relevant, yet even if correct, those cost-benefit considerations aren't decisive. What may properly motivate compliance and eventual acceptance of a law needn't be enough reason to advocate and promote enactment of the law. People may properly protest imposition of a restriction when most of those to be bound by it don't want it and would reject it despite being well informed of its consequences. The fact (if it is a fact) that they would not resent the restriction if they had been raised under that regime does not warrant imposing it against their current well informed wishes. They may wish to remain as they are and may have no compelling reason to acquire a different constellation of concerns.

You may wish for the institution of the proposed animal protectionist reforms and sincerely advocate them without being energized to make significant sacrifices of time, effort, or assets. But could you cleanly hope for the reforms without caring about animals in something like the ways and degrees animal protectionists do? Animal protectionists frame their complaint as a moral indictment, that our harming animals is wronging them.[8] For speciesists the issue is whether we should care about animals in the ways and degrees that animal protectionists do. Some speciesists do care; others don't. To fairly earn the current majority's support of the protectionist agenda they must be shown not just that the costs would be bearable but that people now have some compelling reason to acquire the concerns so they'd welcome the reforms. For many of us that would mean becoming someone else.

Many people have no wish to care more about animals, or to care about them as a matter of principle. They prefer not to linger over the bleak life and brutal death of the cow whose flesh they are savoring. They are vulnerable to reflexes of distressed compassion, horror, and revulsion when an animal is writhing and wailing under their nose. They may acknowledge that their emotional life would be significantly blunted, blocked, or twisted if they couldn't be touched by an animal's agony before their own eyes and ears. Still, they may ask: what's the evidence that we are missing something by not being much moved by a bare report of animal suffering, or that we should prefer living a life in a world of principled caring for animals in something approximating the way we care about human beings? It's a given that (barring our bungling) the animals we care for would be better off. Our question, however, is whether we should care about that. Would our own lives be better? Is there something wrong with us not much caring about other animals beyond those we are personally involved with?

Most of us live in a human world structured by speciesism, an environment objectively and experientially focused on inter-human relationships. What kinds and degrees of concern for animals would work out to be compatible with or promote the flourishing of our species seems a canyon-wide open question. Again, we may have reason to agree on some particulars, for example, to condemn sadistic intentions and to prohibit the conduct. Perhaps a deficiency below some minimal susceptibility to compassion for animal suffering is objectively deplorable, and a life devoted to relieving animal suffering may be admirable. Still, there might be nothing wrong with having only some minimal concern. Perhaps sometimes there is: some human lives are emotionally enriched by greater attention to animals. But is it likely that humans generally would be better human beings by acquiring a more generous concern for animals? At this point in the evolution of human knowledge, we seem far from really knowing any such thing.

Again, there need not be anything wrong with someone who wants a world where humans are more mindful of animal well-being. I agree that our world and we humans in it would be better off if we cared more about our world's well-being. The well being of my own world, however, is a richer condition than a reduction of sheer somatic pain. Somatic pain is the cost of the biological benefits of an organism's injury signal system, hardly the worst thing in the world. What matters fundamentally is the injury suffered, not the "experience" of the suffering expressing the injury. As things are, what actually most stimulates many people (animal protectionists and others) is all the wailing and writhing, the behavioral expressions, the perceptible signals of the occurrence of an experience that signals somatic injury. Here our natural sentimentality and lazy love of simplicity may get the best of us unless reined in by reason. Our real concerns can be misdirected by fixing tightly on animal pains, the transient sensations, rather than the injuries or death sustained.[9]

So there may be some ways of going wrong about these matters. Yet much of it all seems undetermined and likely indeterminable. Consider: We do have reason to think something is wrong with racists who wish for a world where others felt as they do. But I'm not disposed to suppose that there must be something wrong with anyone dedicated to prohibiting abortion because she is gripped by a vivid sense of an embryo living a human life. I rather doubt that every fervent anti-abortionist is a defective human being, in some way a bad example of my kind. I see no prospect of real proof that she is flat out mistaken about some fact, empirical or moral or whatever. I know of no evidence that her horror of abortion must be abnormal, unhealthy, ill-informed, due to some corrupting prejudice, or anything of the sort. Would our kind be better off by our all sharing (or being devoid of) a sense of an embryo as living a human life? Here it may be helpful to consider whether there has been any advantage in humans acquiring some predisposition on this matter. With the pre-technological limitations of our interactions with prenatal life there seems little room for any selectional pressures here. (Presumably the world of rational marsupials would differ in this.)

We don't have reason to think something is wrong with anti-abortionists or pro-abortionists who wish for a world where everyone else felt as they do. The abortion clinic vigils of anti-abortionists may merit some respect and sympathy and not just grudging tolerance—if the protesters can reciprocally sympathize with the resentment and indignation they stir when their protest turns retributive, insensitive, vicious. It's not just their conduct but also their attitudes that get unreasonable and objectionable when they bethink themselves entitled to force themselves upon those who don't share their feelings.

Something like that may be true of animal protectionists. They may, quite understandably, feel that their deeply felt compassion is normative and those who lack it are somehow defective human beings. They may be right, but their literature hasn't yet identified and compellingly refuted any essential assumptions of the indifference and unconcern they condemn. In any case, all their railing against speciesism does nothing to advance their argument.

NOTES

1. Nature needn't be anthropomorphicized or deified for it to favor some things over others. A reason why something happens is a fact favoring its happening, a factor contributing to its happening. The facts of nature, including the fact of the operation of principles of natural selection, may favor the development and perpetuation of biological species with certain traits.

2. See my "Philosophy on Humanity" in R. L. Perkins, ed., *Abortion: Pro & Con* (Schenkman, 1974); reprinted in E. Manier, et al., *Abortion* (Notre Dame, 1977.)

3. See my "Understanding the Abortion Argument," in *Philosophy & Public Affairs*, 1971, Fall (I,1).

4. See my "Applying Ethical Theory: Caveats from a Case Study," in D. Rosenthal and F. Shehadi, eds., *Applied Ethics and Ethical Theory*, (Utah, 1988).

5. "Speciesism" in Rosalind Hursthouse, *Ethics, Humans and Other Animals*, (Routledge, 2000).

6. I know no reason to suspect that something must be wrong with someone who is not race blind in his/her private life (e.g., in his/her sexual preferences and mate selection) or more broadly in those matters not properly regulated by political coercion (e.g., the beneficiaries of one's charity.)

7. I document this elsewhere with ample quotations. See my 2000, "Understanding Speciesism," presented at the 32nd Conference on Value Inquiry and Lund University.

8. This formulation is analytically problematic and may be morally misdirected. A conception of harming an animal as wronging it seems to call for talk of animal rights, a metaphor

preferred by those devoted to making moral matters litigable. Talk of failure in one's duties to (and not just regarding) an animal is less worrisome but it needn't suggest that the animal is thereby wronged. Utilitarianism is challenged to capture the notion of being wronged as something more than being the biggest loser in the sea of an act's ill consequences.

9. Strict hedonists have a hard time making sense of a concern for animal death as the cessation of life and organic experience, and not itself another experience. We watch and feel an animal's agony and are moved by those feelings to protest. What we feel at a creature's death is not a somatic replication of writhing or cringing or gritting or the like. We experience sheer loss, blank terror, or horror or bewilderment: It's gone! That life is no longer there. It's the absence of the life that shocks and dismays, not the pleasures forgone.

BIBLIOGRAPHY

Ryder, Richard. 1975. "Speciesism," in 2000, Rosalind Hursthouse, *Ethics, Humans and Other Animals*, Routledge.

Wertheimer, Roger. 1971. "Understanding the Abortion Argument," in *Philosophy & Public Affairs*, Fall (I,1).

———. 1974. "Philosophy on Humanity," in *Abortion: Pro & Con*, ed. R. L. Perkins, Schenkman, reprinted in 1977, E. Manier, et. al., *Abortion*, Notre Dame.

———. 1988. "Applying Ethical Theory: Caveats from a Case Study," in *Applied Ethics and Ethical Theory*, ed. D. Rosenthal and F. Shehadi, Utah.

———. 2000. "Understanding Speciesism," presented at the 32nd Conference on Value Inquiry and Lund University.

ETHICS AND THE LIFE SCIENCES

VEGETARIANISM, TRADITIONAL MORALITY, AND MORAL CONSERVATISM

DAVID DETMER
PURDUE UNIVERSITY CALUMET

ABSTRACT: "Moral vegetarianism," the doctrine that it is immoral to eat meat, is widely dismissed as eccentric. But I argue that moral vegetarianism is thoroughly conservative—it follows directly from two basic moral principles that nearly everyone already accepts. One is that it is morally wrong to cause unnecessary pain. The other is that if it is wrong in one case to do *X*, then it will also be wrong to do so in another, unless the two cases differ in some morally relevant respect. Since everyone agrees that it is wrong to kill humans for food, this principle entails that defenders of meat eating must find some morally relevant difference between eating humans and eating other animals if they are to justify their practice. I argue that this burden cannot be met. Finally, I offer four arguments against the claim that the moral permissibility of eating meat is intuitively evident.

Traditional morality draws an extremely sharp distinction between human beings and all other animals: while there is a complete consensus that it would be morally wrong to hunt, kill, and eat any humans, or to raise them in the conditions found on modern factory farms for the purpose of subsequently butchering and eating them, such treatment of nonhuman animals is routine. For those who consider tradition and custom to constitute their own warrant, there is no need to inquire as to whether, and, if so, on what basis, this distinction is justified. But for the rest of us, mindful of the many atrocities that have at one time been practiced as traditional and customary, there is a need for justification: why do we consider it to be morally acceptable to do to other animals what we would never do to humans?

Many philosophers have responded to this question by noting some of the ways in which humans differ from other animals. These thinkers have often gone on to claim that human superiority in such areas as rationality, linguistic ability, and capacity for reflective thought fully justifies granting to humans a unique moral status, one to which no other member of the animal kingdom can legitimately aspire.[1]

But I would argue that this defense of our current beliefs and practices cannot succeed, since it is unable to overcome two powerful and fundamental objections. The first is simply that there appears to be no feature or set of features which distinguishes all humans from all nonhuman animals. For, notice that if one appeals to rationality, intellect, linguistic competence, and the like, any standards that would be rigorous enough to exclude all nonhuman animals would also have to exclude some humans—those who are severely retarded or brain-damaged, for example. Alternatively, standards lax enough to include all humans would have the undesired consequence of failing to exclude a good many of the nonhuman animals that we regularly butcher and eat.

As this objection has a long history, several attempts to meet it have been advanced in recent years. For example, it has been argued that the features characteristically possessed by members of the group to which an individual belongs are of greater moral relevance than are the features possessed by the individual in question,[2] or that we can supplement a high, individually-oriented, standard of inclusion into the realm of full moral status with an appeal to our natural sympathy for less fortunate members of our own species.[3]

I find these attempts implausible. With regard to the first, surely it is not true that features characteristically held within a group are of greater moral relevance than features actually held by the specific individuals in question. For example, while it is undoubtedly true that tall people, as a group, are better at basketball than short people, that does not justify giving a basketball-playing job to a tall but hopelessly inept player when a shorter but vastly more talented player is available. And with regard to the second attempt to escape this objection, do we really want to let our natural sympathies play such an important role in our ethics? We would never tolerate such a move in connection with justifying our differing treatment of individuals on the basis of their race, ethnicity, or sex. Why should we do so on the basis of their species-membership?

Let us turn, then, to the other basic objection to the strategy of trying to identify characteristics that distinguish all humans from all nonhuman animals in such a way as to justify granting full moral status to the former while denying it to the latter. The objection is that, when we turn to specific cases, the proposed feature or set of features which is supposed to justify a double standard in our treatment of humans and nonhuman animals is typically irrelevant to the treatment in question. Consider, for example, the claim that the special moral status of human beings derives from their unique linguistic abilities. The obvious reply is that such abilities are morally relevant only in some situations, and are of no apparent relevance in others. After all, while no one could reasonably complain about a decision to deny a cow admission to study at a university (since its linguistic incompetence would render it incapable of benefiting from such admission), it is far from clear what bearing the cow's inability to read or write could possibly have on the question of the moral permissibility of causing it to suffer unnecessary pain. Indeed, the utterance, "it's OK to hurt her—she can't talk," appears to be every bit as much a *non sequitur* as would "it's OK to hurt her—she's from Pittsburgh (or weighs two hundred pounds, or has fair skin)." Thus, even if we were to waive the first objection and concede, contrary to fact, that there are capacities shared by all humans but by no nonhuman animals, it still wouldn't automatically follow that these differences would justify any specific difference in treatment between people and animals. Still less would it directly justify splitting the two groups into two separate moral universes, with one receiving a long list of inalienable rights denied the other, including the right to torture, maim, and kill members of the other group, often for rather trivial reasons. Rather, in order to invoke a factual difference between humans and other animals as a basis for justifying a difference in how the two groups are to be treated, one would still have to show that the different feature or ability in question is genuinely relevant to the proposed difference in treatment in such a way as to render the double standard defensible. But when the difference in treatment concerns the infliction of physical pain, it would appear that Bentham had it right: "the question is not, Can they *reason*? Nor, Can they *talk*? But, Can they *suffer*?"[4]

It does not follow, however, that sentience is the characteristic to which we should always turn in attempting to determine which individuals are entitled to moral consideration, as if the main mistake of the approach I am now criticizing were merely that it had fastened on the wrong characteristic (linguistic competence or rationality, say, rather than the capacity to feel pain) as the one that is always decisive for determining how we might justly treat others. The point, rather, is that no single feature, or single cluster of features, is always relevant (let alone exhaustively relevant) in determining how a given individual should be treated in a given situation. What is relevant in one situation frequently is not relevant in another, so there is no one feature or set of features that distinguishes those worthy of moral consideration from those who are not in a global way, covering all situations. Musical ability, not knowledge of theoretical physics, is relevant when we are deciding which applicant to hire for the position of first violinist in a symphony orchestra. But when we are deciding which candidate to hire as a professor of physics, the identities of the relevant and irrelevant considerations are reversed. And if any of these applicants for jobs at the symphony orchestra or university physics department show up at the emergency room, any considerations concerning their specialized knowledge and ability fade away to complete irrelevance, to be replaced exclusively by considerations regarding the nature of their illnesses or injuries. Similarly, because Jack and Jill, unlike Fido, engage in moral reflection, and have a capacity for moral autonomy that Fido lacks, it is morally harder to justify interfering with their freedom to act as their conscience dictates than it is to restrict Fido's freedom to do as he pleases. But, as Tom Regan points out, "just because moral autonomy is morally relevant to the assessment and weighting of *some* interests, it does not follow that it is relevant to the assessment and weighting of *all* interests. And one interest to which it is not relevant is the interest in avoiding pain. Logically, to discount Fido's pain because Fido is not morally autonomous is fully analogous to discounting Jill's intelligence because she does not live in Syracuse."[5] What these examples show, in James Rachels's words, is that "before we can determine whether a difference between individuals is relevant to justifying a difference in treatment, we must know what sort of treatment is at issue," since "there is no *one* big difference between individuals that is relevant to justifying *all* differences in treatment."[6]

In response to arguments of this sort, some defenders of meat-eating have abandoned the fruitless search for features which (a) all humans possess, (b) no nonhuman animals possess, and (c) are relevant to justifying our use of animals as food. These philosophers have instead boldly asserted that it is intuitively evident that human beings should be granted a privileged moral status. They claim that our moral superiority to other animals, and consequent right to use them for food, is so deeply obvious as to render argument unnecessary. Or, at the very least, they insist that this widespread belief exhibits a greater degree of intuitive obviousness than do any of the proposed arguments attempting to overturn it. For example, Jan Narveson objects that Tom Regan's defense of animal rights relies on "intuitions taken from the very set of received beliefs from which he is so considerably dissenting." Thus, since Regan's conclusions (e.g., that animals have rights and that vegetarianism is morally obligatory) are "rather unintuitive," Narveson states that we are free to draw upon "another selection" of intuitions, which would support our current practices.[7] Or again, consider the words of C. A. J. Coady:

> It will be said that the idea that there is something specially morally important about human beings needs justification . . . but I'm not sure that it has to be [justified]. There are various ground floor considerations in ethics as in any other enterprise—for an animal liberationist such things as the "intrinsic good" of pleasure and the "intrinsic evil" of pain are usually ground floor. No further justification is given for them, or needed. It is not clear to me that membership in the human species does not function in a similarly fundamental way in ethics so that there is as much absurdity, if not more, in asking "Why does it matter morally that she is human?" as in asking "Why does it matter morally that she is in pain?"[8]

What can be said in reply to Narveson and Coady? I, for one, will not deny that "moral vegetarianism," the doctrine that vegetarianism is morally obligatory, relies on intuited fundamental moral principles. One of these is that it is morally wrong to cause unnecessary pain. Another is a principle of equality or consistency: the rightness or wrongness of doing *X* will be the same in different cases unless the two cases differ in some morally relevant respect. This principle is important because all parties to the debate over moral vegetarianism agree that it would be wrong to kill humans for food. Therefore, if the principle of equality or consistency is to be respected, defenders of meat eating must find some morally relevant difference between eating humans and eating other animals if they are to justify their practice.

But Narveson and Cody wish to turn the tables on the moral vegetarians by claiming that they can no more defend their moral first principles (those having to do with the wrongness of gratuitous cruelty and of arbitrary double standards) than the meat-eaters can defend theirs (having to do with the moral superiority of humans over other animals). If I read them correctly, it is not that they wish to deny that the vegetarians' first principles are generally sound, but rather their points are (a) that moral vegetarians cannot demonstrate the soundness of their basic moral principles, and thus they should be willing to cut some slack to the meat-eaters who similarly cannot justify their most basic intuitions, and (b) that the axioms concerning the moral superiority of humans over other animals and the moral legitimacy of eating nonhuman animals are at least as intuitively evident as are the two basic moral principles which seem to stand in the way of meat eating, so that it is arbitrary to insist on upholding one set of basic intuitions over another.

I agree with Narveson and Cody that moral vegetarians cannot provide proof for their most basic principles. That pain is generally a bad thing, that it hurts, is an elementary *datum* of experience, not something that we conclude on the basis of something even more fundamental. And the principle of consistency appears to be a prerequisite to logic or coherent thought, as opposed to something that might be derived from them. But Narveson and Coady claim that their principles, while admittedly incapable of being proved, are every bit as intuitively obvious. We appear to have arrived at a standoff. Is there a principled way to choose between these two competing sets of (allegedly) intuitive claims? Are they really equally obvious? Is there any non-arbitrary basis for choosing between them other than on non-moral grounds, such as those of self-interest?

I think there is. The principles to which moral vegetarians appeal are, in my judgment, plausible candidates for obvious moral truths. In that respect they contrast starkly with the principles cited by Narveson and Cody. Accordingly, I do not regard my upholding of the former set of claims over the latter to be the slightest bit arbitrary. By way of explanation, I offer four points.

First, notice that the basic principles underlying moral vegetarianism are general and simple, whereas those to which meat eaters appeal are specific and complex. While specificity and complexity are no barriers to truth, they are considerable obstacles to intuitive obviousness. It is difficult to assess a many-sided thing at a glance. Rather, one has to consider the ways in which the different parts are relevant to the question at issue, measure the degree to which the different parts point in one direction or another, and attempt to estimate where the weight of the relevant evidence lies. For example, in thinking about the issue of whether or not capital punishment is a morally acceptable practice, one must consider a great number of controversial subsidiary issues: Do murderers deserve to die (which raises further questions about the nature of freedom, responsibility, and desert)? Is the carrying out of executions, even if they are deserved, a proper function of the state (which raises all sorts of basic questions about what states are for, and what their limitations should be)? How reliable is the criminal justice system in convicting and executing only those who are genuinely guilty? Is the death penalty a more effective crime deterrent than is an alternative, such as a long prison sentence? How expensive is a capital punishment system, and are its alleged benefits worth its monetary cost? One could go on in this fashion at some length. Consequently, no matter what one thinks about the morality of capital punishment, it is highly implausible to claim that the truth about its moral permissibility or impermissibility could be intuitively evident. This is an issue that, as a direct result of its complexity, must be settled by examining lots of evidence and many arguments. The same, I would contend, holds for the question of the moral permissibility of eating meat. For, that question raises issues concerning (a) our moral status versus that of other animals, (b) our nutritional need, or lack thereof, for meat, (c) the degree of pain and hardship visited on animals in order to meet the human demand for meat, and many other issues as well. By contrast, the claim that pain is generally bad, and that it is wrong to inflict it on others unnecessarily or without a countervailing reason, is simple. It is not made up of a number of sides, some of which point in one direction and others in another. Consequently, it appears to be the kind of thing that cannot be known by weighing and considering arguments, but by direct insight, that is to say, by intuition.

Similar remarks apply when we consider the generality or specificity of the competing "intuitions." That pain is bad and double standards require justification are both perfectly general principles. That human beings have a unique moral status and that it is morally permissible for human beings to kill animals for food are much more specific. The problem with claiming intuitive obviousness for such specific moral claims is that the logic of moral justification seems to move from the specific to the general. Typically one shows that a particular action is wrong by showing that it offends against a general moral principle. It might then be necessary to defend that moral principle by appealing to an even more general one, and so on. But clearly this process cannot go on forever. The most general of our moral principles, if they can be known at all, will have to be known directly, for there will be no more general principles to support them. They will have to be self-evident or intuitive. On this basis, the claims that pain is bad and that we should be consistent in our moral judgments are plausible candidates as self-evident principles. But the claim that human beings have a special moral status is not, since it seems legitimate to ask for some theory of moral status, some defensible or intuitive principle, from which such a conclusion would follow. And the same holds for the claim that it is legitimate for human beings to kill other animals for food. Surely we can ask for some more general moral principle from which this follows. Or at least we should be able to make this demand when the specific claim in question clearly goes *against* moral principles that defenders of that specific claim otherwise uphold. For surely Narveson and Coady would agree, as a general principle, that one should not inflict pain on others unnecessarily, and that one should be consistent in one's moral reasoning and practice. (Indeed, Narveson seems to endorse the latter principle explicitly, as when he argues that since the first principles of moral vegetarians and meat eaters are equally intuitive, it is unreasonable to demand that the latter set of principles be sacrificed to the former; in other words, since they are relevantly alike, they should be treated alike). So, even if it is perhaps too much to ask that we be able to justify all of our specific moral judgments in terms of general moral principles, surely it is not too much to ask for such justification *when we wish to carve out exceptions, especially self-serving ones, to general principles that we otherwise and generally uphold.* Appeals to intuition, in such contexts, are utterly implausible.

My second point in favor of the first principles of moral vegetarianism, as opposed to those underlying the practice of eating meat, is phenomenological. When I examine my own experience it seems to me powerfully evident that happiness is intrinsically better than misery. But I have no comparably strong experientially-grounded sense that the most vital of animals' interests should be sacrificed for comparatively trivial human concerns. For one thing, it is unclear how the worth of animals' interests could be a *datum* of my experience. Such a matter seems, rather, to belong to that large class of issues about which one would have to reason and evaluate evidence and arguments. But perhaps it is relevant to point out that my experience does quite clearly testify that the badness of my own misery flows directly from its nature as misery, rather than from the fact that I am the one who is experiencing it. That the misery is mine explains why I am the one who is aware of this particular case of badness, but the misery appears clearly to be bad as such—something that would be bad for whoever experienced it, including, for example, a cow. Of course, my experience might be illusory, and so perhaps the truth is that it is not such a bad thing for a cow to be miserable. But then, an *argument* would be needed to establish the truth of this claim. Intuition, in the absence of argument, is insufficient, and, if anything, points in the opposite direction.

My third point concerns what I take to be the biggest single cause of faulty "intuiting"—the common and entirely understandable confusion of the genuinely self-evident with the merely familiar or the culturally given. Narveson shows himself to be aware of this problem, as he points out that "received moral views . . . don't come stamped with 'Genuine' and 'Culturally Biased' labels," and concludes, on that basis, that moral vegetarians should leave him alone to "continue to eat meat in good conscience" until they can explain why they favor one set of intuitions over another.[9] It seems to me that Narveson goes wrong here by resting content with his observation that cultural factors can distort our sense of what is and is not intuitively obvious. What he fails to take into account is that we might, while keeping this possibility vividly present to our minds, through thoughtful inquiry make some progress in distinguishing "genuine" intuitions from "culturally biased" counterfeits. The first step in such an inquiry might be to ask, quite simply, "could it be that my intuition in this case is really nothing more than an uncritical acceptance of what is culturally familiar to me?" With regard to my conviction that happiness is intrinsically better than misery, I think this question can safely be answered negatively. Indeed, I think this conviction would flourish, not only in the absence of cultural support, but even in the face of cultural opposition. I base this conclusion on the phenomenological observation that happiness and misery do not present themselves with the kind of plasticity or flexibility necessary for their evaluation to be strongly susceptible to cultural manipulation. But this cannot plausibly be said about our judgments concerning the relative weight to which the interests of animals should be given in our moral deliberations. To see this, consider the following thought experiment. How would we react if an anthropologist, upon returning from some isolated culture that had not previously been studied in the West, were to claim that the people of that culture prized misery as an end in itself, and shunned happiness, not because they thought that their happiness was undeserved, or would spoil them in some way, or would result in punishment by the Gods, or anything of that sort, but rather because they thought happiness to be a bad thing in itself? I think we would not believe the anthropologist. We would assume that the anthropologist was either lying or (more likely) mistaken, and we would require quite a bit of evidence before we would think otherwise, so powerful a *datum* of human experience is it that happiness is intrinsically better than misery. Conversely, it does not even mildly surprise us to read of cultures which assign to animals a radically different moral status, whether higher or lower, than that which we assign to them.

Finally, I would point out that while the opposed sets of fundamental principles that we have been considering are both widely held, it is possible only with regard to those underlying meat eating to explain the popularity of the principles by appealing to something other than their evidentness. I refer to the simple point that these principles are self-serving, benefiting the relatively powerful humans who rely on them, at the expense of the relatively powerless animals, who have no say in the matter and who suffer the consequences.

In conclusion, if my arguments are on target, our agreement that it would be wrong to slaughter and to eat human beings establishes that we should also refrain from treating other animals that way. For it turns out both that no set of morally relevant features can be found which would distinguish humans from other animals in such a way as to provide a justification for using one of these groups for food while scrupulously refraining from so using the other group, and that intuition cannot justify our current practices, since the principles which would justify them are clearly less intuitively evident than are those which generate the demand for a justification of meat-eating (e.g., the theses that the infliction of unnecessary pain is wrong and that double standards require justification). Moreover, such meager intuitive force as the former principles do possess can, unlike those, which oppose them, be explained away as the result of cultural bias, familiarity, and self-interest.

Moreover, even if one were to grant, just for the sake of argument, and in opposition to everything I've been saying, that humans, because of their superiority to animals in the areas of rationality, self-conscious, linguistic capacity, and the like, possess a kind of moral superiority to animals in all circumstances, and in connection with any area of possible conduct, it still would not follow that meat eating is morally defensible. For notice that even though most people do hold that humans are of vastly greater moral worth than other animals, they also tend to think that cruelty to animals, defined either as the deliberate infliction of pain on animals or the infliction of significant pain on them for trivial or ignoble reasons, is morally wrong. That they do not draw the conclusion that meat-eating is wrong probably stems from either or both of two mistaken factual beliefs: that meat-animals lead relatively happy lives prior to being slaughtered (so that significant pain is not inflicted on them) and that a healthy diet requires the consumption of meat (so that such pain as is inflicted on animals is not done for a trivial reason). But if, as is typically (though not always) the case, cows, pigs, and chickens, are forced to lead miserable lives in order merely to satisfy our taste for their flesh (rather than to sustain us in good health), how can this be justified, even on the premise (granted only for the sake of argument) that animals' interests count for significantly less than do our own?

Finally, notice that if intelligence and higher cognitive function are as morally important as we are presently assuming them to be, it is far from clear how this would help the defense of meat-eating. For, while humans are admittedly somewhat higher up the scale of intelligence than are cows and pigs, their margin of superiority is dwarfed by the huge chasm separating cows and pigs from plants of all kinds. So if intelligence has great moral value, we must by all means, if given a choice between them, kill and consume beings that are utterly lacking in intelligence (or even sentience) instead of those that can think (and feel). That's an argument for vegetarianism.

In spite of such arguments as these, moral vegetarianism is decidedly unpopular. Indeed, the claim that it is immoral, at least for relatively wealthy North Americans and Europeans, to eat meat, is still widely regarded as eccentric, at best, or dangerously radical, at worst.

From a historical and sociological standpoint, such an attitude is entirely understandable. For many people, the mere fact that a practice is common, customary, and traditional, is sufficient to justify it, and to reveal all those who would overturn it to be dangerous, unreasonable radicals. But I have attempted to show that, while moral vegetarianism is indeed radical in that it calls for sweeping changes in our eating practices, it is in fact thoroughly conservative from the standpoint of both logic and moral principle—it follows directly and logically from basic moral principles that nearly everyone already accepts.[10]

NOTES

1. See, for example, the selections by Aquinas, Descartes, and Kant in Tom Regan and Peter Singer, eds., *Animal Rights and Human Obligations*, 2nd ed. (Englewood Cliffs, N.J.: Prentice Hall, 1989).

2. One who argues this way is Carl Cohen, in his "The Case for the Use of Animals in Biomedical Research," in Robert M. Baird and Stuart E. Rosenbaum, eds., *Animal Experimentation: The Moral Issues* (Buffalo, N.Y.: Prometheus, 1991), pp. 103–114.

3. Specimens of this argument can be found in Mary Anne Warren's "Difficulties with the Strong Animal Rights Position," in Baird and Rosenbaum, pp. 89–99, and in Bonnie Steinbock's "Speciesism and the Idea of Equality," in Jeffrey Olen and Vincent Barry, eds., *Applying Ethics*, 3rd ed. (Belmont, Calif.: Wadsworth, 1989), pp. 412–418. Note also the following irony in the history of the debate over the moral status of animals. Whereas it was common, as recently as thirty years ago, for those who call for radical changes in our treatment of animals to be derided as irrational sentimentalists, now it is common for defenders of the status quo to admit that logic, scrupulously and dispassionately adhered to, does indeed favor such changes. But these contemporary advocates for meat eating go on to chastise the friends of animals for being too ruthlessly logical, and for failing to attach sufficient weight to our nature as emotional and sentimental beings, who should not be asked to refrain from acting on behalf of the special feeling we have for our own kind!

4. Jeremy Bentham, from *The Principles of Morals and Legislation*, chapter 17, section 1, in Regan and Singer, p. 26.

5. Tom Regan, "Reply to Carl Cohen," in Carl Cohen and Tom Regan, *The Animal Rights Debate* (Lanham, Md.: Rowman & Littlefield, 2001), p. 296.

6. James Rachels, *Created From Animals* (New York: Oxford University Press, 1990), pp. 177–178.

7. Jan Narveson, "On a Case for Animal Rights," *The Monist*, Vol. 70, No. 1 (January 1987), p. 48.

8. C. A. J. Coady, "Defending Human Chauvinism," in Sylvan Barnet and Hugo Bedau, eds., *Current Issues and Enduring Questions*, 2nd ed. (Boston: Bedford Books, 1990), p. 265.

9. Narveson, pp. 47–48.

10. And this is so even if we ignore, as I have done in order to keep this presentation to a manageable length, arguments for moral vegetarianism that are based on considerations of human hunger, of the environment, and of human health.

ETHICS AND THE LIFE SCIENCES

A RATIONAL DEFENSE OF ANIMAL EXPERIMENTATION

NATHAN NOBIS
MOREHOUSE COLLEGE

ABSTRACT: Many people involved in the life sciences and related fields and industries routinely cause mice, rats, dogs, cats, primates and other non-human animals to experience pain, suffering, and an early death, harming these animals greatly and not for their own benefit. Harms, however, require moral justification, reasons that pass critical scrutiny. Animal experimenters and dissectors might suspect that strong moral justification has been given for this kind of treatment of animals. I survey some recent attempts to provide such a justification and show that they do not succeed: they provide no rational defense of animal experimentation and related activities. Thus, the need for a rational defense of animal experimentation remains.

Each *hour* of each *day* in the United States, many people involved in the life sciences and related fields kill approximately three to six thousand mice, rats, dogs, cats, primates, and other non-human animals.[1] These animals are killed for education and training,[2] product safety testing and medical and psychological experimentation, among other uses. These animals typically experience at least some significant pain and suffering in the course of such experimentation and procurement. In the unlikely event that they do not, they are still killed. And this killing is a harm: their early deaths are *bad* for them even if they are killed painlessly: they are deprived of experiencing whatever goods they would have experienced, had they lived. A recent scientific review even suggests that just *being* in a lab is harmful for animals: the data suggests "significant fear, stress, and possibly distress are predictable consequences of routine laboratory procedures."[3]

Thus, lab life and death is bad for animals: it harms them, and harms them greatly in that everything is taken from them and they gain nothing in the process. Harms, however, require moral justification, *good reasons* for why they are morally permissible. Animal experimentation thus requires moral justification. Most animal experimenters agree. We might suspect that they tend to think that a strong justification has been articulated: after all, we tend to think that scientists think that we should have good reasons for what we believe and good reasons for what we do. So we might suspect that they believe there are such good reasons that explain why animal experimentation is morally justified.

In this paper I survey some recent attempts by both philosophers and scientists to provide such moral justification, i.e., the good reasons that explain why such harms are permissible to inflict. I argue that these defenses do not succeed: they do not show why animal experimentation is morally permissible. Moral and intellectual integrity thereby requires, at least, the development of such a defense. My hope here is to encourage advocates of animal experimentation to accept this challenge and provide some insight into how they might better respond to it.

1. EMPIRICAL MOTIVATIONS, MORAL RESPONSES

Before I survey the recent discussion, let me briefly provide some empirical information about what happens to animals and then sketch a common kind of reasoning given in favor of the conclusion that it is morally wrong to treat animals in these harmful ways.

Experimental procedures that animals routinely endure include drowning, suffocating, starving, and burning; blinding animals and destroying their ability to hear; damaging their brains, severing their limbs, crushing their organs; inducing heart attacks, ulcers, paralysis, seizures; forcing them to inhale tobacco smoke, drink alcohol, and ingest various chemicals, poisons and drugs, such as heroine and cocaine.[4]

For some readers, this list of procedures may seem a bit "pale" as mere words often fail to convey enough of the relevant information. For those for whom this is true, I encourage them to view some of the readily available photos and video documentary footage that documents this treatment; many are now available online. A picture often does speak a thousand words in the sense that more information is conveyed that way. Nobody can responsibly discuss these issues unless they have seen such footage, and a lot of it, including the small amounts of footage made available by industry groups: if they haven't, they are simply missing essentially relevant information.[5]

Mere descriptions of these actions do not entail a moral evaluation, however. Are these actions of harming animals morally permissible, or are they wrong? To answer this we need to do some philosophy. A helpful methodology in philosophy is to start with clear cases and see if the insights gained in understanding them can help us understand a not-so-clear case. This methodology is useful here. To discern the morality of treating animals these ways, we might first ask whether it would be wrong to treat *us* these ways and, if so, *why*. Our most carefully reasoned and defensible responses about *us*, i.e., our best hypothesis about which properties we possess that *makes* it such that it is (or would be) wrong to treat us these ways, *might* have implications for some animals.

A RATIONAL DEFENSE OF ANIMAL EXPERIMENTATION

But who is "us"? Who is the "we" or "ourselves" here that we might first think about? This is an important question and the answer is not obvious since there are so many ways "we" can be grouped. Historically, for example, "we" often only included members of our own race, or ethnic group, or religion, or sex. But if one considered "us" to be "conscious, sentient beings," then many animals *are* like us, certainly the ones mentioned above, mammals, birds, and other vertebrates. So if it's wrong to treat "us" in those ways, then it's wrong to treat many animals those ways also.

If one thinks "us" is "humans," problems arise, due to ambiguity in the term "human": is the suggestion that *anything that is biologically human* is wrong to treat those ways? If that's the suggestion (and the method of treatment here we might describe as "destruction") that would seem to imply that it's seriously wrong to destroy (living) cells, tissues, and organs that are *biologically* human, and that all abortions—even very early ones—are seriously wrong.

At least some of these implications force us to be more precise in who we are initially referring to when we think about whether it would be wrong to treat "us" these ways. Let us begin by considering those who are hearing or reading this paper: I presume that nobody reading this paper would want to be treated the ways animals are treated in labs. I suppose that you think it would be *wrong* for one of your colleagues to treat you the way animals are treated, and you think that's true *even if* you were killed so called-"humanely," i.e., painlessly.[6]

Let's suppose that's true, but let's ask *why* it's true: what *makes* it true? Philosophy is similar to science at least in that they both involve hypotheses or explanations. So what *best explains* the fact that it would be wrong to treat you these ways? If the explanation is along the lines of "I am rational" or "intelligent' or "autonomous," that would suggest a theory about the basis of what I'll (loosely) call "moral rights" to the effect that a being has moral rights—especially the right to be treated with respect and not harmed, against its will, for the benefit of others—*only if* that being is rational, or autonomous, or meets some other highly intellectualized condition.

But then we need to think again about who "we" are. Most of us count (some) humans who are not rational, intelligent or autonomous among "us." Severely mentally challenged individuals, the senile and seriously demented, newborn babies, and even babies that—due to some damage—lack the potential to have sophisticated mental lives; they are all considered to be among the morally significant "us."

If they are, however, then the "bar" for basic moral rights must be set rather low. It cannot be set at "being biologically human," however, since, again, human cells and organs alone don't meet it: they have no rights. The most plausible place seems to be at consciousness, the ability to feel pleasure and pain, and having a perspective on the world that can go better and worse from one's own point of view. It's these psychological features that put us in "the moral ballgame," so to speak.

But since this is the case, since many non-human animals are like that, many animals are in the ballgame as much as *comparably-minded* humans are. Since like cases should be treated as like cases, unless there is a morally relevant difference that would justify failing to give animals' interests equal consideration to *comparable* humans' interests, it follows that—since it would be wrong to treat any humans these ways—it is wrong to treat comparably minded animals these ways also. What is morally relevant, in itself, is not the species the individual is a member of, but rather the mental life of the individual; comparable mental lives deserve equal respect and equal consideration and thus, nearly all animal experimentation is wrong, since such experimentation on comparably-minded beings that are biologically human is also wrong.

This kind of reasoning has been developed and defended by *many* philosophers, from a wide range of theoretical perspectives: utilitarianism and other consequentialisms, rights-based deontology, ideal-contractarianisms and golden-rule ethics, virtue ethics, common-sense morality, religious moralities, feminist ethics, among others.[7] As a matter of fact, not many professional *philosophers* have criticized it: most criticisms come from public relations organizations that are well-paid to paint advocates of such reasoning in a bad light. But that is sophistry, not science or good reasoning, and it remains to be explained why so few philosophers—people trained in formal logic and the identification and evaluation of arguments for (and against) moral conclusions—have disagreed with this reasoning. One explanation is this: serious faults have not been found since the reasoning is sound.

Here, however, I wish to consider some recent objections to this kind of reasoning and show how these objections are unsuccessful. I hope to encourage people to take these issues seriously and engage the debate in a more intellectually serious and responsible manner.

2. SO WHY DOES ANIMAL EXPERIMENTATION MATTER?

The first work I wish to discuss is a collection from 2001 entitled *Why Animal Experimentation Matters: The Use of Animals in Medical Research*. The authors of these eight essays attempt to defend animal research on both moral and scientific grounds. The book flap says that its authors "mount a vigorous and long-overdue defense of animal experimentation," show "that the case for animal rights—in both its philosophical and activist guises—is deeply flawed" and provide a "much-needed corrective to an extremist cause that has up until now been too rarely challenged." But advocates of animal experimentation should find the book a serious disappointment. I will explain why; perhaps these explanations will lead to stronger work on these issues, work that avoids the flaws of this book.

Philosopher R. G. Frey's essay, "Justifying Animal Experimentation: The Starting Point," should have been at the start of the book. However, his contribution is hidden as the last chapter, perhaps because it is the strongest contribution, philosophically, in that it accepts much of the basic reasoning given in defense of animals above. Frey notes that most supporters of vivisection attempt to justify it by appealing to its benefits for humans. But, he argues, this defense is subject to serious objections, suggested above. He notes:

A RATIONAL DEFENSE OF ANIMAL EXPERIMENTATION

> Whatever benefits animal experimentation is thought to hold in store for us, those very same benefits could be obtained through experimenting on humans instead of animals. Indeed, given that problems exist because scientists must extrapolate from animal models to humans, one might think there are good scientific reasons for preferring human subjects.[8]

Thus, Frey sees what far too few experimenters seem unable to see, namely, that the premises of their defenses of animal experimentation obviously and straightforwardly imply that experimentation on vulnerable humans is permissible also. Frey thus sets the moral challenge for the other authors: to explain why, morally, no humans can be subject to the kinds of experiments that animals are subject to, and to do this by identifying the morally relevant properties that animals lack and humans have. He also raises the scientific challenge to explain how researchers can reliably use animal models to understand and cure human disease. He thinks that the first challenge has not been met; the second important *scientific* challenge was, unfortunately, not directly addressed in this book. Important scientific details on *exactly* how "animal models" are *reliably* used to gain information applicable to humans, by reliably predicting humans' responses, are unfortunately not provided.

In responding to philosophers like Frey, Scientist Adrian Morrison states that he "abhors" Frey's position and those of other philosophers who accept the kind of reasoning developed in favor of animals above. He asserts that all "human beings stand apart in a moral sense from all other species" and that all are worthy of "special consideration."[9] Such a view *might* be true, but regrettably Morrison does not defend his claims by identifying the morally-relevant characteristics that *all* humans (even those with less intelligence, sentience, and autonomy than animals) possess and all animals lack that might make his claim true. That omission prevents him from *rationally* criticizing opposing views; he does not provide any plausible reason to think that what animal advocates say is false or their arguments unsound: he merely states his "opinion," in the worst sense of the term.

To defend animal experimentation, Morrison appeals to "self-preservation," he explains (quite, oddly), "in the larger sense, of helping the weak and the helpless from those who consider themselves competent to decide the fate of others."[10] Of course, "self preservation" might justify experimenting on humans: he never explains why that would be wrong, i.e., explain what it is *about* such humans that makes it wrong, what their morally-relevant properties are. But animal advocates partially agree with Morrison as they hold that *all* who are "weak and helpless" should be protected from those who, like Morrison, deem themselves competent to decide their fate. Morrison divines that animal experimenters have "God's blessing"; one wonders how he would respond to the many theologians who argue otherwise.

Elsewhere in the book, appeals to evolution are made to attempt to justify a moral view that is poorly disguised as a "biological perspective." In another essay, biologists Charles Nicholl and Sharon Russell state (falsely) that, "Evolution has endowed us with a need to know as much as we can."[11] This is false because it is simply not true that *if we evolved, then we need to know as much as we can*.

(Interestingly, one never hears attempts to motivate students to do their schoolwork by proclaiming that evolutionary theory implies that they must learn their lessons; furthermore, there is no section on the 'moral implications of evolution' to be found in science textbooks: this is because evolution neither suggests nor supports any particular moral principles.) So appeals to this false imperative to defend the propriety of learning things by harming animals are a mistake.

Morrison also appeals to evolution. He claims that, "to refrain from exploring nature in every possible way would be an arrogant rejection of evolutionary forces."[12] On his view is it therefore "arrogant" to *not* perform painful and lethal experiments on humans since that is a *possible* way to explore nature and satisfy our alleged "need to know"? Obviously not, and, in fact, contrary to Morrison's suggestion, we are often morally required to resist "evolutionary forces." And the way things *are* or *have been* never entails how things *ought* to be: no evolutionary facts ever, *in themselves*, justify any moral views. So appeals to evolution do not settle any moral questions about animal experimentation.

Nicholl and Russell, however, think moral arguments are "beside the point in terms of providing justification for our exploitation of animals" since, they claim, implausibly, this is necessary for our species' survival: they think, implausibly, that if there were no animal experimentation, humans would likely all go extinct.[13] They also claim that, "it is an evolutionary necessity to regard one's own kind as more important than other species," but, again, this is false: if anything is an "evolutionary necessity," it is that *one's own* genes get passed down. Nicholl and Russell again are unable to see that no moral imperative follows from that and no moral constraint against using humans follows either.

They cry "double standard"[14] when animal advocates criticize human cruelty to animals but are silent on animal's eating other animals. In effect, they try to take moral advice from animals, claiming that *since animals act some way, it's perfectly okay for humans to act that way also*. Thankfully, they do not encourage humans imitating any animals by, e.g., eating our offspring or excrement. So their arguments rely on a principle that even they realize is false. They also claim that since animals cannot understand the concept of rights or claim that they have rights, animals cannot have rights, but they forget that some humans can do neither. They somehow don't realize, as Frey pointed out, that their argument against animal rights implies that these kinds of humans lack rights also.

Nicholl and Russell make sociological observations about animal advocates and conclude that they are "adaptively unfit" because they tend to have fewer children than average,[15] but this is simply irrelevant to the moral issues. Are the pro-animal philosopher's arguments unsound *because* animal advocates have fewer children? On their moral theory, unnecessary animal suffering should be avoided, but just sentences before this pronouncement they claim that theorizing about duties to animals is a "pointless enterprise."[16] Thus, their and Morrison's attempt at doing moral philosophy is a disaster. Stronger and more careful criticisms of philosophers who argue in defense of animals are found elsewhere, but not in this book. That's unfortunate, given that the book flap promised that the book would show cases for animal rights to be flawed.

Scientist Jerrold Tannenbaum worries about what might happen if animal experimenters were required to ensure animals not only "freedom from unnecessary or unjustifiable pain or distress, but to well-being, pleasure, and even happy lives."[17] This is especially worrisome, according to him, because calls for this moral consideration come from within the research community itself. He worries that more scientists will see animals as "friends," not "research tools," and then animal research will stop. These worries, of course, provide no reason to think that animal experimentation is morally permissible: these worries are founded on that assumption, but do not defend it.

Scientist Stuart Zola notes that the distinction between "basic" and "applied" animal research is not clear. He expresses worries about restrictions on projects "devoted simply to increasing knowledge" that *might* have "serendipitous" results.[18] However, he provides no calculations of serendipitous results that *might* result from non-animal research, so we are given no basis on which methods of research should be supported. Again, however, these concerns do not, in themselves, answer the question of whether animal experimentation is morally permissible, even from perspectives where the *only* moral concern is optimizing human health.

Philosopher Baruch Brody suggests that there are special obligations between humans (e.g., parents to children), and so there might be special obligations to humans that require discounting comparable animal interests.[19] This analogy is flawed, however: no special obligations to our friends or family allow us to discount strangers' and even enemies' interests so much that, to try to benefit ourselves, we deliberately inflict pain, suffering, and death on them and treat them as animals in laboratories are treated. Thus, the analogy is poor. He criticizes impartial moral thinking but, fairly, notes that his partialist approach requires further reflection. To avoid begging the question, i.e., merely *assuming* that such harmful treatment of animals is permissible, it is clear that much more reflection and argument is needed.

One dissenter, philosopher H. Tristam Engelhardt, defends animal "rights," of sorts, in a chapter surprisingly entitled "Animals: Their Right to Be Used." These "rights," however, include "the right to be skinned" and "transformed into fur coats [and] trimmings on hats," used in bullfights and cockfights, and "used to produce knowledge of interest to humans, even if it will not have any practical application."[20] Animals even have a "special right to be the object of the culinary arts of Chinese and French chefs." Furthermore, he claims, "*pace* Singer" (actually, Henry Spira[21]) that it *is* "appropriate to blind rabbits for beauty's sake." Sometimes he advocates human enjoyment as a criterion for the rightness of causing animal suffering, other times it is necessity or usefulness. His moral principles are logically inconsistent; given his tone, readers might think he isn't even addressing the issue seriously.

There is too little discussion of the scientific issues. Remarks are scattered and, typically, underdeveloped. An important series of articles and books, which some authors surely were aware of, is ignored.[22] One chapter is primarily a historical presentation of various human benefits allegedly gained from animal experimentation, but as one biologist has argued:

> Wind- and steam-powered vessels were certainly vital in the exploration of much of the globe, but this fact in no way indicates that they should continue to be seen as useful even as newer and more efficient technologies develop. In this regard, scientific arguments for and against the use of vivisection are best addressed in the context of modern medical research.[23]

Historical discussion, therefore, may have little to do with contemporary scientific debates. And those who advance modern medicine through clinical and *in vitro* research, computer and mathematical modeling, epidemiology and other methods will be shocked by Morrison's false claim that "medicine cannot progress without animal experimentation."[24] Those who make medical progress by simply providing people, especially needy people, with existing medical treatments will be even more astounded.

To better evaluate this book's defense of animal experimentation, readers should carefully identify the scientific objections to trying to use animals to understand and cure human disease and the case for non-animal-based research methods and see if this book provides an adequate response and an independent, positive case for animal use.[25] The book's value might consist in spurring others to articulate stronger reasons why vivisection matters and is morally justified, despite its high costs for animals and, perhaps, humans as well. There are costs for humans *if* there are better things that could be done for (more) humans that would yield greater benefits and funds, effort and talent could be diverted from animal experimentation toward meeting those goals.

Thus, while it is admirable that these authors at least attempted to give reasons why animal experimentation is morally permissible, if not morally obligatory, we see that their arguments are very weak. Those who seek a rational defense of animal experimentation need to look elsewhere.

3. PUTTING HUMANS FIRST?

A more recent attempt at providing such a defense is found in Tibor Machan's *Putting Humans First: Why We Are Nature's Favorite*.[26] Although Machan is a philosopher, his attempt to defend the status quo regarding animal use fares no better than the authors' attempts above.

Arguments from *moral rights* (in a strict, philosophers' sense of the term) provide one basis for arguments against uses of animals that harm animals. However, they are not the only basis, so it's a mistake to think that animal experimentation is permissible if animals have no moral rights, strictly speaking. There can be other, non-rights-based reasons why it is impermissible.

However, in his first chapter, "A Case for Animal Rights?," Machan argues that animals have no moral rights: he argues that the claim that some animals possess moral rights is "a fiction" and "a trick." This is because, on his view, a

A RATIONAL DEFENSE OF ANIMAL EXPERIMENTATION

being has moral rights, rights that presumably *make* it wrong to harm that being for pleasure or even serious benefits, only if that being has a "moral nature," i.e., a "capacity" to see the difference between right and wrong and choose accordingly (pp. xv, 10). Machan says humans are the "kind" of beings that have this capacity, that animals are not of this "kind," and this is why humans have moral rights and animals have none.

But the premises of this argument are imprecise: true, *only* humans have this capacity, but only *some* humans, not *all*. Many humans lack the "capacity" to see the difference between right and wrong and choose accordingly. Machan's theory of rights therefore seems to provide no protection for vulnerable humans who are not moral agents and so lack the moral nature he describes. They lack what he claims is logically necessary to have any moral rights, and so they have no rights on his view. It would, on Machan's view, apparently be morally permissible to perform painful, lethal experiments on them, especially if they promise benefits for those with the "moral nature" he describes.

But Machan disagrees with this implication and with this objection to his theory of moral rights. He argues that, contrary to appearances, human babies and severely mentally challenged individuals do not "lack moral agency altogether" (p. 16) and so they have rights on his theory. This is puzzling, because it is not clear what it would be to "lack moral agency altogether." Is this to say that babies can "sort of" or "sometimes" make moral decisions? If this is the concept, then babies do indeed lack moral agency altogether.

In response, Machan claims that to see why such humans are moral agents of sorts, we must consider them as they would exist "normally, not abnormally" and focus on the "healthy cases, not the special or exceptional ones" (p. 16; cf. pp. 38, 40). Apparently, Machan thinks that since "normal" human beings are moral agents, abnormal humans are moral agents as well. But this inference is clearly illegitimate: while exceptional humans' characteristics include some properties they share with normal humans (e.g., being biologically human), it is not true that, *in general*, all features of normal beings are shared by abnormal beings. For example, quadriplegics and cancer patients are in their unfortunate conditions even though normal, healthy humans—whom they share much with—are not: it is obviously unsound to reason that *since normal humans have four limbs, all humans have four limbs*, but this claim is parallel to Machan's defense of why humans who are not moral agents still have moral rights.

So, in the absence of arguments to the contrary, the fact that normal humans are moral agents does not *make* abnormal humans moral agents. Since it's just not true that *any property "normal" humans have, "abnormal" humans will have also*, this reasoning seems entirely *ad hoc*, deemed valid for this occasion but not others. Thus, vulnerable humans do not meet Machan's necessary condition for rights; his defense of their rights fails and thereby so does his argument that animals have no moral rights. Since many advocates of animal experimentation accept an argument like Machan's, this error is useful to observe and learn from.

Machan's other main argument against animal rights is surprising. He claims that if animals have the right to not be harmed at the hands of moral agents, then they also have that right against "politically incorrect" animals who, as he repeatedly observes throughout the book, are *not* moral agents (p. 12). He argues that since they don't have that latter right (i.e., animals don't have rights against other animals), they don't have the former right (i.e., they have no rights against us). Basically he suggests that—when it suits our pleasure—it is morally permissible for us to act like some animals and kill other animals. Thankfully (like Nicholl and Russell, above) Machan also does not endorse our imitating some animals by our eating our offspring (or our excrement), but since chickens, pigs, cows, rats, mice, and most primates are primarily vegetarian, they would surely welcome our imitation in that regard.

In the final chapter, "Putting Humans First," he objects to those who argue that, in our relations with animals, humans often should *not* be put first. Rather, on their view, *animals'* interests in avoiding harms like pain, suffering, and death should come before our interests in eating, wearing, and experimenting on them. Machan responds that, "Humans are more important, even better, than animals, and we deserve the benefits that exploiting animals can provide" (p. 116). He calls the kind of altruism that would deny this "insidious" and "perverse" (p. 118). These are interesting *assertions*, but unfortunately, strong arguments were not given to justify this sense of radical human superiority.

Since most philosophical work on ethics and animals is generally pro-animal (and there are psychological if not philosophical explanations why this is so), it is exciting when any anti-animal work is published. Machan's case against animals and their philosophical advocates, however, is a philosophical disappointment. Unanswered, merely rhetorical questions (which could be answered) too often take the place of arguments and, when given, arguments are not carefully and precisely developed or defended, as we have seen.

The deepest problem with Machan's case, however, might be that his position on the use of animals is unclear and, it seems, ambivalent. On the one hand, he recognizes that most uses of animals are merely for sport (i.e., entertainment, presumably including *culinary* entertainment) and convenience, not necessity (p. 19). And he says, even though he thinks animals don't have—strictly speaking—moral rights, it is still "quite likely" wrong to use them for "certain nonvital purposes" (p. 21). These insights *seem* to justify the conclusion that the vast majority of uses of animals—clearly in the food, fashion and product "testing" industries—are quite likely wrong, especially in light of the direct and indirect harms for human health and indirect harms through environmental contamination. This response is similar to the response that fits those who claim that causing "needless" or "unnecessary" animals harms are wrong: since most of these harms are indeed needless, as we can live healthy lives without these various products, even this "animal welfarist"-type perspective provides strong condemnation to many actual uses of animals. Machan does claim, however, that developing some "human potentials" *may* justify inflicting suffering on animals, as might other "rational purposes" (pp. 20, 118). What sort of purposes and potentials might justify such harms? We aren't told, so Machan's moral verdicts seem rather arbitrary and *ad hoc*.

Thus, while Machan's book is one of the few book-length defenses of harming animals, it is very weak. Again, those who seek a rational defense of animal experimentation need to look elsewhere.

4. CONCLUSIONS: UTILITARIANISM AND ANIMAL USE

Before concluding, let me make some general remarks about utilitarianism and arguments from it in favor of animal use.

While few advocates of animal experimentation actually accept utilitarianism, i.e., indeed, they believe utilitarianism is *false*, they often appeal to it to try to justify their actions. However, no one has done anything close to the conceptual and empirical work that would be needed to make a *serious* attempt at justifying any animal research on utilitarian grounds: no advocate of vivisection has provided any method for calculating and comparing (actual) animal harms to (merely possible) human benefits, calculated direct human harms that are the consequence of vivisection, calculated indirect harms and opportunity costs that result from funds being directed towards vivisection and not towards producing other benefits (and utilitarianism has no bias for medical benefits), and somehow added it all up to reasonably conclude that the calculation favors using animals.

Even if *some* benefits were lost and some or all vivisection stopped, that's not enough to justify it on utilitarian grounds, since one has to show that there is no alternative course of action that would yield greater benefits. Nobody has seriously tried to show that some specified amount of vivisection is (likely) indispensable for bringing about the greatest possible overall medical benefits. Nobody has argued that, despite *all* the other research methods available (and, more generally, methods of improving people's health, most of which are just the implementation and distribution of existing medical knowledge anyway, not new research), no other possible use of funds, time, and talent could (or likely would) bring about a greater improvement in health for humans than animal research. *Perhaps* these cases can be made, but until this is done, the most reasonable attitude might be a skeptical one.

In conclusion, I have discussed some recent attempts to defend the status quo regarding animal use, especially in scientific research. There are more attempts that I have not discussed here—in particular, Carl Cohen's (which I have discussed elsewhere and have argued amusingly misfires because his strategy implies that animals actually have rights and humans have none—since they can be grouped into the same and different "kinds" or groups, which Cohen makes rights dependent on—as well as the inconsistent negations of those conclusions[27]). I have shown that, in each case, the reasoning given in favor of some anti-animal perspective is faulty because it *either* depends on false and/or rationally indefensible premises (some of which are not explicit) *or* is of a pattern that is exceedingly *ad hoc*, with nothing *in general* to recommend its cogency.

Given that this is the case, and especially given the clear harms for animals who are victims, it is imperative that those who harm them develop a plausible justification for doing so. If they cannot do so, moral and intellectual integrity requires simply that they stop. Given the unlikelihood that such a defense can be developed, it is likely morally obligatory that those who use animals in harmful manners simply stop doing so. We should think that anything less is rationally indefensible, unless and until good reasons show otherwise.

NOTES

1. Based on a variety of sources of information, Tom Regan estimates that between twenty-five and fifty million animals are killed in US laboratories each year. See Tom Regan, *The Animal Rights Debate* (Rowman & Littlefield 2001), p. 144.

2. For discussion of the ethics and science of using animals in education, see my "In Defense of 'How We Treat Our Relatives,'" *American Biology Teacher*, November/December 2004, pp. 599–600, and "Animal Dissection and Evidence-Based Life-Science and Health-Professions Education," *Journal of Applied Animal Welfare Science*, 2002, vol. 5, no. 2, pp. 155–159. Text available online at www.NathanNobis.com.

3. J. P. Balcombe, N. D. Barnard, and C. Sandusk, "Laboratory Routines Cause Animal Stress," *Contemporary Topics in Laboratory Animal Science*, 2004, November, 43 (6): pp. 42–51. Here is the article's abstract (emphasis mine):

> Eighty published studies were appraised to document the potential stress associated with three routine laboratory procedures commonly performed on animals: handling, blood collection, and orogastric gavage. We defined handling as any non-invasive manipulation occurring as part of routine husbandry, including lifting an animal and cleaning or moving an animal's cage. Significant changes in physiologic parameters correlated with stress (e.g., serum or plasma concentrations of corticosterone, glucose, growth hormone or prolactin, heart rate, blood pressure, and behavior) were associated with all three procedures in multiple species in the studies we examined. The results of these studies demonstrated that animals responded with rapid, pronounced, and statistically significant elevations in stress-related responses for each of the procedures, although handling elicited variable alterations in immune system responses. Changes from baseline or control measures typically ranged from 20% to 100% or more and lasted at least 30 min or longer. We interpret these findings to indicate that laboratory routines are associated with stress, and that animals do not readily habituate to them. The data suggest that significant fear, stress, and possibly distress are predictable consequences of routine laboratory procedures, and that these phenomena have substantial scientific and humane implications for the use of animals in laboratory research.

4. For details, see Tom Regan's *Empty Cages: Facing the Challenge of Animal Rights* (Landham, Md.: Rowman & Littlefield, 2004).

5. For further discussion, see Kathie Jennie's "The Power of the Visual: The Role of Images in Moral Motivation" (unpublished).

A RATIONAL DEFENSE OF ANIMAL EXPERIMENTATION 61

6. In public talks on these issues, I have heard more than one animal experimenter ask the audience if they would like to be "the guinea pig" to be the first one to undergo some surgery or some treatment. Audiences respond, no, and the experimenter somehow takes this to show that animal experimentation is permissible. This is a perverse reversal of any "golden rule" type reasoning, as the suggested false principle is that since *we* wouldn't want to be harmed these ways, it is *therefore* permissible to harm others!

7. For representatives of these various perspectives, see, among others, Peter Singer's *Animal Liberation*, 3rd Edition (New York: Ecco, 2002), Tom Regan's *The Case for Animal Rights*, 2nd Edition (Los Angeles: University of California, 2004), Mark Rowlands' *Animals Like Us* (London: Verso, 2002), Rosalind Hursthouse's *Ethics, Humans and Other Animals* (New York: Routledge, 2000), David DeGrazia's *Taking Animals Seriously: Mental Life and Moral Status* (New York: Cambridge University Press, 1996), and Andrew Linzey and Tom Regan's *Animals and Christianity: A Book of Readings* (New York: Crossroads, 1988).

8. R. G. Frey, "Justifying Animal Experimentation: The Starting Point," in E. F. Paul and J. Paul, eds., *Why Animal Experimentation Matters: The Use of Animals in Medical Research* (N.J.: Transaction Publishers, 2001), p. 200.

9. Adrian Morrison, "Making Choices in the Laboratory," in E .F. Paul and J. Paul, eds., *Why Animal Experimentation Matters: The Use of Animals in Medical Research* (N.J.: Transaction Publishers, 2001), p. 50, p. 51.

10. Ibid., p. 51.

11. Charles Nicholl and Sharon Russell, "A Darwinian View of the Issues Associated with the Use of Animals in Biomedical Research," in E. F. Paul and J. Paul, eds., *Why Animal Experimentation Matters: The Use of Animals in Medical Research* (N.J.: Transaction Publishers, 2001), p. 164.

12. Morrison, "Making Choices in the Laboratory," p. 56.

13. Charles Nicholl and Sharon Russell, "A Darwinian View of the Issues Associated with the Use of Animals in Biomedical Research," p. 150.

14. Ibid., p. 62.

15. Ibid., p. 166.

16. Ibid., p. 168.

17. Jerrold Tannenbaum, "The Paradigm Shift toward Animal Happiness," in E. F. Paul and J. Paul, eds., *Why Animal Experimentation Matters: The Use of Animals in Medical Research* (N.J.: Transaction Publishers, 2001), p. 93.

18. Stuart Zola, "Basic Research, Applied Research, Animal Ethics and an Animal Model of Human Amnesia," in E. F. Paul and J. Paul, eds., *Why Animal Experimentation Matters: The Use of Animals in Medical Research* (N.J.: Transaction Publishers, 2001), p. 90.

19. Baruch Brody, "Defending Animal Research: An International Perspective," in E. F. Paul and J. Paul, eds., *Why Animal Experimentation Matters: The Use of Animals in Medical Research* (N.J.: Transaction Publishers, 2001).

20. H. Tristam Engelhardt, "Animals: Their Right to Be Used," in E. F. Paul and J. Paul, eds., *Why Animal Experimentation Matters: The Use of Animals in Medical Research* (N.J.: Transaction Publishers, 2001), p. 178.

21. See Peter Singer, *Ethics Into Action: Henry Spira and the Animal Rights Movement* (Lanham, Md: Rowman & Littlefield, 2000).

22. See, e.g., H. LaFollette and N. Shanks, *Brute Science: Dilemmas of Animal Experimentation* (New York: Routledge, 1996); H. LaFollette and N. Shanks, "The Intact Systems Argument: Problems with the Standard Defense of Animal Experimentation," *The Southern Journal of Philosophy*, 1993, 31(3): pp. 323–333; H. LaFollette and N. Shanks, "Animal Models in Biomedical Research: Some Epistemological Worries," *Public Affairs Quarterly*, 1993, 7(2): pp. 113–130; H. LaFollette and N. Shanks, "Chaos Theory: Analogical Reasoning in Biomedical Research" *Idealistic Studies*, 1994 (24) 3: pp. 241–254; H. LaFollette and N. Shanks, "Two Models of Models in Biomedical Research." *Philosophical Quarterly*, 1995, 45(179): pp. 141–160.

23. T. R. Gregory, "The Failure of Traditional Arguments in the Vivisection Debate," *Public Affairs Quarterly*, 2000, 14(2): pp. 159–182.

24. Morrison, "Making Choices in the Laboratory," p. 58.

25. See, e.g., C. R. Greek and J. Greek, *Sacred Cows and Golden Geese: The Human Cost of Experiments on Animals* (New York: Continuum, 2000), *Specious Science: How Genetics and Evolution Reveal Why Medical Research on Animals Harms Humans* (New York: Continuum, 2002), *What Will We Do if We Don't Experiment on Animals: Medical Research for the Twenty-First Century* (Victoria, B.C.: Trafford, 2004).

26. Tibor Machan's *Putting Humans First: Why We Are Nature's Favorite* (Lanham, Md.: Rowman & Littlefield, 2004). References to this work will be in the main body of the text.

27. Nathan Nobis, "Carl Cohen's 'Kind' Argument *For* Animal Rights and *Against* Human Rights," *Journal of Applied Philosophy*, March 2004, vol. 21, no. 1, pp. 43–59. Text available online at www.NathanNobis.com.

ETHICS AND THE LIFE SCIENCES

AT THE EDGE OF HUMANITY: HUMAN STEM CELLS, CHIMERAS, AND MORAL STATUS

ROBERT STREIFFER
UNIVERSITY OF WISCONSIN, MADISON

ABSTRACT: Experiments involving the transplantation of human stem cells and their derivatives into early fetal or embryonic nonhuman animals raise novel ethical issues due to their possible implications for enhancing the moral status of the chimeric individual. Although status-enhancing research is not necessarily objectionable from the perspective of the chimeric individual, there are grounds for objecting to it in the conditions in which it is likely to occur. Translating this ethical conclusion into a policy recommendation, however, is complicated by the fact that substantial empirical and ethical uncertainties remain about which transplants, if any, would significantly enhance the chimeric individual's moral status. Considerations of moral status justify either an early-termination policy on chimeric embryos, or, in the absence of such a policy, restrictions on the introduction of pluripotent human stem cells into early-stage developing animals, pending the resolution of those uncertainties.

Some people object to human embryonic stem cell research (hES cell research) because of their beliefs about the moral status of the human embryos that are destroyed when the stem cells are derived. Some people object to using animals in biomedical research because of their beliefs about the moral status of animals. Most biomedical research advisory and regulatory bodies, however, believe that both types of objections can be overcome and are generally supportive of both hES cell research and biomedical research on animals. It is

Reprinted by permission of publisher and author. Robert Streiffer. 2005. "At the Edge of Humanity: Moral Status, Human Stem Cells, and Chimeras." *The Kennedy Institute of Ethics* 5(4): pp. 347–370.

therefore somewhat surprising that many such bodies have expressed serious concern regarding a use of hES cells that combines both kinds of research: the creation of chimeras, organisms with parts from different species, through the xenotransplantation of hES cells into animals that are in the embryonic or early fetal stages of their development. Although the focus has been on the creation of chimeras using human embryonic stem cells, similar concerns might arise with respect to transplants involving other types of human stem cells, as well as their more specialized progeny.

The emerging bioethics literature on the creation of such chimeras has analyzed several possible moral issues. Jason Robert and Françoise Baylis (2003) explore the possibilities that such research is unethical because of its unnatural results, because it violates species boundaries, or because it might harm society by leading down a slippery slope that undermines the categories presupposed by desirable legal and cultural practices. Phillip Karpowicz and colleagues (2004; 2005) also discuss the unnaturalness objection and look at whether chimeras might be problematic because they violate moral taboos, violate species integrity, or undermine human dignity.

With two notable exceptions (Karpowicz, Cohen, and van der Kooy 2004; 2005), the literature has neglected to address issues arising out of concern for the individual most directly affected by the research, namely the chimeric research subject itself. After outlining the relevant scientific and regulatory background, I argue that the effect that certain transplants could have on the moral status of chimeric research subjects raises novel and significant ethical issues. I distinguish between the two different views of moral status that are generating most of the controversy surrounding hES cell research and argue that on each of them certain kinds of human stem cell transplants could significantly enhance the chimeric individual's moral status. Given that the moral evaluation of research normally presupposes a fixed moral status for the subject, this raises novel ethical issues that are just now beginning to receive attention. I therefore construct a taxonomy of principles for evaluating moral status enhancements. I then argue that on the most plausible principle, the introduction of human stem cells into a nonhuman animal in a way that would substantially enhance its moral status is wrong, not because of the fact that the research subject's moral status is enhanced, which is a prima facie good, but rather because of the fact that the subsequent treatment of the subject likely will fall far below what its new moral status demands. Translating that ethical conclusion into a policy recommendation, however, is complicated by the fact that substantial empirical and ethical uncertainties remain about which transplants, if any, would significantly enhance the chimeric individual's moral status. I conclude by discussing various policy options, and I argue that the moral status framework justifies either an early-termination policy on chimeric embryos, or, in the absence of such a policy, restrictions on introducing human pluripotent stem cells into early-stage developing animals, pending the resolution of those uncertainties.

SCIENTIFIC AND REGULATORY BACKGROUND

A chimera is a single individual composed of cells that have different embryonic origins. Intraspecific chimeras, created when the cell donor and the cell recipient belong to the same species, have been an important research tool for decades (Nagy and Rossant 2001). Chimeras commonly are created by transplanting, or injecting, stem cells from one animal into another. Stem cells are cells that renew themselves and also give rise to more specialized kinds of cells. Although much of the research on chimeras involves injecting mouse stem cells into mouse blastocysts, I restrict my use of the term here to animal/human chimeras, by which I mean the individual that results from injecting human stem cells or their derivatives into a nonhuman animal.

Even though the term "chimera" evokes negative connotations for some, chimeras are often no more than animals with some human blood cells inside them. Such chimeras have been created for some time by transplanting stem cells derived from the bone marrow of adult humans, adult hematopoietic stem cells, into postnatal animals. By injecting cells that have been tagged with markers, researchers can observe where the cells and their progeny migrate, how they specialize, and how they interact with other tissues and systems in the animal's body. This provides researchers the opportunity to learn about how stem cells specialize in response to cues from their surrounding cellular environment, to explore their potential for repairing or replacing damaged tissue, and to explore their potential for creating animal models that more closely mimic humans (Okarma 2001; Thomson 2001). Although such research involves biomedical research on animals and, like all biomedical research on animals, raises important concerns within traditional animal ethics, it has not been regarded as especially or distinctively problematic by researchers or by bioethicists.

There is special interest in transplanting human stem cells into prenatal animals. Doing so helps to minimize the risk that the transplant will be rejected by the animal's immune system and provides the opportunity to learn about how stem cells act in a developing organism, which is of interest to developmental biologists. For future clinical therapies involving human stem cells, transplants might need to take place early in fetal or even embryonic development to help prevent complications before they start (Flake and Zanjani 1999). Early in utero transplants of human stem cells into animals would have to precede human clinical trials for such therapies.

Because adult hematopoetic stem cells are believed to be merely multipotent, restricted to differentiating only into types of blood cells, whereas hES cells are believed to be pluripotent, capable of differentiating into any kind of cell (Wagers et al. 2002), there is now a growing interest in differentiating hES cells in vitro, and then transplanting these derivatives into prenatal animals. This would offer a wider range of opportunities to study early human development and potential therapies, not just for blood cells, but for any kind of tissue (Svendsen 2002). Researchers at the University of Wisconsin-Madison, for example, are interested in using chick embryos as a model for studying hES cell derived neural precursor cells (Basu 2005)—precursor cells differentiate into more specialized cells, but do not renew themselves as stem cells do. And there is growing interest in introducing undifferentiated hES cells into prenatal animals. One group has reported inserting hES cells into 1.5- to 2-day-old chick embryos to explore whether the chick embryo "may serve as an accessible and unique experimental system for the study of in vivo development of human ES cells" (Goldstein et al. 2002). Also, one test for pluripotency is to inject cells into a blastocyst and then see whether those cells contribute to the development of all the other tissues of the resulting organism (Kaiser Daily Reproductive Health Report 2002; Dewitt 2002; NAS 2004).

However, as I mentioned, transplanting hES cells into early-stage developing animals has been flagged for special concern by several regulatory and advisory bodies, bodies which view the typical concerns in both animal ethics and hES cell research as answerable. Geron, the company that funded James Thomson's original derivation of hES cells, has an ethics advisory board (EAB) that issued guidelines for hES cell research (Geron Ethics Advisory Board 1999). These guidelines are generally supportive of hES cell research but include a provisional prohibition on research involving "any creation of chimeras" until the EAB has undertaken more extensive analysis of the issues involved in doing so. The initial report on hES cell research of the University of Wisconsin's Bioethics Advisory Committee (1999) recommended that hES cells "not be used for introduction into a uterus without further University of Wisconsin Review and approval." WiCell Research Institute, the not-for-profit company that manages the University of Wisconsin's hES cell lines, went even further in their memorandum of understanding to which recipients of their cells must agree: "Recipient agrees that its research program will exclude (i) the mixing of Wisconsin Materials with an intact embryo, either human or nonhuman; (ii) implanting Wisconsin Materials or products of Materials in a uterus; and (iii) attempting to make whole embryos with Wisconsin Materials by any method" (WiCell 2001). The Clinton administration's proposed guidelines prohibited funding for "research in which human pluripotent stem cells are combined with an animal embryo" (NIH 2000). (Interestingly, there appears to be no similar restriction in Bush's guidelines.)

Two consensus groups have taken up the task of establishing voluntary guidelines for hES cell research. The first was organized by the New York Academy of Sciences, but agreement was stymied because of disagreement over the creation of human/mouse embryonic chimeras (DeWitt 2002). The second was the National Academy of Sciences' Committee on Guidelines for Human Embryonic Stem Cell Research (NAS Committee), which published its guidelines in April 2005. Presentations to the NAS Committee on research involving the introduction of hES cells into embryonic animals ranged from supporting no special review whatsoever to supporting an outright ban, and although there seemed to be no inclination among the presenters to ban the introduction of hES cells into all fetal animals, there was disagreement as to where to draw the line between embryonic and fetal stages (NAS 2004; Weiss 2004). The final guidelines include a prohibition on research "in which hES cells are introduced into nonhuman primate blastocysts," and special review for all research "involving the introduction of hES cells into nonhuman animals at any stage of embryonic, fetal, or postnatal development" (NAS 2005, pp. 47–48).

HUMAN APPEARANCES, HUMAN EXPERIENCES, COGNITIVE CAPACITIES, AND MORAL STATUS

Some have suggested that the problematic aspect of chimeras is the aesthetics involved. William Hurlbut, a member of the President's Council on Bioethics, says that "visible chimeras," animals with visible parts that appear to be human, are unethical because "human appearance is something we should reserve for humans" (Shreeve 2005). The image of animals with human body parts, or even animals with the parts of other species of animals, is surely part of what is motivating the public's reaction to chimeras. It should go without saying that this view is a non-starter. It should go without saying, but evidently it does not, and so it is worth considering here.

Consider work by Yilin Cao and colleagues (1997), in which researchers evaluated whether a polymer template could be used to grow cartilage in the shape of a 3-year-old child's auricle. In order to provide a hospitable environment for the cartilage to form, the template was inserted under the skin on the back of a mouse. Pictures from this experiment showed a small mouse in a Petri dish with what appears to be a fully-formed human ear on its back. These pictures have been used by such anti-biotechnology organizations as the Turning Point Project to elicit negative aesthetic reactions to biotechnology, which are then treated as if they were reactions that carried moral significance. But the ability to grow cartilage in the right shape is one step in the important process of being able to provide functional and aesthetically correct replacements for children who, due to deformity or accident, need total external ear reconstruction. Although such research on animals may be unethical because of traditional concerns in animal ethics, it is not remotely plausible to think that the mere visual appearance of the mouse makes such research wrong. The mere fact that one would be conferring a human appearance on a nonhuman animal is of no consequence.

Others who have discussed the introduction of human stem cells or their derivatives into an embryonic or fetal animal have focused on possible effects on the animal's neural tissue. As the NAS Committee stated, "Perhaps no organ that could be exposed to hES cells raises more sensitive questions than the animal brain, whose biochemistry or architecture might be affected by the presence of human cells" (NAS 2005, p. 41).

The possibility that transplantation of human stem cells or their derivatives could alter neural tissue is already well documented. Two groups have reported that they differentiated hES cells in vitro into neural precursor cells, which they then transplanted into the brains of neonatal mice (AAP 2001; Zhang et al. 2001).

Su-Chun Zhang and colleagues (2001, p. 1129) report that the cells were then "incorporated into a variety of brain regions, where they differentiated into both neurons and astrocytes." Irv Weissman, a Stanford researcher, injected human neural stem cells into the brains of neonatal mice, with the result that "every part of the brain was populated with human cells" (Krieger 2002), although presumably only to a very small degree since Weissman also told the press that the human cells made up only 1 percent of the cells in the mice brains. Ronald Goldstein and colleagues (2002, p. 80) report that the hES cells they transplanted into chick embryos differentiated into neurons and penetrated into the developing central nervous system. Oliver Brüstle (1999, p. 537) reports that human neural precursor cells transplanted into embryonic rats differentiated into "all three major cell types of the nervous system" and generated "an extensive axonal network encompassing large areas of the host brain."

The focus on alterations of neural tissue might be justified in different ways. In its second report on hES cells, the University of Wisconsin Bioethics Advisory Committee (2001) explores the possibility that the introduction of human stem cells could result in a chimera capable of "human experiences":

> Mixing human stem cell lines with experimental animals early in the animal's fetal development may ... result in the development of human neural tissue in the experimental animal, which raises at least the theoretical possibility that such tissue could become integrated in a way that human experiences become possible. After consulting with biologists, the Committee concluded, based on current knowledge of developmental biology, that this risk is extremely remote unless such mixing occurred very early in embryonic life. It is for this reason that introducing human stem cells into developing animals very early in embryonic life raises greater concerns about the creation of chimeras with human-like characteristics, and such experiments should receive careful ethical and scientific scrutiny.

This view appears to single out human experiences, and to require additional review for research that would provide animals with such experiences. The underlying moral principle appears to be something like the following:

The Human Experience Principle: It is always morally problematic to enable a nonhuman individual to have human experiences.

The principle uses the phrase "human experience." How might this be defined? If a human experience is any experience that some humans are capable of having, then the experience of seeing red is a human experience. The Human Experience Principle then implies that it is always morally problematic to enable an animal to see red. This seems plainly false.

Perhaps a human experience is one that some humans are capable of having and no nonhumans are capable of having. That is, human experiences are experiences that are distinctively human. But what if it were true that only humans could see red? The Human Experience Principle would imply that, in those circumstances, it would be morally problematic to enable an animal to see red. Again, this is clearly false. It is not clear that the phrase "human experiences" picks out a morally relevant class of experiences.

A second possible justification for focusing on transplants that alter neural tissue is that neural tissue is the physical basis for those cognitive capacities that themselves form the basis for the robust moral agency and rational autonomy of which normal adult humans are, so far as we know, distinctively capable (DeGrazia 1996, pp. 199–210). Call such cognitive capacities "high-level cognitive capacities." Then, the underlying moral principle would be something like the following:

> The Cognitive Capacity Principle: It is always morally problematic to enable a nonhuman individual to have high-level cognitive capacities.

Because being able to experience red is not a high-level cognitive capacity, the Cognitive Capacity Principle does not imply that it would be morally problematic to enable a nonhuman to experience red, and thus it avoids the aforementioned problem with the Human Experience Principle.[1]

Nonetheless, the Cognitive Capacity Principle looks dubious. If one were to discover beings of another species, the normal adults of which had high-level cognitive capacities, it would not be morally problematic to cure one of them of severe brain damage—i.e., to enable *that* nonhuman individual to have high-level cognitive capacities.

Another possibility, which I think gets at something deeper than the previous principles, focuses on the possible impact of transplanting human stem cells into early-stage developing animals on the moral status of the resulting chimera:

> The Moral Status Principle: It is always morally problematic to cause an individual that would otherwise have a lower moral status to have the moral status of a normal, adult human.

Because being able to experience red is not what gives humans our moral status, the Moral Status Principle does not imply that it is morally problematic to enable an individual to experience red. Because the hypothetical beings of another species cognitively similar to our own presumably have the same moral status as we do, and retain their moral status even when they are brain damaged, the Moral Status Principle does not imply that it is morally problematic to return their cognitive capacities to normal.

In the remainder of the paper, I focus on the implications of the Moral Status Principle for the moral evaluation of chimeric research.

THE MORAL STATUS FRAMEWORK

Which effects would a transplant need to have to confer upon a chimeric research subject the moral status of a normal, adult human being, and which transplants, if any, would produce those effects? The answers depend on why normal human adults have the comparatively high moral status that they do.

On cognitive capacity views of moral status, an individual's cognitive capacities give it its moral status, and the high-level cognitive capacities that normal adult humans have is what gives them their relatively high moral status (VanDeVeer 1979). Although Karpowicz and colleagues (2005, p. 120) do not use the language of "moral status," they articulate one attractive view of these cognitive capacities in their discussion of human dignity:

> Human dignity is a widely shared notion that signifies that humans typically display certain sorts of functional and emergent capacities that render them uniquely valuable and worthy of respect. It is not only the capacities for reasoning, choosing freely, and acting for moral reasons, as Kant argues, or for entertaining and acting on the basis of self-chosen purposes, as Gewirth holds, that are at the core of what we mean by human dignity. The notion also encompasses such capacities as those for engaging in sophisticated forms of communication and language, participating in interweaving social relations, developing a secular or religious world view, and displaying sympathy and empathy in emotionally complex ways.

Given that high-level cognitive capacities are intimately related to the individual's neural tissue, this view obviously justifies focusing on transplants that could affect neural tissue in a way that enhances cognitive capacities.

As Karpowicz and his colleagues point out (2005, pp. 124–26), there are many constraints on the ability of differentiated human stem cells, particularly retinal and neural stem cells, to significantly enhance cognitive functions. In many cases, such enhancements likely would be prevented by the animal's smaller skull size and shorter gestation period, as well as by the surrounding nonhuman cellular environment that would provide developmental cues to transplanted cells. This is also the view that Dr. Fred Gage, a neuroscientist at the Salk Institute who specializes in neuroplasticity and neural stem cells, presented to the NAS Committee (NAS 2004).

It may be that many human stem cell xenotransplants would not confer any high-level cognitive capacities onto the resulting chimeric subjects, as seems to be the case with hematopoetic stem cell xenotransplants late in an animal's fetal development. But even with stem cells that are merely multipotent, there are still two uncertainties. First, the mechanism by which oocyte cytoplasm de-differentiates cells, a procedure involved in somatic cell nuclear transfer, is still unclear (NAS 2005, p. 35). Thus, there is the possibility that introducing multipotent stem cells into an embryonic environment might de-differentiate the cell, restoring it to a pluripotent state. Second, it is not yet clear what kinds of cognitive enhancements might be possible even within a constrained environment. For example, it is possible to use genetic engineering to enhance the learning and memory of mice without modifying skull size or gestational period (Tang et al. 1999).

Moreover, it would be premature to claim to know that the introduction of hES cells very early in development will not dramatically affect cognitive capacities. If a large enough quantity of hES cells were introduced, the cells themselves could induce changes that would eliminate some of the constraints mentioned above. As the NAS Committee concluded, "it is not now possible to predict the extent of human

HUMAN STEM CELLS, CHIMERAS, AND MORAL STATUS

contribution to such chimeras" (NAS 2005, p. 34). And Gage, when asked what the effects would be of introducing hES cells or neural stem cells into an animal early in development, said, "We don't know the answer to [that] question because the experiment hasn't been done, that I know of" (NAS 2004). Although some of the experiments cited above do involve the introduction of hES and neural progenitor cells early in development, experience in this area is limited. It therefore seems premature to place much confidence in our ability to draw a precise line between those introductions of human stem cells that will, and those that will not, confer high-level cognitive capacities.

It is also important to note that some researchers will be interested in the bases of the restrictions on cognitive development and in whether they can be overcome through the use of human stem cells. They will be interested in designing experiments that seek to overcome existing limitations on cognitive development in nonhuman animals. Other researchers will be interested in creating chimeras in which such limitations are overcome so that the chimeric individuals can be used as models that more closely mimic human beings. Such chimeras might be created to study diseases or injuries that impair high-level cognitive functions in humans, or to do basic research on the neurological development involved in language acquisition, mathematical concept acquisition, moral development, or any number of other cognitive capacities that are now limited to normally functioning human beings. As noted by the NAS Committee,

> [T]he idea that human neuronal cells might participate in "higher-order" brain functions in a nonhuman animal, however unlikely that may be, raises concerns that need to be considered. Indeed, if such cells are to be used in therapeutic interventions, one needs to know whether they could participate in that way in the context of a treatment. (NAS 2005, p. 33)

On anthropocentric views of moral status, normal human adults have the moral status they do simply because they are human beings, that is, because they are members of the species *homo sapiens* (Noonan 1970; Devine 1978; Schwarz 1990). As has often been noted, anthropocentric views seem to suffer from an explanatory gap: it is hard to see how being a member of a certain species could give an individual its moral status (Regan 1978; Feinberg 1980; DeGrazia 1996, pp. 56–61).[2] Nonetheless proponents of these views argue that they provide the only way to explain adequately the equal moral status of all human beings, even human beings who lack high-level cognitive capacities, and thereby to avoid the so-called "marginal humans" problem that afflicts cognitive capacity views.

Karpowicz and colleagues (2005, p. 120), for example, maintain that human's special moral status is "attributable equally to all human beings," but, quite clearly, the capacities they cite as the basis of human's special value are not held equally by all human beings, and some human beings lack them altogether. The authors recognize that their view may exclude infants and seriously disabled individuals and respond:

[W]e tend to ascribe [human dignity] to all humans, no matter how seriously impaired or ill they may be, because there is no clear agreement about just how many dignity-associated capacities a person must possess to be said to have human dignity. To avoid the possibility of mistakenly failing to treat those with severe disabilities as ends in themselves, human dignity proponents ascribe dignity to all humans. (Karpowicz, Cohen, and van der Kooy 2005, pp. 121–122)

But an appeal to uncertainty and disagreement seems implausible given that there is no real uncertainty or disagreement that a newborn fails to have the capacities they cite and so would, on their view, *clearly* lack the special moral status that accompanies individuals with human dignity.

Anthropocentric views raise difficult questions about how much human material an individual needs in order to be a human being. Since normal human embryos are both human and organisms, they are human beings, albeit ones at the earliest stages of development (Feinberg 1980, pp. 288–291). But when faced with an organism that has some human cells and some nonhuman cells, how is one to decide whether the organism is human, and hence, whether it is a human being? It is not plausible to suppose that the individual in question has to have a human brain: an anencephalic infant is a human being and would possess human moral status on an anthropocentric view. Thus, on anthropocentric views, the focus on alterations of neural tissue is overly narrow, and the focus on alterations of neural tissue that affect cognitive capacities even more so. Presumably, replacing the entire inner cell mass of an animal blastocyst with hES cells would suffice to make the resulting individual a human being since, in normal human development, the inner cell mass is what goes on to form the fetus. In such cases, one could, at least in principle, end up with a human being surrounded by a nonhuman trophectoderm. On the other hand, having only a few nonhuman cells in the final individual would not suffice since a human with a porcine heart valve is still a human being. But where to draw the line is unclear, as it is with other objects, such as the ship of Theseus, that have vague identity conditions (Parfit 1984, pp. 231–243; Thomson 1987; Thomson 1997).

It seems, then, that on both anthropocentric and cognitive capacity views of moral status—views that, in some form or other, generate most of the controversy about hES cell research—the transplantation of human stem cells or their derivatives into developing animals could, at least in principle, significantly enhance the chimeric research subject's moral status. The question remains, though: why think that significantly enhancing an individual's moral status is always morally problematic?

EVALUATING ENHANCEMENTS OF MORAL STATUS

What are the moral principles regarding enhancements in moral status? To focus the discussion, I shall concentrate primarily on issues that arise from concern for the altered individual, concerns from the perspective of that individual itself.

There would seem to be the following possibilities. First, an enhancement in moral status might always be an unequivocal good from the individual's perspective. Any deleterious effects an enhancement might have on other factors that are relevant from the individual's perspective are always outweighed by the enhancement itself. I will call this view the Millian View since it echoes Mill's remark that it is better to be Socrates unsatisfied than a pig satisfied.

Second, whether an enhancement in moral status is good or bad from the individual's perspective might be entirely derivative upon its effects on other, independently relevant factors. For example, if the chimeric research subject suffers more than it would have, and if the research has no other morally relevant effects, then the enhancement in status would be bad from its perspective just to the degree that its suffering was bad. Because the relevant baseline in this case is how the individual would have fared had it not received the enhancement, I will call this view the Instrumentalist View with the Non-Moral Baseline.

The third view is the Instrumentalist View with the Moral Baseline: how good an enhancement is from the individual's perspective depends entirely upon how the individual's life compares, in terms of other, independently relevant factors, to the life it would have were its new moral status fully respected. To the extent that the individual's life meets this moral baseline, the enhancement is good from the individual's perspective; but to the extent that its life falls short of the moral baseline, the enhancement is bad from the individual's perspective. The two instrumentalist views disagree in cases where the chimeric research subject's life is better than it would have been had its status not been enhanced, but given its new moral status, it deserves to have its life be even better.

The fourth and fifth possibilities are mixed views that attach some positive moral weight to the fact that the individual's status has been enhanced, but allow that this improvement might be outweighed by deleterious effects on other factors that are independently relevant from the individual's perspective. According to the Mixed View with the Non-Moral Baseline, the relevant question is how the individual's life, taking into account both the *prima facie* good of the enhancement and its other effects, compares to the life it would have had had it not received the enhancement. And according to the Mixed View with the Moral Baseline, the relevant comparison is to the life it would have if its new moral status were fully respected.

Finally, a sixth possibility holds that conferring an enhanced moral status on an individual is always bad from the individual's perspective. I will call this the No-Enhancing View. There are other logical possibilities, but these are the most interesting ones.

If any of these views is to underwrite a general moral objection to status-enhancing research, such as the one expressed in the Moral Status Principle, it would have to be the No-Enhancing View. The others allow that, in some circumstances, an enhancement could be good from the individual's perspective. But of the views, the No-Enhancing View is the least plausible. Imagine conferring an enhanced moral status on an entity, and then ensuring that it lives a life in which it receives much better treatment than it would have gotten otherwise and in which it is given everything it is owed in virtue of that enhanced moral status. It is hard to see how this outcome could be bad from the individual's perspective. I conclude, then, that if the Moral Status Principle is to be sustained at all, it must be by appeal to some factors other than the ones relevant from the individual's perspective.

It is worth considering briefly whether the Moral Status Principle might be justified on such grounds. Perhaps bringing new individuals into existence that have the distinctive moral status of normal human adults somehow lessens the value of that status for extant humans, as expanding membership in an exclusive club might lessen its value to its already existing members. That thought, however, is belied by the fact that every time people reproduce, they engage in just such an activity, and they produce far more individuals with human moral status than would ever be produced by chimeric research.

Another thought might be that the extension of human moral status to "lesser animals" somehow diminishes the value of that status for humans, as extending a university diploma to those who do not deserve it lessens the value of that diploma for those who do. If, however, a transplant truly has enhanced the moral status of the chimeric research subject, then in whatever sense we humans "deserve" our special status, it now does as well, and so the value placed on human moral status will not be lessened.

Undoubtedly, there are other possible arguments that support the No-Enhancing View—e.g., enhancing is unnatural, enhancing is playing God, and so on. Although I have seen no conclusive refutation of all such arguments, I am dubious that any of them are sound and maintain that many would be unacceptable grounds for public policy even if they were (Streiffer 2003; Streiffer and Hedemann 2005).

The opposite extreme of the No-Enhancing View is the Millian View, according to which moral status enhancements are always an unequivocal good from the perspective of the enhanced individual. This view is also implausible: what kind of life an individual with an enhanced status will lead surely matters. My life is better than the life of even a *very* satisfied pig, but if my life were filled with enough pain and misery, and with extremely limited prospects, it arguably would be worse.[3]

The Instrumentalist Views might be motivated by a hedonistic view of how good an animal's life is. Since an enhanced moral status it not itself pleasurable or painful, an Instrumentalist View must be true. But there are two problems.

First, it is doubtful that hedonism is the correct view regarding animals. Hedonism about animal welfare seems to presuppose that all nonhuman animals have an exceedingly limited mental life, incapable of being interested in anything other than experiencing pleasure and avoiding pain. There is substantial empirical research against this idea (see DeGrazia 1996, pp. 97–257, for extensive discussion of the mental life of animals and its relationship to welfare).

There is also a lively debate in the agricultural biotechnology literature about reducing an animal's suffering by genetically altering it so as to eliminate its natural desires and capacities (Thompson 1997, pp. 96–99; Cooper 1998; Rollin 1998). Consider the following example. In the crowded conditions of industrial agriculture, chickens frequently hurt and kill one other (Cheng and Ali 1985). Since this is economically inefficient, producers often cut off the chickens' beaks, combs, and toes to minimize the damage the chickens can do (Duncan 2001). Although this measure mitigates the problem, it does not eliminate it. It is also economically costly to producers and painful to the chickens. Scientists have known for some time about genetically blind chickens, the result of a chemically-induced genetic mutation (Smyth, Boissey, and Gawron 1977). These chickens are of interest to producers because their blindness makes them less mobile, which means that they use less feed, and makes them unable to cannibalize their eggs, which means that they have increased egg productivity (Cheng et al. 1980). Their blindness also limits their ability to injure or kill their cagemates, and flocks of blind chickens suffer less feather damage and fewer injuries (Cheng and Ali 1985). Even assuming that blind chickens experience substantially less suffering than they would have experienced without the alteration, it is arguable that such alterations are still morally problematic from the individual's perspective. (For a discussion of similar cases, see Gavrell Ortiz 2004; Comstock 2000, pp. 95–138.) If so, then hedonism with respect to animals cannot be correct.

Second, whatever plausibility hedonism might have when it is applied to animals, it surely has even less plausibility when applied to humans. Since status-enhanced chimeric research subjects may be similar to humans in the morally relevant respects, it is dubious to suppose that hedonism about animals extends to hedonism about chimeric research subjects in the kind of research at issue.

The Mixed Views have the advantage of accommodating the Millian intuition that status enhancement could be good from the individual's perspective even if it had some harmful consequences. Also, the plausible idea that status diminishments are *prima facie* bad from the individual's perspective lends plausibility to the idea that status enhancements are *prima facie* good from the individual's perspective. I thus conclude, albeit tentatively, that one of the Mixed Views is correct.

All that remains, then, is to determine which baseline, the moral or the nonmoral, is the relevant one. Consider an example from the literature on exploitation in which a transaction provides someone with a benefit, but with far less benefit than they deserve: an employer who pays an employee a wage that is beneficial compared to the alternatives, but is still substantially less than what justice requires. In such cases, the relevant baseline for the evaluation of the transaction is the moral baseline: given that the employee deserves more, it is no defense of the employer's behavior to say that the employee is better off than he would have been without the job (see Wertheimer 1996 for extended discussion). Similarly, the relevant question for evaluating status-enhancing transplants is surely whether the subject's new entitlements are respected. I therefore conclude that the Mixed View with the Moral Baseline is correct.

ETHICAL AND POLICY IMPLICATIONS

If a Mixed View is correct, then transplants that confer the moral status of a normal human being onto a chimeric research subject are *prima facie* good. But because the relevant baseline is the moral baseline, transplants that enhance an animal's moral status to that of a normal human adult raise the following problem. The view institutionalized by animal research oversight committees is that almost any valid research objective justifies sacrificing even the most fundamental interests of animals (Francione 1995). In contrast, the view institutionalized by human subjects research oversight committees is that humans have a moral status which provides them with substantial moral protections, including a very stringent prohibition on harmful research without informed consent. So long as experiments that involve the xenotransplantation of human stem cells into animals are overseen by animal research oversight committees, or by human subjects committees only attentive to concerns of those who provided the gametes or embryos from which the stem cells were derived, the wrong, or an incomplete, set of moral protections is likely to be afforded to status-enhanced chimeric research subjects. If the relevant baseline were the non-moral baseline, then transplants that enhanced moral status probably would be no more problematic than other kinds of biomedical research on animals. But because the relevant baseline is the moral baseline, sacrificing the fundamental interests of the chimeric research subject as they would have been sacrificed in any other animal research is the moral equivalent of sacrificing the fundamental interests of a fully functional adult human being. On all but the most extreme animal rights views, this makes status-enhancing chimeric research much worse than other biomedical research on animals, and on any plausible view, makes it absolutely unacceptable.

Alternatively, if researchers guaranteed adequate protections for any chimeric research subject whose status had been enhanced to that of a normal adult human, then at least from that individual's perspective, there would be no objection to the research. It is difficult to see, however, how researchers could do that without undermining their research objectives, since most biomedical research on animals involves procedures that plainly would be unacceptable if performed on individuals with the moral status of a normal human adult. And if it were acceptable to do the research on something with the moral status of a normal human adult, then actual humans presumably would provide a better model in which to learn about human development and in which to test possible therapies intended for human beings. Why then go to the trouble of creating a chimera and introducing the need to ascertain whether the results obtained from the chimera are generalizable to humans?

If adequate research protections cannot be guaranteed, then the Mixed View with the Moral Baseline implies that status-enhancing transplants are unethical. But which transplants run an unacceptably high risk of enhancing the status of the chimeric research subject? The epistemological difficulties here are daunting, especially in light of the moral stakes. How does one know whether the harms being

HUMAN STEM CELLS, CHIMERAS, AND MORAL STATUS

imposed are no more morally problematic than those usually imposed in biomedical research on animals, or, whether despite outward appearances, the harms being imposed amount to the moral equivalent of Nazi-style research? This question is especially problematic on cognitive capacity views of moral status because high-level cognitive capacities do not manifest themselves without substantial care and education, treatment unlikely to be provided to most research animals.

The empirical uncertainties regarding the effects that various kinds of xeno-transplants would have and the moral uncertainties regarding which effects would be status-enhancing seem to me to be the crux of the practical problems about how to set an acceptable policy. I conclude by evaluating some of the existing policy proposals from the perspective of the moral status framework.

Karpowicz and his colleagues (2004; 2005) focus their attention on transplants of disassociated retinal and neural stem cells and citing the constraints mentioned above—smaller skull size, shorter gestation period, and nonhuman environment—conclude that such cells would "not be able to achieve human brain size and the human brain organization needed to give rise to human neural functions and behaviors [that form that basis of human dignity] when transplanted to nonhuman hosts" (2005, p. 26). Presumably, such transplants also would not result in a human being and so would not enhance moral status on an anthropocentric view either. The introduction of disassociated multipotent, but not pluripotent, stem cells, then, looks promising as a class of research that could be permissible, but there are two issues that need to be resolved.

The first is the empirical issue already mentioned, namely whether retinal and neural stem cells could revert to a more pluripotent state by being introduced into an embryonic environment. The second has both an empirical and an ethical component. As previously discussed, the list of robust cognitive capacities that Karpowicz and his colleagues say are necessary for human dignity and its associated moral status looks excessively demanding. So even if a transplant would not confer all of those cognitive capacities, the question remains whether it nonetheless might confer cognitive capacities that, although less robust, are still sufficient for significantly enhancing the moral status of the chimeric research subject. If the answer to both of those questions is no, then such transplants would seem to be acceptable within the moral status framework. Transplants of other non-neural, multipotent stem cells, such as hematopoetic stem cells, presumably would be even easier to justify.

More difficult questions arise for transplants of hES cells, which are pluripotent and not merely multipotent, and still more difficult questions arise for transplants of hES cells during the embryonic or early fetal stages of development. As already discussed, such transplants might induce changes in the animal that would alter features of the animal that otherwise would have constrained the transplant's effects.

Regarding the introduction of hES cells into developing animals, the NAS Committee proposes special review by a newly instituted committee, the Embryonic Stem

Cell Research Oversight (ESCRO) Committee. The ESCRO Committee would be an important institutional mechanism for assuring that the kinds of considerations raised in this article, which fall outside the types of concerns normally addressed by animal care and use committees or human subjects committees, would have an opportunity to be addressed. The ESCRO Committee's special review would address the following:

> the number of hES cells transferred, what area of the animal body will be involved, and whether the cells might migrate through the animal's body. The hES cells may affect some animal organs rather than others, raising questions about the number of organs affected, how the animal's functioning would be affected, and whether some valued human characteristic might be exhibited in the animal, including physical appearance. (NAS 2005, p. 41)

These are quite general considerations, and, with the exception of physical appearance, which is irrelevant to moral status, the moral status framework offers a constructive way to sharpen them. What effects would a transplant have to have in order to significantly enhance the moral status of the chimeric research subject? What is the likelihood that a given transplant would have those effects? And would the researcher be able to provide adequate research protections for the resulting chimeric individual, were the research to proceed?

With respect to the introduction of hES cells into embryonic animals, the NAS Committee's guidelines include a ban on the introduction of any hES cell into a nonhuman primate blastocyst (NAS 2005, pp. 47–48). (The restricted focus on hES cells appears not to represent a substantive claim that transplants of more specialized human stem cells are unproblematic, and instead appears to be an artifact of the Committee's restricted mandate (NAS 2005, p. 4). The Committee explicitly says that other kinds of human stem cells can raise issues similar to those raised by hES cell transplants.) And if pluripotent human stem cells were to become available from another source besides human embryos, these too would surely raise the issues highlighted by the moral status framework.

According to the moral status framework, this ban is both overly permissive and unnecessarily restrictive. It is overly permissive because the moral status framework would not sharply distinguish between primate and nonprimate blastocysts or between blastocyst stages and slightly later developmental stages. If one introduces enough pluripotent human stem cells into an animal embryo, primate or otherwise, one could, in principle at least, end up with a human inner cell mass surrounded by a nonhuman trophectoderm, affecting both its species and its potential to develop robust cognitive capacities. Furthermore, because brain development occurs after the blastocyst stage, it seems likely that even a transplant that occurred after the blastocyst stage still could affect the characteristics relevant to the individual's cognitive capacities.

From the perspective of the moral status framework, the ban on introducing pluripotent human stem cells into a nonhuman primate blastocyst is also unnecessarily restrictive. From an anthropocentric view of moral status, a transplant would not significantly enhance the individual's moral status so long as two conditions are met. First, the number of cells introduced into the animal blastocyst is sufficiently low so that the original transplant itself does not constitute the creation of a human being. Second, the chimeric individual is terminated before the human cells increase in sufficient proportion to result in the entity's being deemed a human being. From a cognitive capacity view of moral status, early termination prior to the onset of consciousness would ensure that a transplant would not significantly enhance the individual's moral status. So on either view, a general ban on the introduction of pluripotent human stem cells into a nonhuman primate blastocyst is overly broad. The Committee's position cannot reflect any general concern about early termination policies and potential entanglement in the abortion debate since it requires early termination of *human* embryos used in research at 14 days of development, or the appearance of the primitive streak, whichever comes earlier (NAS 2005, p. 46). If an early termination policy is acceptable with respect to human embryos, it surely is acceptable with respect to chimeric embryos, even those with some human pluripotent stem cells in them.

Any early termination policy that permits the creation of human beings but requires termination prior to the onset of any cognitive capacities, as the Committee's policy does with respect to human embryos, will be acceptable according to cognitive capacity views of moral status, but not according to anthropocentric views. Such policies therefore will directly entangle the chimera policy debate with the abortion policy debate. This is surely an unwelcome result for proponents of this research, but perhaps is unavoidable if certain lines of research are to be pursued.

At any rate, in the absence of an early termination policy, a general ban on the introduction of pluripotent human stem cells into nonhuman primate blastocysts is a reasonable response to the present empirical and ethical uncertainties. One might object to such a ban that, in the face of uncertainty, potentially beneficial research should be allowed to proceed rather than be restricted, but in other areas of basic and therapeutic biomedical research, experiments that pose risk of serious harm to individuals with the moral status of normal human adults can only be carried out once they have been shown to be reasonably safe. That is, there is a clear moral requirement to perform such experiments in animals first, even if doing so slows down research and the provision of medical benefits. Given the uncertainties as to which human stem cell transplants into embryonic or early fetal animals would result in research on something that has the moral status of a normal adult human, it seems reasonable to require, in a similar fashion, that further research be done on the transplantation of *animal* pluripotent stem cells into embryonic or early fetal animals before similar work is done with human pluripotent stem cells. If such research confirms the view that some transplants of pluripotent human stem cells into early-stage developing animals will not substantially enhance the individual's moral status, then the moral problems discussed here will be laid to rest regarding those transplants. And if such research disconfirms the view, then it is surely best to know that before the research goes any further.

NOTES

For their many helpful comments on this topic, I thank the University of Wisconsin Bioethics Advisory Committee, Antonio Rauti, Alan Rubel, Brad Majors, Mark Brown, Justine Wells, Sara Gavrell Ortiz, Christopher Ciocchetti, Norm Fost, Alta Charo, Su-Chun Zhang, Clive Svendsen, Ian Duncan, Dan Hausman, Elliott Sober, the chimera discussion group at Bioethics Retreat 2004, and anonymous reviewers for the *Kennedy Institute of Ethics Journal*. This material is based upon work supported by the University of Wisconsin Graduate School.

1. Karpowicz and colleagues (2005, p. 121) seem to adopt something close to the Cognitive Capacity Principle. They claim that it is always wrong to confer the physical basis for high-level cognitive capacities onto an individual unable to exercise those capacities to a significant degree because, in so doing, the researcher "would diminish or eliminate the very capacities associated with human dignity." But if conferring the physical basis also confers the capacities, then the individual's capacities are enhanced in such cases, not eliminated or diminished, compared to what they would have been. And if the physical basis is present without the high-level cognitive capacities being present, then it is still not true that the researcher eliminated or diminished the capacities, since they were never there to begin with.

2. One way to try to avoid being blatantly anthropocentric and yet still accord equal moral status to all human beings is to hold a view that attributes equal moral status to all individuals that are members of species, the normal adult members of which have high-level cognitive capacities (Fox 1978, p. 110). Such a view, however, suffers an explanatory gap of its own: why should mere membership in the same species as individuals who have high-level cognitive capacities confer equal moral status on those members who do not have high-level cognitive capacities?

3. Indeed, the Millian view is even too extreme for Mill (1979 [1861], p. 9), who agrees that some people's lives are filled with "unhappiness so extreme" that they would be better off exchanging "their lot for almost any other, however undesirable in their own eyes."

BIBLIOGRAPHY

AAP. Australian Associated Press. 2001. "Scientists to Use Stem Cells to Fix Brain Damage." *AAP Newsfeed*, 1 December.

Basu, Paroma. 2005. "Scientists Grow Critical Nerve Cells." Available at http://www.news.wisc.edu/packages/stemcells/10648.html. Accessed 25 October 2005.

Brüstle, Oliver. 1999. "Building Brains: Neural Chimeras in the Study of Nervous System Development and Repair." *Brain Pathology* 9: pp. 527–45.

Cao, Yilin, Joseph P. Vacanti, Keith T. Paige, et al. 1997. "Transplantation of Chondrocytes Utilizing Polymer-Cell Construct to Produce Tissue-Engineered Cartilage in the Shape of a Human Ear." *Plastic and Reconstructive Surgery* 100: pp. 297–302.

Cheng, Kimberly M., and Ahmed Ali. 1985. "Early Egg Production in Genetically Blind (rc/rc) Chickens in Comparison with Sighted (RC+/rc) Controls." *Poultry Science* 65: pp. 789–794.

Cheng, K. M., R. N. Shoffner, K. N. Gelatt, et al. 1980. "An Autosomal Recessive Blind Mutant in the Chicken." *Poultry Science* 59: pp. 2179–2182.

Comstock, Gary. 2000. *Vexing Nature? The Ethical Case against Agricultural Biotechnology*. Boston: Kluwer Academic Publishers.

Cooper, David. 1998. "Intervention, Humility and Animal Integrity." In *Animal Biotechnology and Ethics*, ed. Alan Holland and Andrew Johnson, pp. 145–155. London: Chapman and Hall.

DeGrazia, David. 1996. *Taking Animals Seriously: Mental Life and Moral Status*. Cambridge: Cambridge University Press.

Devine, Philip. 1978. *The Ethics of Homicide*. Ithaca, NY: Cornell University Press.

DeWitt, Natalie. 2002. "Biologists Divided Over Proposal to Create Human-Mouse Embryos." *Nature* 420: p. 255.

Duncan, Ian. 2001. "Animal Welfare Issues in the Poultry Industry: Is There a Lesson to Be Learned?" *Journal of Applied Animal Welfare Science* 4: pp. 207–221.

Feinberg, Joel. 1980. "Abortion." In *Matters of Life and Death*, 2d ed., ed. Tom Regan, pp. 256–293. New York: Random House.

Flake, Alan, and Esmail Zanjani. 1999. "In Utero Hematopoietic Stem Cell Transplantation: Ontogenic Opportunities and Biologic Barriers." *Blood* 94: pp. 2179–2191.

Fox, Michael. 1978. "'Animal Liberation': A Critique." *Ethics* 88: pp. 106–118.

Francione, Gary L. 1995. *Animals, Property, and the Law*. Philadelphia: Temple University Press.

Gavrell Ortiz, Sara. 2004. "Beyond Welfare: Animal Integrity, Animal Dignity, and Genetic Engineering." *Ethics and the Environment* 9 (1): pp. 94–120.

Geron Ethics Advisory Board. 1999. "Research with Human Embryonic Stem Cells: Ethical Considerations." *Hastings Center Report* 29 (2): pp. 30–36.

Goldstein, Ronald S., Micha Drukker, Benjamin E. Reubinoff, and Nissam Benvenisty. 2002. "Integration and Differentiation of Human Embryonic Stem Cells Transplanted to the Chick Embryo." *Development Dynamics* 225: pp. 80–86.

Kaiser Daily Reproductive Health Report. 2002. "Stem Cell Researchers at New York Forum Disagree About Creation of Mixed-Species Embryos." Available at http://www.kaisernetwork.org/daily_reports/print_report.cfm?DR_ID=14743&dr_cat=2. Accessed 25 October 2005.

Karpowicz, Phillip, Cynthia Cohen, and Derek van der Kooy. 2004. "Is It Ethical to Transplant Human Stem Cells into Nonhuman Embryos." *Nature Medicine* 10: pp. 331–335.

———. 2005. "Developing Human-Nonhuman Chimeras in Human Stem Cell Research: Ethical Issues and Boundaries." *Kennedy Institute of Ethics Journal* 15: pp. 107–134.

Krieger, Lisa. 2002. "Scientists Put a Bit of Man into a Mouse." *Mercury News*. Available at http://www.mercurynews.com/mld/mercurynews/4698610.htm. Accessed 24 August 2004.

Mill, John Stuart. 1979 [1861]. *Utilitarianism*, ed. George Sher. Indianapolis: Hackett Publishing Company.

Nagy, Andras, and Janet Rossant. 2001. "Chimeras and Mosaics for Dissecting Complex Mutant Phenotypes." *International Journal of Developmental Biology* 45: pp. 577–582.

NAS. National Academy of Sciences. Committee on Guidelines for Human Embryonic Stem Cell Research. 2004. "Workshop on Guidelines for Human Embryonic Stem Cell Research," 12–13 October. Washington, DC.

———. 2005. Guidelines for Human Embryonic Stem Cell Research. Washington, DC: National Academies Press.

NIH. National Institutes of Health. 2000. National Institutes of Health Guidelines for Research Using Human Pluripotent Stem Cells. Available at http://www.bioethics.gov/reports/stemcell/appendix_d.html. Accessed 25 October 2005.

Noonan, John T., Jr. 1970. *The Morality of Abortion*. Cambridge, MA: Harvard University Press.

Okarma, Thomas. 2001. "Human Embryonic Stem Cells: A Primer on the Technology and Its Medical Applications." In *The Human Embryonic Stem Cell Debate*, ed. Suzanne Holland, Karen Lebacqz, and Laurie Zoloth, pp. 3–13. Cambridge, MA: MIT Press.

Parfit, Derek. 1984. *Reasons and Persons*. Oxford: Clarendon Press.

Regan, Tom. 1978. "Fox's Critique of Animal Liberation." *Ethics* 88: 126–33.

Robert, Jason Scott, and Françoise Baylis. 2003. "Crossing Species Boundaries." *American Journal of Bioethics* 3 (3): pp. 1–13.

Rollin, Bernard E. 1998. On *Telos* and Genetic Engineering." In *Animal Biotechnology and Ethics*, ed. Alan Holland and Andrew Johnson, pp. 156–171. London: Chapman and Hall.

Schwarz, Stephen. 1990. *The Moral Question of Abortion*. Chicago: Loyola University Press.

Shreeve, Jamie. 2005. "The Other Stem-Cell Debate." *New York Times Magazine* (10 April): pp. 42–47.

Smyth, J. R., Jr., R. E. Boissey, and M. F. Gawron. 1977. "An Inherited Delayed Amelanosis with Associated Blindness in the Domestic Fowl." *Poultry Science* 56: p. 1758.

Streiffer, Robert. 2003. "In Defense of the Moral Relevance of Species Boundaries." *American Journal of Bioethics* 3 (3): pp. 37–38.

Streiffer, Robert, and Thomas Hedemann. 2005. "The Political Import of Intrinsic Objections to Genetically Engineered Food." *Journal of Agricultural and Environmental Ethics* 18: pp. 191–210.

Svendsen, Clive. 2002. "Stem Cells. In *A Companion to Genethics*, ed. Justine Burley and John Harris, pp. 7–17. Malden, MA: Blackwell Publishing Ltd.

Tang, Ya-Ping, Eiji Shimizu, Gilles R. Dube, et al. 1999. "Genetic Enhancement of Learning and Memory in Mice." *Nature* 401: pp. 63–69.

Thompson, Paul. 1997. *Food Biotechnology in Ethical Perspective*. London: Blackie Academic and Professional.

Thomson, James. 2001. "Human Embryonic Stem Cells." In *The Human Embryonic Stem Cell Debate*, ed. Suzanne Holland, Karen Lebacqz, and Laurie Zoloth, pp. 14–26. Cambridge, MA: MIT Press.

Thomson, Judith Jarvis. 1987. "Ruminations on an Account of Personal Identity." In *On Being and Saying: Essays for Richard Cartwright*, ed. Judith Jarvis Thomson, pp. 215–240. Cambridge, MA: MIT Press.

———. 1997. "People and their Bodies." In *Reading Parfit*, ed. Jonathan Dancy, pp. 202–229. Oxford: Blackwell Publishers.

University of Wisconsin, Bioethics Advisory Committee. 1999. "Final Report of the Bioethics Advisory Committee on Human Stem Cell Research at the University of Wisconsin-

Madison." Available at http://www.news.wisc.edu/packages/stemcells/bac_report.html. Accessed 25 October 2005.

———. 2001. "Second Report of the Bioethics Advisory Committee on Human Stem Cell Research at the University of Wisconsin-Madison." Available at http://www.news.wisc.edu/packages/stemcells/bac_report2.html. Accessed 25 October 2005.

VanDeVeer, Donald. 1979. "Interspecific Justice." *Inquiry* 22: pp. 55–70.

Wagers, Amy, Richard Sherwood, Julie Christensen, and Irving Weissman. 2002. "Little Evidence of Developmental Plasticity of Adult Hematopoietic Stem Cells." *Science* 297: pp. 2256–2259.

Weiss, Rick. 2004. "Of Mice, Men and In-Between: Scientists Debate Blending of Human, Animal Forms." *Washington Post* (20 November): p. A01.

Wertheimer, Alan. 1996. *Exploitation*. Princeton, NJ: Princeton University Press.

WiCell. 2001. "Memorandum of Understanding." Available at http://www.news.wisc.edu/story.php?get=6457. Accessed 25 October 2005.

Zhang, Su-Chun, Marius Wernig, Ian Duncan, et al. 2001. "*In Vitro* Differentiation of Transplantable Neural Precursors from Human Embryonic Stem Cells." *Nature Biotechnology* 19: pp. 1129–1133.

ETHICS AND THE LIFE SCIENCES

CLIMATE CHANGE AND THE CHALLENGE OF MORAL RESPONSIBILITY

STEVE VANDERHEIDEN
UNIVERSITY OF MINNESOTA DULUTH

ABSTRACT: The phenomenon of anthropogenic climate change—in which weather patterns and attendant ecological disruption result from increasing concentrations of greenhouse gases released into the atmosphere through human activities—challenges several conventional assumptions regarding moral responsibility. Multifarious individual acts and choices contribute (often imperceptibly) to the causal chain that is expected to produce profound and lasting harm unless significant mitigation efforts begin soon. Attributing responsibility for such harmful consequences is complicated by what Derek Parfit terms "mistakes in moral mathematics," or failures to correctly assess the various individual contributions to collectively produced harm. Combined with the difficulties in attributing responsibility to agents for spatially and temporally distant harmful effects and that of holding agents culpable for effects (resulting from socially-acceptable acts) about which they may be ignorant, this paper attempts to sort out several ethical problems surrounding the identification of responsible parties contributing to climate change.

The phenomenon of anthropogenic climate change—in which weather patterns and attendant ecological disruption result from increasing concentrations of greenhouse gases released into the atmosphere through human activities—challenges several conventional assumptions regarding moral responsibility. Multifarious individual acts and choices contribute (often imperceptibly) to the causal chain that is expected to produce profound and lasting harm unless significant mitigation efforts begin soon. Persons, that is, through various individual acts, *cause* climate change (i.e., they are *causally* responsible for the phenomenon), but are they *morally* responsible for the harm that is expected to result? Anthropogenic climate change is a problem that cries out for ethical evaluation, but poses two serious challenges to standard accounts of moral responsibility, which this paper surveys.

Attributing responsibility for these predicted harmful consequences is complicated by what Derek Parfit terms "mistakes in moral mathematics," or failures to correctly assess the various individual contributions to collectively produced harm. In addition, agents causally responsible for contributing to climate change are often ignorant of both their individual contributions and the larger effects they cause. Whether or not this ignorance about causal responsibility diminishes or excuses moral responsibility shall be the second ethical challenge examined within the paper.

Given the costs inherent in an effective international climate change mitigation regime, a necessary (but insufficient) condition for the willing participation of relevant parties (e.g., nation-states) is the perceived fairness of the allocation of greenhouse gas abatement burdens, and the most obvious criterion of fairness in distribution involves some conception of responsibility. To wit: insofar as some nation-states bear greater responsibility for creating the problem, fairness demands that they bear proportionally greater responsibility for its remedy. Thus, assessing moral responsibility in a philosophically defensible manner is a necessary first step in designing an effective remedy. Given the complexities involved in establishing causation and disaggregating individual contributions from epiphenomenal problems like anthropogenic climate change, the devil is, so to speak, in the details. The necessary first step in designing a response to climate change is to sort out these details.

INDIVIDUAL RESPONSIBILITY AND "IMPERCEPTIBLE EFFECTS"

Supposing that anthropogenic climate change will, as is predicted, cause significant harm to persons (in both the near and distant future), to what extent can that harm be attributed to individual offenders (that is, to individual emissions of greenhouse gases, through such mundane activities as driving, using electrical appliances, and ea ting beef raised in deforested subtropical areas)? Connecting effect to cause is complicated by the countless tiny point sources of GHG emissions, threshold considerations, and problems of disaggregating large cumulative consequences into apparently negligible discrete acts. Given the nature of an aggregative harm like air pollution, there exists a kind of paradox of small effects: it appears to be true of no one that their acts (by themselves) cause any palpable harm to anyone, and yet the combined acts of many cause significant harm. In view of consequences, it seems paradoxically true that a morally significant harm has resulted from a series of morally insignificant acts. That is, some bad outcomes have been caused by entirely blameless acts.

The problem of attributing responsibility to individual polluters is similar in form to the *voter's paradox*: given the vanishingly small chances of a single vote altering the outcome of a national election, participation by each is thought to be irrational (that is, given the expected costs of voting outweigh any expected benefits from the activity); yet, in combination individual votes may make a difference. As Parfit has shown, however, such accounting involves not so much a paradox as a "mistake in moral mathematics"—it consists (he argues) in "ignoring small chances" of a single act. The *effects* of altered electoral outcomes may not be small (as in the case of climate change), but the *chances* of a single vote altering those effects certainly are. Nonetheless, we must consider those very small chances against the large potential effects. Parfit here argues by analogy: we would ordinarily treat a one-in-a-million chance of killing a person as an acceptable level of risk, but would not so readily dismiss the same odds (faced by a nuclear engineer) of killing a million people. "When the stakes are very high," he suggests, "no chance, however small, should be ignored."[1]

The case of minute and multifarious individual contributions to climate change is not identical, however, in that it is not the small *chances* of individual acts causing significant harm that creates the apparent paradox, but instead the very small (or imperceptible) *effects* of individual actions causing (in combination with other like actions) significant harm. A great many actions have small, almost immeasurable effects on environmental quality, and yet no single one of them may have a perceptible effect on any person. Taken collectively, though, billions of tiny point sources of pollution do cause perceptible harm.

May we conclude from this aggregative effect that each contribution to pollution is an act which indeed causes harm (albeit a tiny fraction of a significant problem, and one that may be both spatially and temporally distant from the agent), and which can be so treated by a regulatory regime? Must we treat these micro-offenses as constituting palpable harm, therefore (following Mill) warranting state interference?[2] If so, the likely result is a ban on nearly all human activity, including breathing, insofar as it requires the emission of greenhouse gases into the atmosphere. Such a conclusion, obviously, would be absurd. Does the rejection of this absurd conclusion justify its opposite: that we should not regard any small act of pollution or resource degradation as an act of harm? It doesn't follow necessarily (although this conclusion is often adopted by policymakers nonetheless), and such an inference would at least require further argument before its implication for environmental regulation could be accepted.

One obvious mistake in the above inference concerns threshold effects: given a smaller global population, many of the present activities that contribute to increasing concentrations of atmospheric GHGs could safely be assimilated into the environment without any discernable effects upon climate. For 10,000 years, atmospheric GHG concentrations remained remarkably stable, and only began increasing during the early stages of industrialization, and more recently began a more rapid ascent. Thus, some act that—even in combination with other like acts—*once* was entirely benign (and would still be, given smaller global population), *became* harmful once some threshold level of population or emissions was exceeded. Literally, the wrongness of some act may depend upon how many other people are able to benignly commit that same act.

Parfit refers to the difficulties surrounding this threshold problem as the mistake of "ignoring the effects of sets of acts." Since many desirable outcomes require cooperation among groups of people, and in the absence of such cooperation individual acts fail to bring about even a fraction of the desired outcome, some notion of the role of individual participation in a group endeavor, he rightly argues, is needed. Likewise, some kinds of actions don't appear to cause harm individually (small sources of pollution being an example), but taken together with other similar acts they do cause measurable harm. The principle that Parfit offers for overcoming this mistake in moral mathematics is: "Even if an act harms no one, this act may be wrong because it is one of a *set* of acts that *together* harm other people. Similarly, even if some act benefits no one, it can be what someone ought to do, because it is one of a set of acts that together benefit other people."[3] The benefit or harm of a cooperative endeavor must be disaggregated so that individual acts can properly be assigned responsibility for their contributions to the overall outcome, and the above principle does just that.

A similar mistake involves "the belief that imperceptible effects cannot be morally significant." Even if a person's share of a collective act that causes either harm or benefits is so small that the effects of each individual contribution cannot be perceived by a beneficiary or sufferer, those effects, Parfit claims, matter morally. Environmental harm is especially prone to this mistake in moral mathematics, because of the very small contribution that individual point sources of pollution have on aggregate pollution levels. Likewise with the depletion of natural resources, individual contributions to larger aggregate problems appear to be trivial (as in cutting down a single tree), and yet the countless occurrences of such seemingly trivial acts together add up to quite serious harms. As Parfit puts it, "each of our acts may be *very* wrong, because of its effects on other people, even if none of these people could ever notice any of these effects. Our acts may *together* make these people very much worse off."[4]

To illustrate, he proposes a "commuter's dilemma" in which persons commuting into a city must decide whether to drive (thus contributing a small amount of pollution) or take public transportation (yielding a much smaller share). The latter option has a convenience cost that most people are unwilling to bear, at least not in the absence of a significant countervailing benefit. Like many other coordination problems involving groups, the commuter's dilemma leaves each potential driver weighing the trivial cost of pollution (shared by all) against the seemingly more significant convenience cost of using public transportation (borne by each). Even though they may recognize air pollution as a significant public health issue and acknowledge that automobile emissions are a prime cause of such pollution, each commuter regards their own contribution to the problem as insignificant, and further reasons that their own "sacrifice" in taking public transportation will have an imperceptible effect on the overall pollution problem (given the likelihood that others will continue to drive). As game theory predicts in such dilemmas, they all choose to drive. But should they (or, more to the point, can the state legitimately discourage them from doing so)?

Parfit, rejecting the mistaken intuition that holds imperceptible effects to be insignificant, answers that they should not. In the past, he notes, when most people lived in small communities, the consequences of making these mistakes in moral mathematics were far less serious. A relatively small group of persons each making trivial contributions to air pollution may never harm anyone, insofar as that pollution remains dispersed and falls within the capacity of the natural environment to cleanse itself (known as its carrying capacity). With urbanization, these mistakes began to take on far more serious consequences when carrying capacity was exceeded, and many more trivial contributions to aggregate problems like pollution became more insidious.

Increasingly concentrated populations dramatically intensify individual contributions to problems of pollution, especially once carrying capacity is exceeded, but large populations do nothing to disperse those harms or diminish their effects. Each person may contribute only a tiny fraction of the overall levels of pollutants in the environment, but each bears the full cost of that pollution. A single person inhaling carbon monoxide does almost nothing to cleanse the air of that pollutant. While perhaps true that society at the time could safely ignore small or imperceptible harms, such is no longer the case. "It now makes a great difference whether we continue to believe that we cannot have greatly harmed or benefited others unless there are people with obvious grounds for resentment or gratitude," Parfit urges. "If this is what we think, what we do will often be much worse for all of us."[5]

IGNORANCE, NEGLIGENCE, AND CULPABILITY

Given the nature of individual contributions to aggregative problems like climate change, and given also the massive public relations effort designed to discredit the scientific basis for the causal link between fossil fuel combustion and deforestation, increased atmospheric concentrations of GHGs, and their predicted consequences, it should not be surprising that many Americans do not feel at all responsible for the consequences of their GHG emissions-producing acts. That is, people do *cause* climate change through a wide variety of otherwise innocuous acts, but they are largely ignorant about the effects on these acts. Does this ignorance in any way diminish their culpability?

At issue here is not whether or not the many small contributions of GHGs into the atmosphere *cause* the problems of climate change—the Intergovernmental Panel on Climate Change has established this causal link with a high level of certainty[6]—but whether or not (and under what circumstances) persons might be held *morally responsible* for their contributions to the problem. The relevant distinction, then, is between *causal responsibility* (where effects are linked to particular causes) and *moral responsibility* (or culpability, where agents are blamed for consequences for which they are causally responsible. In at least some cases, persons may be held causally but not morally responsible, and the problem of assessing culpability for anthropogenic climate change might begin with two of those.

The first kind of case in which causal responsibility does not entail moral responsibility concerns thresholds: *ought* implies *can*, so it cannot be the case than *any* emission of GHGs triggers moral blame unless persons could plausibly be expected to refrain from exhaling carbon dioxide. This problem is relatively easy to address. Though absurd to blame a person for exhaling (thereby contributing GHGs into the atmosphere), it hardly follows that any individual emission is morally blameless. As suggested above, some threshold (perhaps calculated from *carrying capacity*, or the ability of carbon sinks to absorb or recycle GHGs without deleterious effects upon climate) sets the limit for blameless individual emissions, where further emissions beyond the threshold constitute a morally blameworthy failure to within the bounds of sustainability. It may not be wrong to produce carbon dioxide, but it is wrong to produce an excessive amount of it.

The second problem is more difficult, and involves the reasonableness of ignorance concerning the likely effects of one's intentional acts. For one reason or another, most Americans do not regard their everyday acts of GHG production to be causally related to global climate change, despite the oft-observed fact that U.S. *per capita* emissions well exceed the threshold of sustainable GHG emissions and the widely disseminated reports linking fossil fuel combustion, changing land use patterns, and climate-related problems. Though based on a mistake, it nonetheless complicates attributions of moral responsibility to observe that harmful acts result not from individual malevolence but from adherence to widely accepted social norms, which offer no prohibitions against such acts. Can we attribute moral responsibility to persons (or to entire nations of such persons) for harm that results from acts that they neither think to be wrong nor violate existing ethical norms?

Ordinarily, we do not blame persons for bad consequences that result from accidents, where an accident is defined as a consequence that could not reasonably be anticipated to follow from some act. Notice that accidents involve causal responsibility; people do, in fact, *cause* the consequences of accidents (and so they are responsible for them in one sense), although they are not held to be culpable for those unintended consequences (hence are *not* responsible in another sense of the term). This divergence between causal and moral responsibility, however, only applies to a subset of acts that produce unintended consequences. That is, agents are assumed to be *morally* responsible (as opposed to merely being *causally* responsible) for their acts (or omissions) insofar as they can reasonably be expected to anticipate the consequences of those actions (or inactions), whether or not they do, in fact, anticipate them.

The obvious exception to this gap between causal and moral responsibility involve *negligent* acts, where agents fail to adequately consider the possible bad consequences of their actions, and thus subject others to undue risk. In cases of negligence, we say that although some agent did not in fact anticipate that her act might cause morally significant adverse consequences for someone, a reasonable person would have done so. The problem with defining negligence in terms of what a reasonable person would anticipate, however, is that it relies upon societal norms, and a "reasonable" North American may well commit the same "mistake in moral mathematics" of which our agent stands accused. The common law tradition from which negligence has conventionally been defined is confounded by problems like anthropogenic climate change, in which once-permissible actions become impermissible at some point, once thresholds are crossed and empirical knowledge establishes the causal link between particular acts and future harm. In such cases, social norms lag behind the factual bases of our obligations.

Nonetheless, in assessing the mitigating effects of ignorance concerning our causal responsibility for climate change upon the moral responsibility that might be attributed to us for excessive emissions of GHGs, a conception of reasonable ignorance is instructive. In order to avoid cognitive dissonance and thereby to continue upon our current trajectory of greenhouse gas production, a strong temptation to ignore reports about the effects of everyday actions may be psychologically understandable, but this does not make it morally defensible. Twelve years after the Rio Declaration committing developing nations to GHG abatement, and with three scrupulously researched and widely disseminated IPCC Assessment Reports, claims to reasonable ignorance concerning anthropogenic climate change are fully implausible, despite the uncertainties that remain in climate science. Even if the predictions about the harmful consequences of climate change turn out to be overstated, ignoring the considered recommendations of the vast majority of the world's scientific community can only be described as willful ignorance, and cannot exonerate one from moral responsibility for resultant harm.

As citizens of the nation that continues to both lead the world in per capita greenhouse gas emissions and to lead the effort to subvert or undermine any global effort to reduce these emissions, we cannot any longer invoke ignorance as a defense against acknowledging our causal and moral responsibility for this potentially devastating global problem, nor can we plausibly continue to deny the seriousness of anthropogenic contributions to climate change or commit the fallacy that views individual contributions as trivial and therefore unworthy of moral scrutiny. Insofar as we continue to fail to adequately address this problem, we shall do so irresponsibly.

NOTES

1. Derek Parfit, *Reasons and Persons* (Oxford: Clarendon Press, 1984), p. 75.

2. Mill famously argues for interference in individual liberty if (and only if) the act in question causes harm to others, but rejects the application of his harm principle to acts that produce "imperceptible" effects. John Stuart Mill, "On Liberty," from *Utilitarianism, On Liberty, Considerations on Representative Government*, ed. by H. B. Acton (Rutland, Vt.: Everyman's Library, 1972), pp. 78–83.

3. Parfit, *Reasons and Persons*, p. 70.

4. Ibid., p. 83.

5. Ibid., p. 86.
6. Intergovernmental Panel on Climate Change, *Climate Change 2001: A Synthesis Report*, ed. R. T. Watson and the Core Writing Team (New York: Cambridge University Press, 2001).

BIBLIOGRAPHY

Intergovernmental Panel on Climate Change. 2001. *Climate Change 2001: A Synthesis Report*, ed. by R. T. Watson and the Core Writing Team. New York: Cambridge University Press.

Mill, John Stuart. 1972. "On Liberty." In *Utilitarianism, On Liberty, Considerations on Representative Government*, ed. by H. B. Acton. Rutland, VT: Everyman's Library.

Parfit, Derek. 1984. *Reasons and Persons*. Oxford: Clarendon Press.

ETHICS AND THE LIFE SCIENCES

PERSPECTIVES ON ETHICS AND WATER POLICY IN DELAWARE

GERALD J. KAUFFMAN
UNIVERSITY OF DELAWARE

ABSTRACT: Water is a finite resource held in common by the community yet coveted by individuals and special interests. The water management field is filled with disputes about water allocation, rights, and pollution. Environmental ethics is a basis for equitable water policy making in Delaware. The resource allocation dilemma is examined in relation to conflicting objectives imposed by a market economy between individual self-interests and community environmental well being. Two forms of water law are practiced in the USA—eastern riparian rights and western prior appropriation. Both forms seek an ethical balance to resolve conflicts and protect individual water rights while protecting downstream users (the common good). Delaware Valley case studies discuss how environmental ethics can help the water policy specialist make difficult decisions during conflicts. Surveys polls indicate that 81 percent have values supportive of a balance between the economy and environment, or pro-environment, indicating that an environmental ethic is central to decisions concerning water policy.

INTRODUCTION

Water resources managers in Delaware often recommend water policy based on Federal, State, and local laws and regulations. Water resources decision—making becomes a dilemma because of politics, special interests, and limitations in water law or regulations that do not apply to a particular case. The central theme of this article is that the fundamentals of environmental ethics can provide a basis to make difficult yet equitable water policy decisions in the face of often-conflicting special interests. Environmental ethics is an overall philosophy or set of principles concerned with protecting the earth's natural resources.

PROBLEM STATEMENT

What are the bases for sound decisions regarding water policy in Delaware and other areas? What tools are available where water law or economic principles do not extend far enough to assist a water policy maker in making equitable decisions? The literature and several case studies suggest that the code of environmental ethics is available to the water resources manager to make tough yet fair decisions.

Water is a finite natural resource held in common good by the community yet coveted by the individual and special interest groups. Hence, the field of water resources management is a discipline filled with conflicts and disputes about water allocation, water rights, and water pollution. This article explores the field of environmental ethics as a basis for fair and equitable water policy making in Delaware. The dilemma of resource allocation is examined, given the seemingly conflicting objectives imposed by a market economy between individual self-interest and the environmental well being of the larger community, and seeks concepts and methods to facilitate ethical decision-making. We examine the ethical dimension of water law as practiced in different forms in the eastern and western USA. We explore the dilemma: the economic interests of the individual or the environmental needs of the community. Can one have both?

This article reviews a brief history of the evolution of environmental ethics over the last fifty years. A series of case studies in Delaware and other states discusses how a sense of ethics can help the water policy specialist make difficult decisions in the face of various conflicts.

Lastly, we surveyed several committees of water resources practitioners who work in the watersheds in and near Delaware to gauge their beliefs in an environmental ethic as a basis for decision-making. In an effort to employ a sort of environmental ethics index, the survey responses are grouped according to three categories: (1) responses with a pro-economic viewpoint, (2) responses with a more balanced viewpoint between the environment and the economy, and (3) responses with a pro-environmental viewpoint.

CONCEPTUAL FRAMEWORK

The economic interests of the individual or the environmental needs of the community—can we have both? On the front pages, this ethical dilemma is debated in columns about construction of new reservoirs to address drought, oil drilling in national wildlife refuges, and the effects of sprawl development on the landscape. With a dwindling land base, rising population, and the quest for economic prosperity, water policy makers should reconsider the discipline of environmental ethics when evaluating policies to protect our land, water, and natural resources.

The public manager confronts many difficult decisions in enacting policies to protect natural resources. Should a builder be permitted to cover an aquifer with acres of shopping mall pavement at the expense of the drinking water supply? Should an oil refinery be allowed to operate without adequate air or water pollution control equipment?

Usually ordinances and regulations are designed to protect the environment by limiting the maximum levels of water and air pollution generated by these economically productive activities. Presumably the natural resources laws or regulations were written in an ethical spirit to guide development or progress without appreciably harming the environment. But what do we make of the developments that meet the letter of a regulation but not the ethical intent? In many cases a regulation or ordinance or law is not available or complete enough to guide public manager's decisions in protecting natural resources. For instance, in some states there is no law setting a minimum stream flow to protect fisheries. The ethical public manager would advise a water supplier to leave some portion of the water flow in the stream to support the fishery even if there is no legal requirement to do so.

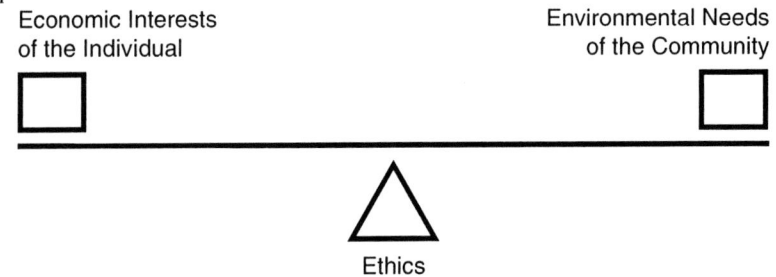

Figure 1. The ethical dilemma: balancing the economic interests of the individual with the environmental needs of the community.

Ethical decisions in water resources management are not black or white but rather shades of gray. Water policy-making is often influenced by the perspectives of various stakeholders in the watershed: the developers, environmentalists, government officials, farmers, and industrialists. Mahatma Gandhi's ethical philosophy emphasized protection of individual rights while preserving the welfare of collective society. Gandhi would advocate that the rights of all the stakeholders should be protected in the watershed not just the rights of a few. But often this is not the case. The viewpoints of special interests in the watershed are often acceded to at the expense of greater society (Cech, 2003). At the March, 2003 Third World Water Forum in Kyoto, the president of the World Water Council announced four key priorities to meet "the greatest challenge of the 21st Century—fresh water." He went on to say that first there needed to be a focus on the ethics of water use namely in the areas of water rights and regulation (Third World Water Forum, 2003). In the ethical spirit, water is described as a basic human right, a common good, not a commodity to be sold by special interests (Brunner et al. 2002).

Perhaps conflicts between upstream and downstream owners are inherent in water management as both the words "river" and "rival" are derived from the Latin *rivalis* meaning "one taking from the same stream as another" (Webster's Dictionary, 1980).

Ethics is a discipline that implies a moral duty or obligation. The ethical public manager evaluates the strengths and weaknesses of a proposal regarding impacts on the environment and makes a decision based on the strength of water law and a set of ecological values. When confronted with difficult water resources management decisions that are clouded with political overtones, we recommend "doing what's right for the water" as the ethically preferable thing to do.

The ethical dilemma inherent in water management is compounded because streams and watersheds know no political boundaries. Most state, county and municipal boundaries do not coincide with watershed boundaries and the polyglot of individual governments is what makes water management so complex. The many governments in a watershed may have different stream standards or different economic goals such as pro-development or pro-preservation. Because the many governments have different agendas, it puts them in dispute with their upstream and downstream neighbors leading to conflicts that must be resolved by public managers usually through the principles of watershed management.

Speaking at *Drinking Water 2001*, a public policy forum organized at the University of Delaware in October 2001, McKay Jenkins described this dilemma when he said:

> What I would like to do today is try and expand our notion of the importance of watersheds to talk about borders and flow in a larger context. Ecologists and drinking water experts have long acknowledged the silliness—not to say utterly counterproductive, and potentially destabilizing—notion of political boundaries when it comes to the flow and distribution of water. What does a county line mean to an aquifer? What does a state line mean to a raincloud? What does a national border mean to a river? . . . The point I want to make here is that any effort to reject the permeability and flow of boundaries, be they natural or psychological, runs against the natural way of things. Water wants to flow—it's in the nature of water. People want to flow—it's in the nature of people. . . . Finally, at least in some places in the country, we are beginning to think in terms not of boundaries, but in terms of watersheds, and flow. (Jenkins, 2001)

Water policy makers strive to make decisions to provide the greatest public good. But how? Water laws and regulations and economic principles have evolved to address this question. Where laws fall short, we turn to ethics as the basis for water resources decision-making to protect the rights of the individual as well as the common good in the watershed.

ETHICAL PRINCIPLES IN WATER LAW

Water laws are based on a set of ethical principles that seek to protect the water resource while trying to balance the needs of various environmental, economic, and social interests. Water law codifies a set of ethical principles designed to protect the water resource and its users.

In the United States, we practice two disparate forms of water law—riparian rights (eastern water law) and prior appropriation doctrine (western water law). Riparian rights law is practiced in the humid, water rich eastern USA in states such as Delaware (Dzurik and Theriaque, 1996). Prior appropriation doctrine is practiced in arid western states such as Arizona. Both forms of law seek to protect the individual water user while trying to protect other users on the stream (the common good). However, there are important differences.

Eastern Water Law

Ethics form a tenet of water law or what are commonly called riparian rights. Riparian rights law is practiced in the humid, water—rich eastern United States and is derived from English common law. Riparianism provides the right to use water as a property right. Owners of land or property in contact with lakes or rivers are granted the right to use the water from that waterway. It's use originally evolved to protect water-powered mill owners and later in pollution prevention to provide downstream users a clean supply. The ethical dimension of eastern water law is that upstream riparian owners may not obstruct the water flow, impair the water quality or otherwise injure the lower or downstream owner. Lower riparian owners may not back-flood an upstream owner. Injured riparian owners may recover damages.

Riparianism dictates that an upstream builder may develop land provided that the construction of new pavement and hard surfaces do not increase downstream flooding or cause more stormwater pollution. Riparian water law provides the basis for most floodplain and stormwater ordinances that have been adopted in many watersheds. The riparian philosophy is the precept of whole basin principles whereby streams are managed on a "watershed" basis with lesser regard for often-conflicting viewpoints expressed by governments separated by political boundaries.

In Delaware, riparian water law is modified by reasonable use doctrine. That is, the upstream owner can take an amount of water as long as the use does not interfere with the "reasonable use" of the downstream owner. The term "reasonable use" is a gray area that is usually defined by an ethical water policy maker and, if not, is decided by a judge and jury.

Western Water Law

Prior appropriation doctrine is practiced in the arid western USA in states such as Arizona. Western water law evolved as a way to protect miners and settlers. Since the federal government owned most of the land in the west, settlers could not claim riparian water rights since they did not own the land. So water rights law in the west grew to resemble the claims system set up by gold and silver miners. Just as the miners who arrived first claimed a grubstake, western water law grew where the first person to divert water is granted the vested right to that water. Prior appropriation doctrine grew to a "first in time first in right" doctrine.

In western water law there evolved two types of users: senior appropriators and junior appropriators. A senior appropriator is the first user on the stream who, since he was there before all the others, has rights to as much water as needed and the full flow in the stream. The junior appropriators settled later and have secondary rights to the water. They receive their allotted supply except during drought when they may have the right to no water at all.

This system normally provides sufficient water to the senior and junior users on the stream. However during drought, the senior appropriator has the right to all the water they need. The junior user has rights to whatever water is left, which could be none at all. Senior rights can be lost if water is not withdrawn over a certain period of time. If the senior owners do not use the available water in the stream, they could forfeit their right to that water.

Comparison of Water Law Doctrines

Table 1 compares and contrasts the two forms of water law practiced in the United States. The ethical forms of eastern and western water law are quite different. Where eastern water law provides an ethical dimension where the upstream user must provide sufficient water to the downstream user, in western water law the senior user may use all of the water in the stream leaving none for the downstream user. In the west the owner can actually forfeit water rights if water is not withdrawn over a certain period. Travelers in the west may observe irrigation running continuously even during periods where watering is not needed. This quirk in western water law actually promotes water waste, which is counter to a water conservation ethic. For this reason many western states such as Colorado have adopted a hybrid form of water law combining riparian and prior appropriation doctrines (Cech, 2003).

Table 1. Forms of Water Law in the United States

Parameter	Riparian Rights	Prior Appropriation
Geography	Eastern USA	Western USA
Climate	Humid, water rich	Arid, dry
States	Delaware	Arizona
Right to Water	Property right provided by adjacency to stream.	First in time, first in right.
Withdrawal	Must return unused water.	Can take all water in stream.
Superior Rights	Upstream owner must allow sufficient flow to downstream owner.	Senior user has rights to all water. Junior user takes leftover water.
Transferability	Can transfer water rights with property.	Can forfeit seniority/water rights if water not used for a certain period.

LITERATURE REVIEW

Philosophers throughout history have considered the triad of ethics, the environment, and the economy and whether mankind can achieve balance with nature. Some philosophers believed in an ethic that supports mankind's dominion over nature. The Book of Genesis in the Old Testament of the *Bible* recorded a biblical belief in a human right to master the earth and it's creatures (Attfield, 1983). Rene Descartes in the seventeenth century wrote that mankind's goal was to become nothing less than the master and possessor of nature (White, 1967). American frontiersmen believed in a nineteenth-century manifest destiny in a right to settle the west, string up barbed wire, and tame nature. More recently the anti-environmentalism views of James Watt, Rush Limbaugh and the rising tide of the "Wise Use" groups contribute accusations that ecological protection hurts the economy (Brick, 1995). These principles of conquering nature and harnessing the environment are still practiced today in the twenty-first century even in the face of a growing environmental movement.

Over the last fifty years, the environmental ethic has evolved to a tradition concerned with balancing the wise use of the earth and its creatures. In 1949, Aldo Leopold in *Sand County Almanac* proclaimed that conservation is a "state of harmony between man and the land (Leopold, 1966). In 1962 Rachel Carson wrote in *Silent Spring* that we have a moral obligation toward nature thus giving rise to the federal water and air pollution laws that followed (Carson, 1962). In April 1970, millions assembled to celebrate the first Earth Day, a watershed moment in the progression of an environmental ethic. The philosophy of the "deep ecology" movement blossomed in 1973 which reflected the inter-relatedness of all mankind as a biotic community (Stark, 1995).

The environmental movement evolved from revolution to regulation when Richard Nixon signed the Federal Clean Water Act in 1972 which set fishable and swimmable standards for waterways in the United States. Later Congress passed the Safe Drinking Water Act Amendments of 1986 and 1996 which set enforceable drinking water standards including requirements for wellhead protection. In 1990, New Castle County, Delaware adopted one of the first water resource protection area ordinances in the country that set thresholds on the amount of new development over an aquifer to protect drinking water supplies (Kauffman and Brant, 2000).

In 1992, the Congress on Renewable Natural Resources "called for our nation and its resources community to develop and adopt a stewardship/sustainability ethic incorporating a long-term perspective to guide both public and private resources decisions" (Renewable Natural Resources Foundation, 1992).

Research over the last few decades supports the feeling of a growing pro-environmental ethic in the USA. A 1989 Harris poll reported that 97 percent of the respondents felt that the country should be doing more to protect the environment and curb pollution (Kempton, Boster, and Hartley, 1999). A 1990 Roper Organization survey of the public indicated that 50 percent of those polled believe that environmental laws and regulations don't go far enough, up from 30 percent with the same belief in 1980. Membership in environmental lobbying organizations such as the Sierra Club and Natural Resources Defense Council exceeded 3,100,000 in 1990, a 30-fold increase from 120,000 members in 1960 (Mitchell, Mertig, and Dunlap, 1991). The survey results indicate that more people care for the environment than they used to.

So the environmental ethic evolved from a state of dominion to a more modern ecological concept of balance with nature. The theme of stewardship was resurrected meaning that people are entrusted with a duty to preserve the earth's beauty and fruitfulness. The modern ethical public manager assumes a stewardship role and maintains an obligation to secure a clean environment for future generations and posterity.

Table 2. Abbreviated Chronology of Watershed Moments in Environmental Ethics

Period	Watershed Moments in Environmental Ethics
1949	Aldo Leopold wrote *Sand County Almanac* and defines conservation as state of harmony between man and land.
1962	Rachel Carson wrote *Silent Spring* emphasizing a moral obligation toward nature.
1970	First Earth Day celebration in April.
1972	Passage of Federal Clean Water Act setting swimmable and fishable standards.
1973	Philosophy of "deep" ecology blossomed.
1989	Harris poll reports 97% feel country should do more to protect environment.
1990	Roper organization reports 50% believe environmental laws do not go far enough.
1990	Membership of environmental groups exceeds 3,000,000 up from 120,000 in 1960.
1992	Congress on Renewable Resources calls for nation to adopt a stewardship/ sustainability ethic.
1996	Congress passes amendments to Federal Safe Drinking Water Act.
1997	New Castle County, Delaware modifies one of first water resource protection area ordinances in USA to protect ground and surface water drinking water supplies.

With knowledge of the evolution of the modern environmental ethic, the public manager has a sturdy foundation upon which to make difficult, but balanced decisions concerning natural resource protection.

CASE STUDIES

The following case studies explore the ethical dilemmas that face the water policy manager.

Water Law versus the Constitutional Takings Issue

The principles of water law can be used by the ethical public manager to address constitutional takings challenges. The Fifth Amendment of the United States Constitution reads that private property shall not be taken for public use without just compensation (Farmer, 2001). According to interpretations of the Fifth Amendment, landowners have the right to develop land and realize reasonable fair market value provided what is done on the land does not unduly harm others or the environment. If the landowner is impeded from realizing an economic return on the land by government ordinance, then the landowner has the right to fair compensation by the government.

Pro-development and/or anti-environmental interests commonly cite the takings issue as the mechanism to overturn environmental regulations such as floodplain or wetlands protection ordinances. The common economic argument provided is that floodplain ordinances monetarily injure the landowners by limiting the number of homes or acres of pavement built in the floodway thus reducing the value of the land. This, in turn, triggers an accusation of a "taking" of land from the owner by the government.

Incidentally, those concerned about losing the value of floodplain and wetland land due to ordinances often oversell the value of the land as it was usually acquired for many thousands of dollars per acre less than adjacent high land. Floodplain land has little economic value in the first place. Why? The function of the floodplain is to flood.

In response to a takings challenge, an ethical public manager might point out from an environmental perspective that the floodplain ordinance does not outright prohibit the development of the land. It just sets a protective threshold limiting the number of structures to protect the ecological value of the floodplain, prevent flood damage to the owner, and minimize downstream flooding. The basis of the floodplain ordinance then is in line with the ethical dimension of riparian water law whereby the upstream owner (developer proposing to put homes in the floodplain) may not injure the downstream owner.

The ethical dimensions of water law may be used to address other economic arguments for the development of land particularly when builders raise the specter of a constitutional taking challenge. Should 100 acres of forest be cut down to accommodate 100 homes to maximize economic return to the builder? This proposal may not meet the ethical principles of water law because the loss of trees and addition of pavement will increase downstream flooding and stormwater pollution thus injuring the riparian rights of the downstream or subservient landowner. A compromise might be to cluster the 100 homes on twenty-five acres of the parcel leaving much of the forest intact thus limiting the possible increase in downstream flooding

An upstream owner has the right under the constitution to develop land and maximize economic return but not if these actions injure the downstream owner which would violate the ethical dimension of riparian water law. The public manager strives to balance the rights of landowners under the constitution so that they do not suffer economic injuries while at the same time ensuring the rights of downstream owners under the protection of water law so that they are not injured by upstream development interests.

Imperviousness and the Drinking Water Aquifer

Consider the case of the ethical public manager and a proposal for a new shopping center over a drinking water aquifer in New Castle County, Delaware (Kauffman, 2001). During the early 1990s, a prominent land development firm filed plans to build a new shopping center with approximately 60 percent impervious roof and pavement area over a limestone aquifer that provides drinking water to 20,000 people. This proposal did not comply with the county water resource protection area (WRPA) ordinance that set a maximum 50 percent limit on new pavement and roof area to protect the sensitive drinking water aquifer.

The shopping center developer circumvented the WRPA ordinance and secured an agreement from the county to build the project at 60 percent impervious cover. This project not only violated the letter of the WRPA ordinance but the intent of the ordinance, which was to protect the quality and quantity of drinking water supply. The developer pursued his financial self-interest which threatened the value of the aquifer hence the choice may not have been in concert with the modern conservation ethic.

During deliberations concerning this matter before the county's Resource Protection Area Technical Advisory Committee, the public manager employed the ethical argument of "doing what's right for the water." The public manager advised against approving this proposal because at 60 percent impervious cover the development was out of harmony with the water resource. The project was not only technically out balance with the ordinance but ethically challenged since the developer cut a deal to circumvent the ordinance.

The public manager lost the ethical battle with this particular project. But the war was won as the WRPA ordinance was toughened in 1997 to prevent further projects of this kind that could harm the drinking water resource (New Castle County Unified Development Code, 1997).

Ethics of Watershed Management

The Christina River Basin in Delaware, Pennsylvania, and a small sliver of Maryland contributes drinking water to over a half million people in these states situated near Wilmington halfway between Philadelphia and Baltimore (Greig, Bowers, and Kauffman, 1998). Streams in the basin such as the Brandywine Creek flow from the headwaters in Pennsylvania and Maryland downstream into Delaware before flowing out to the Delaware River. Pennsylvania has designated the creek as a warm water stream, a less protective designation because the commonwealth is a large state with hundreds of hilly Piedmont streams. In tiny Delaware with three counties and only six Piedmont streams of this type, the state considers the Christina Basin to be of statewide significance. Delaware regulates the same creek just a few yards away over the state line as exceptional resource water with more protective stream water quality standards.

A factory in the Pennsylvania portion of the watershed files a wastewater discharge permit that meets the state's stream water quality standards. However, a few miles downstream over the state line in Delaware, the discharge effluent from the industry violates the Delaware's more protective stream standards. The industry discharge meets the upstream state's water quality standard but does not comply with the downstream state standards. The two states are in dispute.

What course should an ethical watershed manager pursue in this instance? One opportunity is to employ the ethic of riparianism as the principle of watershed management. Since watersheds know no political boundaries, an ethical manager would advocate the regulation of the industrial discharge without regard for political boundaries. Using the principle of the watershed, the public manager advocates unifying the differing water quality standards across state boundaries so that a single water quality standard is employed for the watershed regardless of state status.

Fortunately, the Total Maximum Daily Load provisions of the Federal Clean Water Act provide the opportunity for common regulation of stream water quality standards across state boundaries. In 2001, the two states banded together as part of a joint Christina Basin Watershed Committee Strategy and developed a maximum load that the industry can discharge into the stream to meet both states water quality standards. The ethical water managers from both states averted a lawsuit. Using the principles of watershed management, water managers in both states employed a common environmental ethic to reach across state lines to solve this water pollution problem.

Table 3. Ethical Considerations in Case Studies

Case Study	Ethical Considerations
Case Study 1 Water Law versus the Constitutional Takings Issue	The ethical dimension of riparian water law is that the upstream owner (developer proposing to put homes in the floodplain) may not injure the downstream owner. The public manager should employ an ethical spirit to balance the rights of landowners under the constitution so that they do not suffer economic injuries while at the same time ensuring the rights of downstream owners under the protection of water law so that that they are not injured by upstream development interests.
Case Study 2 Imperviousness and the Drinking Water Aquifer	The public manager employs the ethical argument of "doing what's right for the water." The public manager advised against approving this proposal because at 60 percent impervious cover the development exceeded the 50% impervious cover requirement of the ordinance and was out of harmony with the water resource. The project was not only technically out balance with the ordinance but ethically challenged since the developer cut a deal to circumvent the ordinance.

Case Study	Ethical Considerations
Case Study 3 Ethics of Watershed Management	The ethical water managers from both states in the Christina Basin averted a lawsuit when an industry in upstream Pennsylvania proposed a wastewater discharge that did not meet downstream Delaware's water quality standards. Ultimately a strategy was crafted under the Federal Clean Water Act whereby the wastewater discharge could meet both states' standards. Using the principles of watershed management, water managers in both states employed a common environmental ethic to reach across state lines to solve this water pollution problem.

SURVEY METHODS

Thus far this article discusses the evolution of environmental ethics and through case studies it's availability as a tool for equitable decision making in water resources policy. But what are the current environmental attitudes and values of those who participate in water resources policy making in Delaware? To assess the environmental ethic of the water resources community, we surveyed members of the Delaware Water Supply Coordinating Council and Christina Basin Clean Water Partnership using a survey instrument adapted from the peer-reviewed literature. Research of the published literature indicates that the following survey methods are available as a possible environmental ethic "measuring instrument."

The New Environmental Paradigm summarized responses to twelve questions designed to measure a new Environmental Paradigm Index (EPI) as the state of the worldview and desire to protect the environment (Dunlap and Van Liere, 1978). The survey directed a series of statements to 806 respondents from the general population and 407 respondents from environmental organizations and asked them to strongly agree, mildly agree, mildly disagree, or strongly disagree on their views toward the environment. The survey included the following statements:

1. We are approaching the limit of the number of people the earth can support.
2. The balance of nature is very delicate and easily upset.
3. Humans have the right to modify the natural environment to suit their needs.
4. Mankind was created to rule over the rest of nature.
5. When humans interfere with nature it often produces disastrous consequences.
6. Plants and animals exist primarily to be used by humans.
7. To maintain a healthy economy we will have to develop a steady state economy where industrial growth is controlled.
8. Humans must live in harmony in nature in order to survive.
9. The earth is like a spaceship with only limited room and resources.
10. Humans need not adapt to the natural environment because they can remake it to suit their needs.

11. There are limits to growth beyond which our industrialized society cannot expand.
12. Mankind is severely abusing the environment.

The article concludes that in 1978 the "general public tends to accept the content of the emerging environmental paradigm much more than we had expected." The survey concluded that concepts such as balance with nature were beginning to permeate the consciousness of the public and that the new Environmental Paradigm Index is useful in assessing the attitudes and values of the public and environmental organizations toward the environment.

Public Opinion in the 1980s Clear Consensus, Ambiguous Commitment surveys trends in public opinion and support of the environment (Dunlap, 1991. The thesis of the article is that public support of the environment continues on the upswing. Members of the public were asked eight questions drawn from original surveys by the National Opinion Research Center, Cambridge Reports, the Roper Organization, and New York Times/CBS Polls. For instance, one of the questions asked:

1. Are we (the Government) spending too much, too little, or about the right amount on improving and protecting the environment?

- Don't know   5%
- Too much   4%
- About Right   21%
- Too Little   70%

The results of the survey indicate that in 1990 there was a measurable "proenvironmental sentiment." For instance, 75 percent of those surveyed said the government is spending too little on improving the environment. Overall, more than half of the respondents said that the environment was so important that more should be done to protect it. One could interpret from the survey that a majority of the public surveyed had a relatively strong environmental ethic.

Environmental Values in American Culture included a fixed-form survey to measure the environmental values of selected sectors of the American public (Kempton et al., 1995). One hundred forty-nine questions were answered by members of Earth First, Sierra Club, lay public, dry cleaner, and sawmill worker samples. The respondents were asked to strongly agree, agree, slightly agree, slightly disagree, or disagree with the questions. Several of the 149 survey questions are listed below:

7. People have a right to clean air and clean water.
24. We have to protect the environment for our children, and our grand-children, even if it means reducing our standard of living today.
39. A healthy environment is necessary for a healthy economy.
74. The environment doesn't need as much protection as we imagine.

The results of the survey indicate, "our data demonstrate that environmental values are now closely tied to many other deep valued systems in American culture." At least 93 percent of the respondents agreed that people have a right to clean air and clean water. The authors concluded, "today's environmentalism is unlikely to be a passing fad."

We proposed to utilize one of the above survey instruments to poll members of the Delaware water resources community and determine their level of environmental ethics as a basis for decision-making. Dunlap and Van Liere employed a lengthy twelve-question survey as a measure of the new environmental paradigm that concluded, "concepts such as the environment are beginning to permeate the consciousness of the public." Dunlap utilized a precise eight question survey that indicated there is a measurable pro-environmental sentiment among the American public. Kempton et al. surveyed sectors of the American public in a 149-question survey that concluded that environmentalism is not a passing fad. The surveys concluded that the environment is favored by a majority of the American public thus indicating a rise in environmental values. Table 4 summarizes the surveys examined in the literature:

Table 4. Comparison of Environmental Survey Instruments

Author (s)	No. of Survey Questions	Sentiment
Dunlap and Van Liere (1978)	12	"Concepts such as balance with nature are beginning to permeate the consciousness of the public."
Dunlap (1990)	8	"There is a measurable pro-environmental sentiment among the American public."
Kempton, Boster, Hartley (1995)	149	"Environmentalism is not a passing fad."

Of the three survey methods, we chose the Dunlap (1990) survey instrument to poll the Delaware water resources community about their environmental ethic because:

- The survey is not as lengthy as the other methods (it is eight questions) and can be taken by the respondents in a brief amount of time, which hopefully leads to a higher response rate.
- It is measurable and the results can be analyzed by standard statistical methods.
- It contains general questions about environmental values that should be familiar to the Delaware water resources community.

SURVEY DESIGN

This section describes the design of the survey instrument to assess the environmental ethic level of the Delaware water resources community using questions from Dunlap's 1990 method. Members of two committees, the Delaware Water Supply Coordinating Council and Christina Basin Clean Water Partnership, were

ETHICS AND WATER POLICY IN DELAWARE 107

asked a series of questions to gauge their level of environmental understanding as a basis for decision-making. Each of the committees are composed of public and private stakeholders in water management and are often challenged to make ethically equitable decisions regarding water resources matters.

The Governor and General Assembly appointed the Delaware Water Supply Coordinating Council in July 2000 to develop and recommend water supply policy in Delaware.

The Christina Basin Clean Water Partnership is an interstate initiative between Delaware and Pennsylvania to protect and improve water quality in the streams used for over 50 percent of the drinking water supply in Chester County, Pennsylvania and New Castle County, Delaware.

We employed the following survey methods:

1. Identify Respondents—Survey members of the Delaware Water Supply Coordinating Council and Christina Basin Clean Water Partnership (n = 65) with a series of fixed format questions. The composition of these two water policy bodies are listed below:

Membership of the Delaware Water Supply Coordinating Council

Water Purveyors:	Artesian Water Company
	City of Newark
	City of Wilmington
	New Castle Municipal Services Commission
	Tidewater Utilities, Inc.
	United Water Delaware
Business Owners:	New Castle County Chamber of Commerce
	Delaware State Chamber of Commerce
	Delaware Nursery and Landscape Association
	Grounds Management Society
	Delaware State Golf Association
Government:	Office of the Governor
	Delaware Department of Natural Resources & Environmental Control
	Delaware Department of Public Safety
	Delaware Department of Agriculture
	Public Service Commission
	Delaware Emergency Management Agency
	Delaware Division of Public Health
	Public Advocate
	Delaware River Basin Commission
	New Castle County Executive
Academia:	Delaware Geological Survey
	University of Delaware Water Resources Agency

Non-Profits: Delaware Nature Society
Coalition for Natural Stream Valleys
New Castle County Civic League

Membership of the Christina Basin Clean Water Partnership

Water Purveyors: City of Newark
City of Wilmington
United Water Delaware

Business Owners: URS Corporation

Government: Cecil County Office of Planning and Zoning
Chester County Conservation District
Chester County Health Department
Chester County Parks & Recreation
Chester County Planning Commission
Chester County Water Resources Authority
Delaware County Planning Department
Delaware Department of Natural Resources
Delaware River Basin Commission
Delaware Department of Transportation
U.S. Environmental Protection Agency—Region III
Lancaster County Planning Commission
New Castle Conservation District
New Castle County Department of Land Use
Pennsylvania Department of Environmental Protection
U.S. Army Corps of Engineers
U.S.D.A.—Natural Resources Conservation Service
U.S. National Park Service

Academia: Delaware Geological Survey
University of Delaware, Water Resources Agency

Non-Profits: Brandywine Valley Association
Red Clay Valley Association
Delaware Nature Society

II. Survey Questions—The survey includes a series of global questions adapted from Dunlap (1990) to identify the respondent's overall global environmental ethic (questions 1–5 in the survey) and a second series of questions that pertain locally to the Delaware water policy area (questions 6–10 in the survey). We posed the following survey questions to the members of the Delaware Water Supply Coordinating Council and Christina Basin Clean Water Partnership:

Global Orientation (from Dunlap, 1990)

1. Are we (the Government) spending too much, too little, or about the right amount on improving and protecting the environment?
    a. Don't know

b. Too much
c. About Right
d. Too Little

2. Do you think environmental laws and regulations have gone too far, or not far enough, or have struck about the right balance?
   a. Don't know
   b. Too far
   c. Right balance
   d. Not far enough

3. Do you agree or disagree with the following statement: Protecting the environment is so important that the requirements and standards cannot be too high, and continuing environmental improvements must be made regardless of cost.
   a. No opinion
   b. Disagree
   c. Agree

4. Which of these two statements is closer to your opinion: We must be prepared to sacrifice environmental quality for economic growth. We must sacrifice economic growth in order to preserve and protect the environment.
   a. Don't know
   b. Sacrifice environmental quality
   c. Sacrifice economic growth

5. Do you think that the overall quality of the environment around here is very much better than it was five years ago, slightly better than it was five years ago, slightly worse, somewhat worse, or very much worse than it was five years ago?
   a. Slightly, somewhat, very much worse
   b. About the same/don't know
   c. Very much, somewhat, or slightly better

Local Perspective (from Kauffman, 2003)

6. A developer proposes to construct a shopping center over an aquifer recharge area. The shopping center will provide property tax income. The aquifer is the only source of drinking water. Which of the following options would you recommend?
   a. Build on 100% of the site, $1 million in annual tax income
   b. Build on 50% of site, $500,000 in annual tax income
   c. Build on 20% of site, $200,000 in annual tax income
   d. Deny project, no tax income.

7. A bottling plant is proposed which would require a wastewater discharge upstream from your city. The stream is the sole source of drinking water

for the city. The industry will generate jobs. Which of the following options would you choose:
   a. 100,000 bottles per day, 500 jobs provided, stream water quality reduced by 100%.
   b. 50,000 bottles per day, 250 jobs provided, stream water quality reduced by 50%
   c. 10,000 bottles per day, 50 jobs provided, stream water quality reduced by 10%
   d. Deny projects, no jobs created, water quality remains at existing level.
8. For every dollar of water supply revenue, what percentage would you apply to the following programs?
   a. Improved distribution
   b. Better water treatment
   c. Stream restoration
   d. Profit
9. A reservoir is proposed in a wild and scenic river valley, which of the options would you recommend?
   a. One billion gallon reservoir, sufficient water through 2040, disturbs 20 acres of wetlands
   b. 500 million-gallon reservoir, provides half the needed water through 2040, disturbs 10 acres of wetlands.
   c. 250 million gallon reservoir, provides one quarter the needed water through 2040, disturbs 5 acres of wetlands
   d. No reservoir, water conservation provides 10% of water needed through 2040.
10. During drought, a water purveyor needs to provide 20 mgd of drinking water from a stream which is flowing at 20 mgd. Which of the following options would you prefer?
    a. Draw 20 mgd for drinking water leaving 0 mgd instream to sustain fishery at 0%.
    b. Draw 15 mgd for drinking water and release 5 mgd from new reservoir which disturbed 5 acres of wetlands leaving 5 mgd instream to sustain fishery at 50%.
    c. Draw 10 mgd for drinking water and release 5 mgd from new reservoir which disturbed 10 acres of wetlands leaving 10 mgd instream to sustain fishery at 100%.
    d. Draw 10 mgd for drinking water and conserve 10 mgd by not watering lawns leaving 10 mgd instream to sustain fishery at 100%.

III. Survey delivery and analysis—Questions were delivered to the respondents via email and the respondents were asked to transmit their answers back via email, fax, or US mail. The survey results are compiled in a tabular and graphical manner using statistical methods.

SURVEY RESULTS

Of sixty-five surveys distributed, we received thirty-three responses, a 51 percent response rate. The respondents are classified into five categories. Table 5 summarizes the responses to the survey.

Category	Responses
Water Purveyors	7
Business Owners	4
Government Water Agencies	13
Academia	5
Non-Profit Environmental Organizations	4
Total responses	33

Table 5. Responses to Survey

Question	a.	b.	c.	d.
1. Is the government spending too much, too little, or about the right amount on protecting the environment?	Don't know (12%)	Too much (3%)	About Right (24%)	Too Little (58%)
2. Do you think environmental laws have gone too far, or not far enough, or have struck the right balance?	Don't know (12%)	Too far (9%)	Right balance (30%)	Not enough (48%)
3. Protecting the environment is so important that requirements cannot be too high, environmental improvements must be made regardless of cost.	No opinion (6%)	Disagree (58%)	Agree (33%)	
4. We must sacrifice environmental quality for economic growth. Or, we must sacrifice economic growth to protect the environment.	Don't know (27%)	Sacrifice environment (9%)	Sacrifice economic growth (64%)	
5. Do you think overall quality of the environment 5 years ago is: much better, better, slightly worse, somewhat worse, or much worse?	Slightly, somewhat, very much worse (27%)	About the same/don't know (33%)	Very much, somewhat, or slightly better (39%)	
6. A developer proposes to construct a shopping center over an aquifer recharge area. Which of the following options would you recommend?	Build 100% of site, $1 million in annual tax income (0%)	Build 50% of site, $500,000 in tax income (18%)	Build on 20% of site, $200,000 in tax income (55%)	Deny project, no tax income. (27%)

Question	a.	b.	c.	d.
7. A bottling plant is proposed which would require a wastewater discharge upstream from your city along a drinking water stream	100,000 bottles/day, 500 jobs, water quality reduced 100%. (0%)	50,000 bottles/day, 250 jobs, water quality reduced 50%. (6%)	10,000 bottles/day, 50 jobs, water quality reduced 10%. (64%)	Deny project, no jobs, water quality no change. (30%)
8. For every dollar of water supply revenue, what % would you apply to the following?	Improved distribution. (23%)	Better water treatment (31%)	Stream restoration (29%)	Profit (17%)
9. A reservoir is proposed in a wild and scenic river valley. Which of the options would you recommend?	1 BG reservoir, sufficient water for 2040, disturbs 20 acres wetlands. (39%)	500 MG reservoir, provides half water for 2040, disturbs 10 ac. wetlands. (24%)	250 MG reservoir, provides ¼ water for 2040, disturbs 5 ac. wetlands (6%)	No reservoir,
10. During drought, a purveyor proposes to provide 20 mgd of drinking water from a stream flowing at 20 mgd. Which of the following options would you prefer?	Draw 20 mgd for drinking water, leave 0 mgd instream fishery at 0%. (6%)	Draw 15 mgd for drinking water, release 5 mgd from new reservoir, sustain fishery at 50%. (18%)	Draw 10 mgd for drinking water and release 5 mgd from new reservoir, sustain fishery at 100%. (6%)	Draw 10 mgd for drinking water, conserve 10 mgd, sustain fishery at 100%. (70%)

(%) = percent of respondents in favor

Question 1. Fifty-eight percent of the respondents felt that the government spends too little on the environment. By comparison, the survey conducted by Dunlap in 1990 indicated that 70% believed that too little was spent on the environment. While a majority felt that more could be spent on the environment, several of the respondents believed that too much has been spent. One of the water purveyors answered the question stating: "Too much of what? Time or money? I think government means well but is extremely inefficient. Government should do a lot more with what they have available and work to streamline processes." Since overall 84% said that government spending was about right or too little, this indicates a tilt toward an environmental ethic.

Question 2. Seventy-eight percent of the respondents think environmental laws and regulations have gone not far enough or have achieved the right balance. By

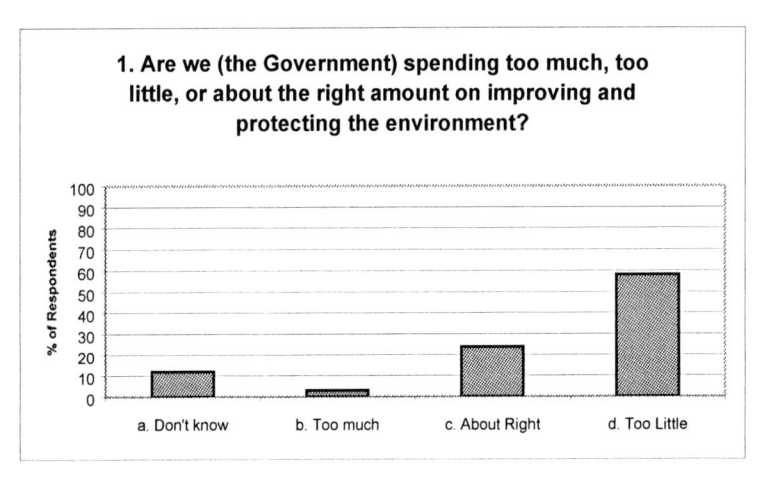

ETHICS AND WATER POLICY IN DELAWARE 113

1. Are we (the Government) spending too much, too little, or about the right amount on improving and protecting the environment?

[Bar chart showing % of Respondents:
a. Don't know: ~13
b. Too much: ~3
c. About Right: ~25
d. Too Little: ~60]

comparison, the survey conducted by Dunlap in 1990 indicated that 77% had the same feeling. While 75% of the respondents have what can be interpreted as a pro-environmental belief, several of the respondents sought to clarify their answers. A water purveyor answered the question stating "if government is truly the will of the people, we must conclude the balance is 'right' for today's people." A member of a government water conservation district did not know how to answer the question stating: "Several laws and regulations are in place but are not enforced. Since laws and regulations have not been enforced, it is difficult to determine whether the right balance has been struck." And a nonprofit environmental group member said that: "Environmental laws and regulations now existing have nearly enough substance if they were fully implemented and enforced."

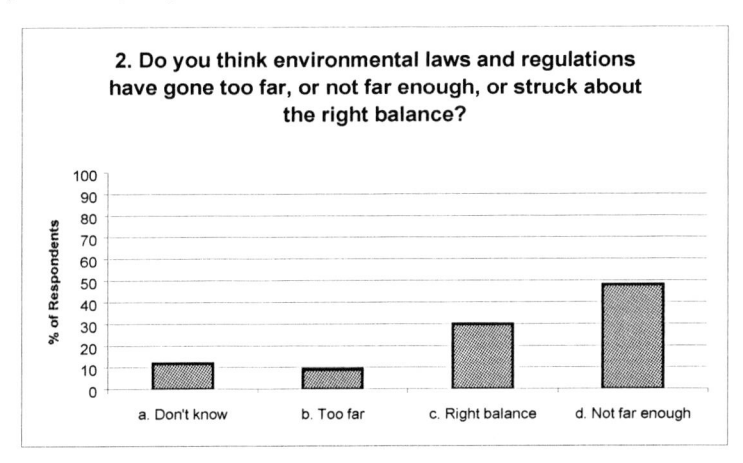

2. Do you think environmental laws and regulations have gone too far, or not far enough, or struck about the right balance?

[Bar chart showing % of Respondents:
a. Don't know: ~13
b. Too far: ~10
c. Right balance: ~30
d. Not far enough: ~50]

Question 3. In contrast to the pro-environmental theme of the proceeding two questions, close to 60% felt that that there must be some cost limits set on spending for environmental improvements. By contrast, the survey conducted by Dunlap in

1990 indicated that only 21% disagreed that environmental improvements must be made regardless of cost. The first two questions indicated that the respondents believed that the environment is important, however, spending limits should be set. Interpretation of the pro-environmental sentiments of the first two questions with the need for spending limits in the third question indicates that the respondents are trying to achieve some sort of balance between the environment and economics. One of the nonprofit environmental group members corroborated this belief stating: "Obviously some balance must come into play. Cost-benefit ratios should not be ridiculous."

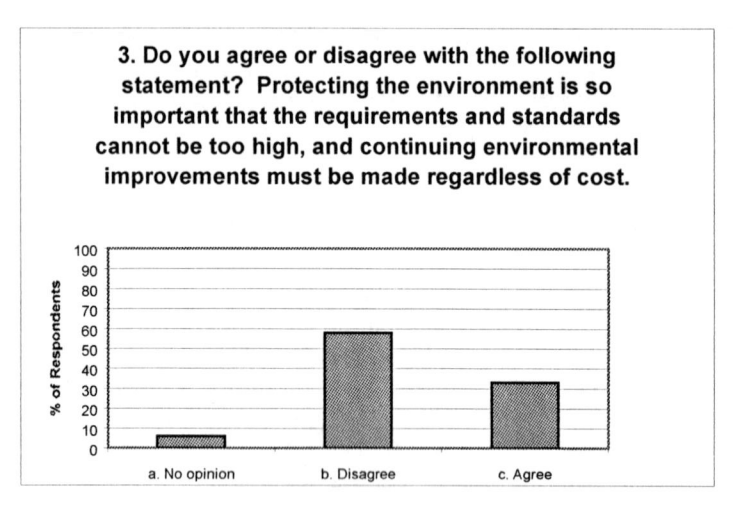

3. Do you agree or disagree with the following statement? Protecting the environment is so important that the requirements and standards cannot be too high, and continuing environmental improvements must be made regardless of cost.

Question 4. While in question 3 the majority of those surveyed believe that spending limits should be set, in this question 65% believed that we must sacrifice economic growth to protect the environment. By comparison, the survey

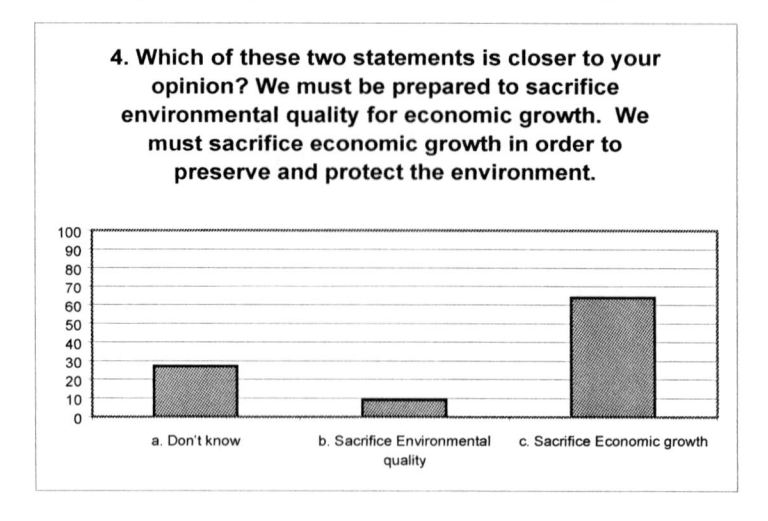

4. Which of these two statements is closer to your opinion? We must be prepared to sacrifice environmental quality for economic growth. We must sacrifice economic growth in order to preserve and protect the environment.

ETHICS AND WATER POLICY IN DELAWARE 115

conducted by Dunlap in 1990 indicated that 64% believed the same. The results of this question indicate that almost 2/3 of those surveyed have a pro-environmental belief when compared to sacrificing economic growth. Several of the respondents clarified their answers as follows. Government water conservation district official: "Both need to be sacrificed in order to reach a balance. The issue is not as black and white as this question implies." Nonprofit environmental group member: "I don't really agree with either. I believe that the economy and the environment are interdependent."

Question 5. No real consensus was achieved here as 39% believed the environment is better than it was, 33% believe it is about the same, and 27% believe it is worse. By comparison, the survey conducted by Dunlap in 1990 indicated that 32% said the environment is better, 13% said it was the same, and 55% said it was worse. One member of a nonprofit environmental group probably summarized the sentiments of those surveyed saying: "On balance environmental quality is worse, but some facets are better, others slightly to considerably worse." It is difficult to determine from this question whether there is a pro-environmental sentiment from the surveyed groups.

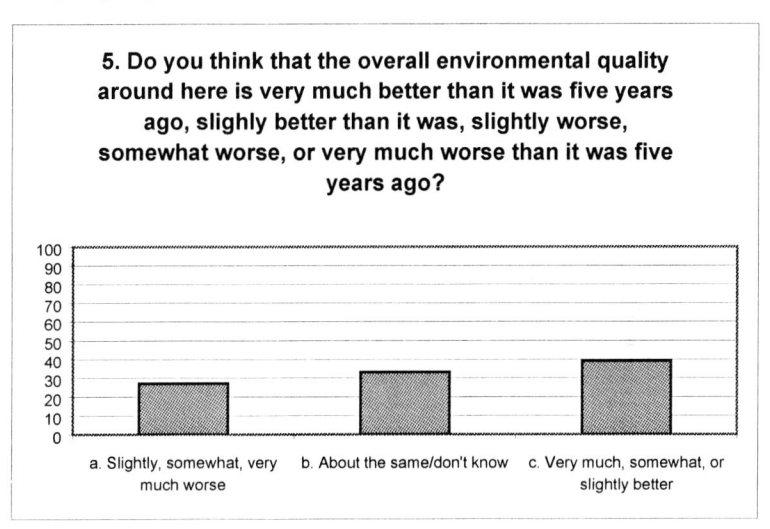

5. Do you think that the overall environmental quality around here is very much better than it was five years ago, slighly better than it was, slightly worse, somewhat worse, or very much worse than it was five years ago?

a. Slightly, somewhat, very much worse
b. About the same/don't know
c. Very much, somewhat, or slightly better

Question 6. Almost 3/4 of the respondents chose a moderate approach in trying to balance the environment with the economy. Seventy-three percent of those surveyed chose options B or C which would allow the developer to build on 20% or 50% of the site over the aquifer, while still allowing for $200,000 or $500,000 in annual tax income. None of the respondents chose option A which represents the greatest economic return ($1 million in annual tax income) but has the greatest environmental impact (100% of the site would be developed). On the other extreme, 27% chose option D which has the least economic return (they voted to deny the project) and would have the least environmental impact to the aquifer.

All the respondents selected an option which would provide at least some form of environmental protection to the aquifer.

6. A developer proposes to construct a shopping center over an aquifer recharge area. The shopping center will provide property tax income. The aquifer is the only source of drinking water. Which of the following options would you recommend?

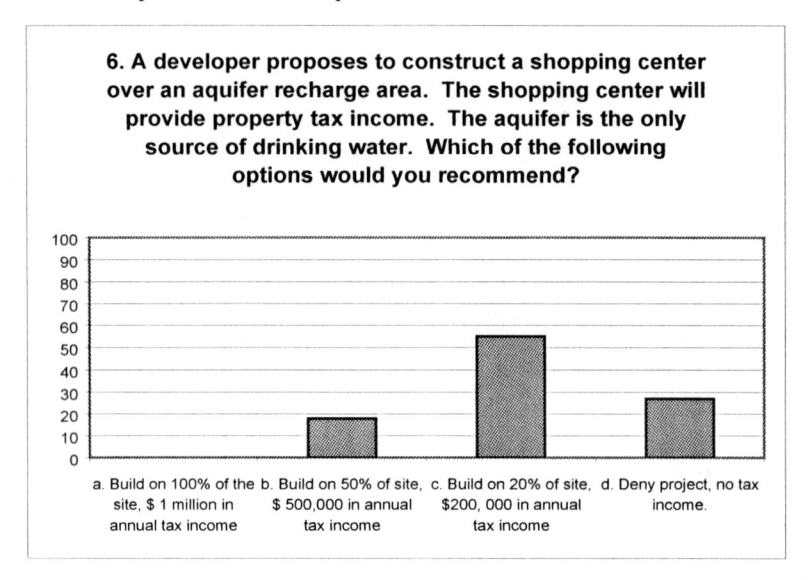

a. Build on 100% of the site, $ 1 million in annual tax income
b. Build on 50% of site, $ 500,000 in annual tax income
c. Build on 20% of site, $200, 000 in annual tax income
d. Deny project, no tax income.

Several provided comments along with their choices. A government water conservation district official recommended "Using BMPs such as porous pavement and filtration systems to collect and treat water from the site, and allowing treated water to recharge the aquifer. A member of a nonprofit environmental group mentioned that the choice between option B and C depended on "How much impervious surface already exists over the aquifer recharge area?"

7. A bottling plant is proposed which would require a wastewater discharge upstream from your city. The stream is the sole source of drinking water for the city. The industry will generate jobs. Which of the following options would you choose?

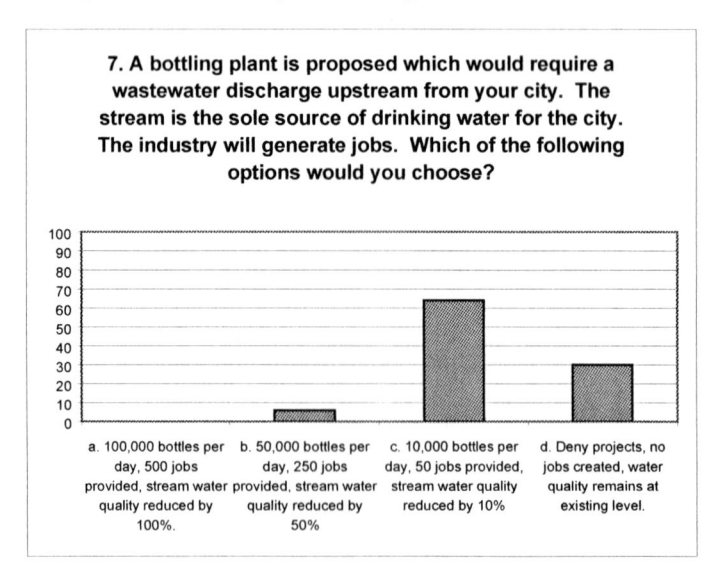

a. 100,000 bottles per day, 500 jobs provided, stream water quality reduced by 100%.
b. 50,000 bottles per day, 250 jobs provided, stream water quality reduced by 50%
c. 10,000 bottles per day, 50 jobs provided, stream water quality reduced by 10%
d. Deny projects, no jobs created, water quality remains at existing level.

Question 7. Seventy percent of the respondents chose options B and C which provide a moderate balance between the environment and the economy. Almost 3/4 of those surveyed would allow a bottling plant wastewater discharge with an economic return of 50 to 250 jobs created but reduces the stream water quality by 10% to 50%. None surveyed chose the most favorable economic option A which would provide 500 jobs but reduce the stream water quality by a factor of 100%. Thirty percent chose the most favorable environmental option D, which would deny the project, no jobs created, at no cost to the environment. Most of those surveyed seemed to employ an ethic which sought to balance the economy (create jobs) and minimize reduction in stream water quality.

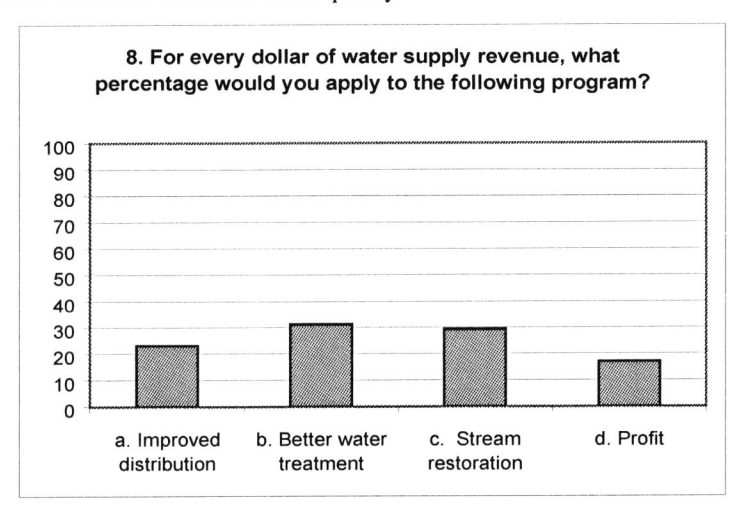

8. For every dollar of water supply revenue, what percentage would you apply to the following program?

a. Improved distribution — ~25
b. Better water treatment — ~32
c. Stream restoration — ~30
d. Profit — ~18

Question 8. When it comes to spending water supply revenue, the members of the Water Supply Coordinating Council and Christina Basin Clean Water Partnership felt that 60% of the money should be spent on environmental projects such as option B (better water treatment) or option C (stream restoration). The respondents felt that 23% of the funds should be reinvested to improve the distribution system (option A) and 17% should be reserved for profit (option D). Those surveyed generally came out in favor of spending the majority of the revenues on the environment while favoring lesser expenditures on reinvestment to the system and profit.

Question 9. Almost 70% of those surveyed felt that some sort of reservoir could be constructed but at differing environmental costs. Forty percent chose option A which would construct the largest reservoir but have the largest environmental cost at 20 acres of wetlands disturbed. Thirty percent chose options B or C which would be smaller reservoirs, which would provide one quarter to half the water needed with less environment impact to wetlands. A little over one quarter chose the least environmentally damaging option D (Water Conservation) which would provide 10% of the water needed meaning other costly projects would need to be implemented. Options B and C represent reservoir alternatives where the members of the Delaware water resources committee chose to balance the economy (the need for water) with the environment (need to minimize wetland impacts).

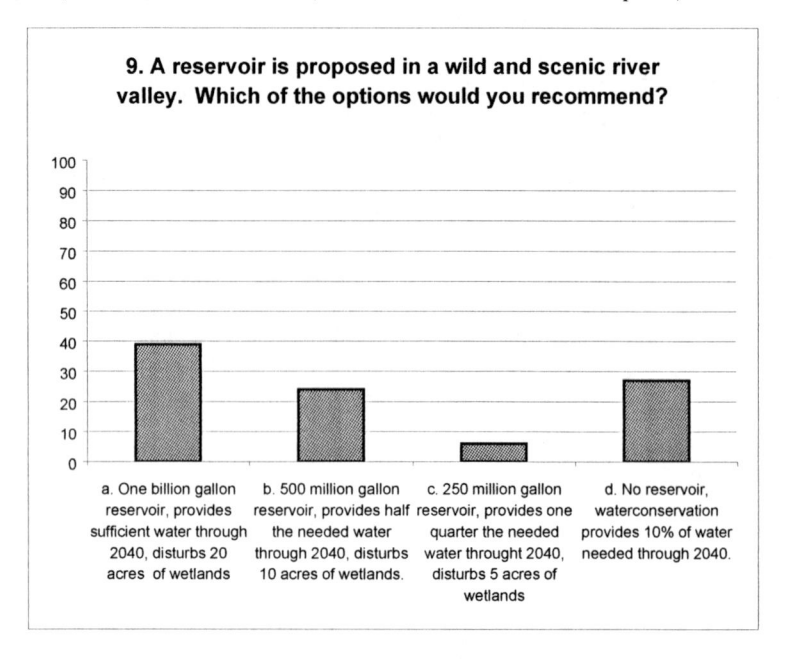

9. A reservoir is proposed in a wild and scenic river valley. Which of the options would you recommend?

[Bar chart showing responses:]
- a. One billion gallon reservoir, provides sufficient water through 2040, disturbs 20 acres of wetlands: ~39
- b. 500 million gallon reservoir, provides half the needed water through 2040, disturbs 10 acres of wetlands: ~24
- c. 250 million gallon reservoir, provides one quarter the needed water throught 2040, disturbs 5 acres of wetlands: ~5
- d. No reservoir, waterconservation provides 10% of water needed through 2040: ~27

Question 10—A majority (70%) of those surveyed preferred the most environmentally favored water withdrawal option D where half the needed water would

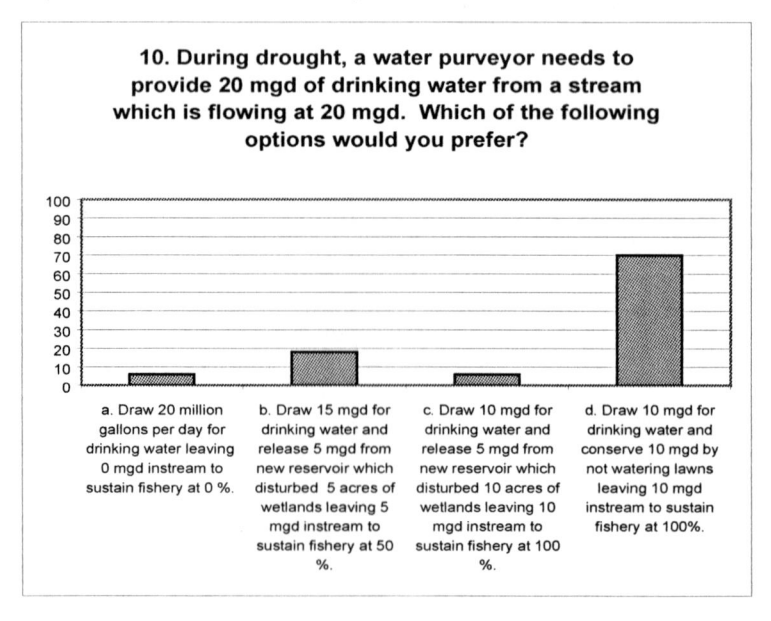

10. During drought, a water purveyor needs to provide 20 mgd of drinking water from a stream which is flowing at 20 mgd. Which of the following options would you prefer?

[Bar chart showing responses:]
- a. Draw 20 million gallons per day for drinking water leaving 0 mgd instream to sustain fishery at 0 %.: ~2
- b. Draw 15 mgd for drinking water and release 5 mgd from new reservoir which disturbed 5 acres of wetlands leaving 5 mgd instream to sustain fishery at 50 %.: ~18
- c. Draw 10 mgd for drinking water and release 5 mgd from new reservoir which disturbed 10 acres of wetlands leaving 10 mgd instream to sustain fishery at 100 %.: ~3
- d. Draw 10 mgd for drinking water and conserve 10 mgd by not watering lawns leaving 10 mgd instream to sustain fishery at 100%.: ~70

be withdrawn from the stream and half would be conserved by not watering lawns thus leaving enough water in the stream to sustain the fishery at 100%. This option indicates that over 3/4 of the respondents would prefer the environmental benefit of sustaining the fishery over the negative economic impacts of sustaining lawns and landscaping during drought. Only 6% prefer option A at the other end of the environment vs. economy spectrum—withdrawing all of the water out of the stream leaving none to sustain the fishery. About one quarter chose more balanced environmental-economic options B and C which sought to withdraw a portion of the needed drinking water from the stream and the balance from a reservoir. These options have forms of environmental impact as the reservoir from which the water was released originally had damaged wetlands. One of the water purveyors preferred a combination of options A and D writing: "I would ask for conservation to save the water in my reservoir—NOT to sustain the fishery at 100%. Damn the fish!"

Table 6 tabulates the responses to the survey by total sample and then disaggregated by each of the following sectors: water purveyors, business, government, academia, and non-profit environment groups. There were discernible differences in the responses depending on the perspective of each sector.

1. Are we (the Government) spending too much, too little, or about the right amount on improving and protecting the environment? Over 50% of the total sample, business, government, academia, environmental groups responded that too little is spent on the environment. In contrast, only 14% of the water purveyors felt the same way. The largest segment of the water purveyors (43%) thought that spending was just right,
2. Do you think environmental laws and regulations have gone too far, or not far enough, or struck about the right balance? Over 47% of the total sample, business, government, academia, environmental groups responded that environmental laws have not gone far enough. In contrast, only 14% of the water purveyors felt the same way. The largest segment of the water purveyors (43%) thought that environmental laws went too far or struck the right balance.
3. Do you agree or disagree with the following statement? Protecting the environment is so important that the requirements and standards cannot be too high, and continuing environmental improvements must be made regardless of cost. The majority (over 50%) of the total sample, water purveyors, business, and government responded that they disagreed that continuing environmental improvements must be made regardless of cost. The majority of the academic sector (60%) and environmental groups (75%) agreed with this statement.
4. Which of these two statements is closer to your opinion? We must be prepared to sacrifice environmental quality for economic growth. We must sacrifice economic growth in order to preserve and protect the environment. Interestingly, at least half of the total sample and the five sector groups

shared a similar environmental perspective responding that we must sacrifice economic growth in order to preserve and protect the environment.
5. Do you think that the overall quality of the environment around here is very much better than it was five years ago, slightly better than it was five years ago, slightly worse, somewhat worse, or very much worse than it was five years ago? The responses to this question varied widely by sector group. The majority of the total sample (39%), water purveyors (72%), and environmental groups (75%) responded that the environment was better. The majority of the business (50%), government (46%), and academia (40%) groups thought the environment was about the same.
6. A developer proposes to construct a shopping center over an aquifer recharge area. The shopping center will provide property tax income. The aquifer is the only source of drinking water. Which of the following options would you recommend? The majority of the business sector (50%) recommended building on 50% of the site. The majority of the total sample (55%) recommended building on 20% of the site. The majority of the water purveyors (72%), government (61%), academia, and environmental groups (75%) recommended denying the project.
7. A bottling plant is proposed which would require a wastewater discharge upstream from your city. The stream is the sole source of drinking water for the city. The industry will generate jobs. Which of the following options would you choose? The majority of the environmental groups (75%) chose the option of a 50,000 bottle per day plant with water quality reduced by 50%. Over 60% of the total sample (64%), water purveyors (86%), business groups (100%), and academia (60%) chose the option of a 10,000 bottle per day plant reducing water quality by 10%. The majority of the government groups preferred to deny the project.
8. For every dollar of water supply revenue, what percentage would you apply to the following program? This question was not disaggregated by sector.
9. A reservoir is proposed in a wild and scenic river valley. Which of the options would you recommend? The majority of the total sample (39%), water purveyors (43%), government (46%) and academia (40%) preferred the largest reservoir (1 billion gallons) at the largest environmental cost. The majority of the business groups (50%) preferred a reservoir at half the size (500 million gallons) and half the environmental cost. The majority of the environmental groups (75%) preferred no reservoir instead preferring water conservation.
10. During drought, a water purveyor needs to provide 20 mgd of drinking water from a stream which is flowing at 20 mgd. Which of the following options would you prefer? There was consensus here. The majority of all the groups preferred the least environmental costly alternative: Draw 10 mgd for drinking water and conserve 10 mgd by not watering lawns leaving 10 mgd instream to sustain fishery at 100%.

ETHICS AND WATER POLICY IN DELAWARE

To conclude the analysis of the survey and employ an environmental ethics index, the responses were grouped according to three categories: (1) responses with a somewhat pro-economic viewpoint, (2) responses with a more balanced viewpoint between the environment and the economy, and (3) responses with a somewhat pro-environmental viewpoint. Table 6 summarizes the criteria for an environmental index derived from the survey results.

Table 6. Environmental Index Criteria

Question	Pro-economic viewpoint	Balanced viewpoint, environment and economy	Pro-environmental viewpoint
1. Are we (the Government) spending too much, too little, or about the right amount on improving and protecting the environment?	Option B (3%)	Option C (24%)	Option D (58%)
2. Do you think environmental laws and regulations have gone too far, or not far enough, or have struck about the right balance?	B (9%)	A and C (42%)	D (48%)
3. Do you agree or disagree with the following statement? Protecting the environment is so important that the requirements and standards cannot be too high, and continuing environmental improvements must be made regardless of cost.	B (58%)	A (6%)	C (33%)
4. Which of these two statements is closer to your opinion: we must be prepared to sacrifice environmental quality for economic growth. We must sacrifice economic growth in order to preserve and protect the environment.	B (9%)	A (27%)	C (64%)
5. Do you think that the overall quality of the environment around here is very much better than it was five years ago, slightly better than it was five years ago, slightly worse, somewhat worse, or very much worse than it was five years ago?	A (27%)	B (33%)	C (39%)

Question	Pro-economic viewpoint	Balanced viewpoint, environment and economy	Pro-environmental viewpoint
6. A developer proposes to construct a shopping center over an aquifer recharge area. The shopping center will provide property tax income. The aquifer is the only source of drinking water. Which of the following options would you recommend?	A (0%)	B and C (73%)	D (27%)
7. A bottling plant is proposed which would require a wastewater discharge upstream from your city. The stream is the sole source of drinking water for the city. The industry will generate jobs. Which of the following options would you choose?	A (0%)	B and C (70%)	D (30%)
8. For every dollar of water supply revenue, what percentage would you apply to the following program?	A and D (40%)		B and C (60%)
9. A reservoir is proposed in a wild and scenic river valley. Which of the options would you recommend?	A (39%)	B and C (30%)	D (27%)
10. During drought, a water purveyor needs to provide 20 mgd of drinking water from a stream which is flowing at 20 mgd. Which of the options would you prefer?	A (6%)	B and C (24%)	D (70%)
Composite Viewpoints	19% Pro-economic	36% Economic—environment balance	45% Pro-environment

The results of the survey indicate that the majority of those polled from the Delaware water resources community (81%) have values that may be interpreted as either (1) supportive of a balance between the economy and the environment or (2) pro-environment, thus indicating that (at least for the questions posed) this ethic is injected into decisions concerning water policy. Thirty-six percent answered the survey questions in a manner that attempts to strike a balance between economic and environmental needs. Forty five percent answered the questions from what could be construed as a pro-environmental viewpoint. The minority viewpoint was pro-economic as 19% of those surveyed preferred economic needs in answering the questions. The survey of a cadre of water resources policy makers on the Delaware Water Supply Coordinating Council and the Christina Basin Clean Water Partnership concludes that (at least for the questions posed) they instill water resources decision-making with a reasonably strong environmental ethic.

CONCLUSIONS

This article explored the field of environmental ethics as a basis for fair and equitable water policy making in Delaware. We examined the dilemma of resource allocation, given the seemingly conflicting objectives imposed by a market economy between individual self-interest and the environmental well being of the larger community, and seek concepts and methods to facilitate ethical decision-making. We have the following conclusions:

Evolution of Environmental Ethic—A literature review indicates that mankind's overall environmental ethic has evolved over the 50 years from a state of dominion to a more modern ecological concept of balance with nature. For instance, membership in environmental lobbying organizations such as the Sierra Club and Natural Resources Defense Council exceeded 3,100,000 in 1990, a 30-fold increase from 120,000 members in 1960. More people care for the environment than they used to thus serving as an underpinning for water law and ethical water resources decision-making.

Water Law—Environmental ethics is a basis of current water law, which is designed to guide water resource allocation and help resolve conflicts. In the USA, two forms of water law are practiced—riparian rights (found in the eastern states) and prior appropriation doctrine (predominately in western states). Both forms of water law seek an ethical balance to protect the rights of the water user (the individual) while trying to protect other upstream and downstream users (the common good).

Case Studies in Water Resources Decision-making—A series of case studies from Delaware and other states discuss how an awareness of environmental ethics can help the water policy specialist make difficult decisions in the face of various conflicts. Witness:

- The ethical dimension of riparian water law is that the upstream owner (developer proposing to put homes in the floodplain) may not injure the downstream owner.
- The public manager advises against approving a shopping center proposal over an aquifer because at 60 percent impervious cover the development exceeds the 50 percent impervious cover requirement of the water resources ordinance and is out of harmony with the water resource.
- The ethical water managers from both states in the Christina Basin averted a lawsuit when an industry in upstream Pennsylvania proposed a wastewater discharge that did not meet downstream Delaware's water quality standards. Ultimately a strategy was crafted under the Federal Clean Water Act whereby the wastewater discharge could meet both states' standards.

Survey of Delaware Water Resources Policy Makers—A survey was conducted of sixty-five water resources policy makers in the water purveyor, business, government, academic, and nonprofit environmental sectors on the Delaware Water Supply Coordinating Council and the Christina Basin Clean Water Partnership. The purpose of the survey is to assess their beliefs in an environmental ethic as a basis for water resources decision-making. The results of the survey indicate that the majority of those polled from the Delaware water resources community (81 percent) have values that may be interpreted as either (1) supportive of a balance between the economy and the environment or (2) pro-environment, thus indicating that this ethic is injected into decisions concerning water policy. Thirty-six answered the survey questions in a manner that attempts to strike a balance between economic and environmental needs. Forty five percent answered the questions from what could be construed as a pro-environmental viewpoint. The minority pro-economy viewpoint was expressed as 19 percent of those surveyed preferred economic needs in answering the questions. The survey of the cadre of water resources policy makers on the Delaware Water Supply Coordinating Council and the Christina Basin Clean Water Partnership concludes that (at least for the questions posed) water resources decision-making in Delaware is instilled with a reasonably strong environmental ethic.

The general public was not part of the survey and therefore it is uncertain what the public's attitude would be toward an environmental ethic. Some perspective may be gleamed from a comparison of Dunlap's survey in 1990 which surveyed the general public and the current survey of policy makers. For instance, 58 percent of the water policy makers felt that the government spends too little on the environment. By comparison, the survey conducted of the public by Dunlap in 1990 indicated that 70 percent believed that too little was spent on the environment. Seventy eight percent of the policy makers think environmental laws and regulations have gone not far enough or have achieved the right balance. By comparison, the survey of the public conducted by Dunlap in 1990 indicated that 77 percent had the same feeling. In contrast to the pro-environmental theme of the proceeding two questions, close to 60 percent felt that that there must be some cost limits set on spending for environmental improvements. By contrast, the survey conducted by Dunlap in 1990 indicated that only 21 percent disagreed that environmental improvements must be made regardless of cost. Sixty-five percent of the policy makers believed that we must sacrifice economic growth to protect the environment. By comparison, the survey of the public conducted by Dunlap in 1990 indicated that 64 percent believed the same.

Summary—When confronted with difficult decisions concerning the protection of natural resources such as water and land, the public manager may find it useful to remember the concept of environmental ethics as a state of harmony between humans and the land. The economic interests of the individual and the environmental needs of the community can be balanced provided it is done on the scale of ethical decision-making.

Quod natura non sunt turpia. What is natural cannot be bad.

BIBLIOGRAPHY

Attfield, A. 1983. *The Ethics of Environmental Concern*. New York: Columbia University Press.

Brick, P. 1995. *Determined Opposition: The Wise Use Movement Challenges Environmentalism. Environment*, 37 (8). Heldreff Publications.

Brunner, R. D., C. H. Colburn, C. M. Cromley, R. A. Klein, and E. A. Olson. 2002. *Finding Common Ground, Governance and Natural Resources in the American West*. Yale University Press.

Carson, R. 1962. *Silent Spring*. Boston: Houghton Mifflin.

Cech, Thomas R. 2003. *Principles of Water Resources History, Development, Management, and Policy*. New York: John Wiley and Sons, Inc.

Dunlap, R. E. October 1991. *Public Opinion in the 1980's Clear Consensus, Ambiguous Commitment. Environment*.

Dunlap, R. E. and K. D. Van Liere. 1978. "The New Environmental Paradigm." *Journal of Environmental Education*. Volume 9: 10—19.

Dzurik, A. A., and D. A. Theriaque. 1996. *Water Resources Planning* (2nd Edition). New York: Rowman and Littlefield Publishers.

Farmer, C. G. 2001. "A Historical/Legal Perspective on Takings." *Forum 7, Oregon's Future*. Spring/Summer 2001.

Greig, D., J. Bowers, and G. Kauffman. May 1998. "Phase I and II Report." *Christina River Basin Water Quality Management Strategy*.

Jenkins, M. 2002. "Water Borders." *Proceedings of Drinking Water 2001: The Issues Concerning Delaware's Most Precious Natural Resource*. University of Delaware, Institute for Public Administration. Public Policy Forum, October 2001.

Kauffman, G. J. July 2001. "The Ethical Dilemma of the Public Natural Resources Manager." *PA Times*. American Society for Public Administration, Volume 4, No. 8.

Kauffman, G. J., and T. Brant. 2000. "The Role of Impervious Cover as a Watershed-based Zoning Tool to Protect Water Quality in the Christina Basin of Delaware, Pennsylvania, and Maryland." *Proceedings of the Water Environment Federation*. Watershed Management 2000 Conference. Vancouver, British Columbia.

Kempton, W., J. S. Boster, and J. A. Hartley. 1999. *Environmental Values in American Culture*. Cambridge, Massachusetts: MIT Press.

Leopold, A. 1966. *A Sand County Almanac*. Oxford University Press, Inc.

Mitchell, R. C., A. G Mertig, and R. E. Dunlap. 1991. "Twenty Years of Environmental Mobilization: Trends Among National Environmental Organizations." *Society and Natural Resources*, 4, 1991.

New Castle County Department of Planning. 1997. *Unified Development Code*, Article 10.

Renewable Natural Resources Foundation. 1992. "Congress on Renewable Natural Resources: Critical Issues and Concepts for the Twenty-First Century." *Renewable Resources Journal*. Autumn 1992.

Stark, J. A. 1995. "Postmodern Environmentalism: A Critique of Deep Ecology." *Ecological Resistance Movements*. Chapter 14. B. R. Taylor, editor. State University of New York Press.

Third World Water Forum. 2003. Kyoto, Japan. http://news.bbc.co.uk/1/hi/sci/tech/2856755.stm. Accessed March 24, 2003.

Webster's New Collegiate Dictionary. 1980. Springfield, Massachusetts: G & C Merriam Company.

White, L. March 1967. The Historic Roots of Our Ecological Crisis." *Science*, Vol. 155, No. 3767."

ETHICS AND THE LIFE SCIENCES

WELL-ORDERED SCIENCE:
THE CASE OF GM CROPS

MATTHEW LISTER
UNIVERSITY OF PENNSYLVANIA

ABSTRACT: The debate over the use of genetically-modified (GM) crops is one where the heat to light ratio is often quite low. Both proponents and opponents of GM crops often resort more to rhetoric than argument. This paper attempts to use Philip Kitcher's idea of a "well-ordered science" to bring coherence to the debate. While I cannot, of course, here decide when and where, if at all, GM crops should be used I do show how Kitcher's approach provides a useful framework in which to evaluate the desirability of using GM crops. At the least Kitcher's approach allows us to see that the current state of research in to, and use of, GM crops is very far from the ideal of a well-ordered science and gives us a goal to work towards if we wish to achieve a more well-ordered agricultural policy.

WHAT IS A "WELL-ORDERED SCIENCE"?

There are, Kitcher tells us, a number of different ways in which the research agendas of science may be set up. Four possible models we might consider are, *Internal Elitism*, which, "consists in decision-making by members of scientific subcommunities," *External Elitism*, which "involves both scientists and a privileged group of outsiders, those with funds to support investigations and the ultimate applications" that is, "paymasters," *Vulgar Democracy*, which, "imagines that the decisions are made by a group that represents (some of) the diverse interests in the society with advice from scientific experts," and finally, *Enlightened Democracy*, which, "supposes decisions are made by a group that receives tutoring from scientific experts and accepts input from all perspectives that are relatively widespread in society."[1]

In most societies the status quo is a variety of external elitism that scientists actively try to turn into internal elitism.[2] Additionally, when the paymasters of external elitism are not governments or universities, but large corporations, there is a strong tendency to pander to the interests of vulgar democracy insofar as this helps sell products made by the corporations in question.[3] It is not clear that this is an improvement. Kitcher's own proposal, "well-ordered science" is meant to be an idealization of the values that he hopes would come from enlightened democracy as a means of setting a scientific agenda. I turn now to the details of this view.

Well-ordered science is not a description of how science is currently structured or practiced. Rather, it is, "Intended as an ideal" towards which we might aim in setting up the governing of science.[4] This ideal, as a variety of enlightened democracy, results from the process of deliberation of a certain sort, loosely based on the idea of deliberative democracy found in such political theorists as Rawls and Gutmann.[5] While vulgar democracy tends to lead to the 'tyranny of the ignorant,' a dismissal of epistemic significance, and an emphasis on the short-term and 'hot topics,'[6] it is hoped that the enlightened democracy favored by well-ordered science will find and use our tutored preferences and so will be able to arrive at the real common good which we should use to order science in a democratic society.

Importantly, however, "The collective good is whatever is identified as such through this ideal democratic deliberation."[7] This follows from Kitcher's rejection of the idea that there is an idea of scientific significance or a goal of and for science independent of our interests. That is to say, the interests of deliberators, or at least ideal deliberators, play a constructive role in the idea of a well-ordered science much like that deliberators play in Rawls's constructive political theory. Through this, Kitcher hopes to avoid appealing to any sort of interest-independent idea of a goal for science.

Kitcher distinguishes three phases in his ideal inquiry. In the first, decisions are made (by the ideal deliberators) to commit resources to particular projects. In the second, projects are pursued in the most efficient way, subject to the moral constraints, also decided on by the ideal deliberators. Finally, in the third step, the results of the investigation are turned into practical applications, again under the guidance of ideal deliberators.[8]

Ideal deliberators, among other traits, are those who have 'tutored preferences.' Without tutored preferences, we get not 'enlightened' but rather 'vulgar democracy.' Tutored preferences arise when each deliberator is informed of the significance, both epistemic and practical, assigned to a project by the other deliberators.[9] Assumedly this process involves the deliberators becoming familiar with the scientific significance of a project, even if they cannot become experts in the science.

WELL-ORDERED SCIENCE: THE CASE OF GM CROPS

The next step in Kitcher's model involves an exchange of tutored preferences among the deliberators. This allows for deliberation, not negotiation, among the parties as to what the goals of society are. Next, probabilities of reaching these goals are assigned. If we cannot agree on a definitive list of goals or probabilities, we may defer to a group-chosen set of experts and arbitrators. Finally, the deliberators vote on budgets for the research programs, which must be followed within ethical guidelines also decided on by the deliberators. The result of this process is taken to define that which is best for the community.[10] It is not, however, strictly necessary that we actually follow this process for science to be well-ordered. Rather, what matters is that we have institutions and practices that will mimic the outcome of such an ideal process.[11] We may well wonder how we can know that we have such institutions and practices without actually following the process, and may further worry that if one group in society has the ability to dominate others, they will be able to insist that the goals of well-ordered science have been reached, even if they have not. We leave these worries, however, for a deeper discussion.

OBSTACLES TO WELL ORDERED SCIENCE

Kitcher makes note of four obstacles to the realization of a well-ordered science. The first of these is the problem of *Inadequate Representation*. He claims,

> A group is inadequately represented when the research agenda and/or the application of research systematically neglects the interests of the members of that group in favor of other members of society. Because of the Nonrepresentational Ratchet an early problem of inadequate representation may be self-perpetuating.[12]

If a group is not adequately represented among the ideal deliberators (or the institutions that serve as proxies for them), then their interests may be systematically ignored. Kitcher elaborates on this idea in two important ways. First, even in a democratic system, we cannot expect the invisible hand to solve this problem. There will always be incentives to ignore the weak and minority groups. Kitcher also insists, however, that the mere lack of numbers of a particular group is not enough to show inadequate representation. This is because it is the interests, and not the mere members, of a group that are important. So long as the interests of a group are represented, it need not matter, Kitcher seems to think, if the actual members represent them.[13] This is perhaps not completely correct. In practice there is substantial reason to think that the interests of groups with little social power will be miss-represented if they are not actually parties to deliberation, even if the intentions of the powerful groups are good. There is good reason to think, for example, that in a patriarchical society, men will tend to systematically misrepresent the interests of women even if they have no intention to do so. To this extent it seems that actual representation by minority groups and others with less social power is more likely to provide for well-ordered science, and to help avoid the problem of false consciousness discussed below, than merely trying to take the interests of everyone into account will do.

The second obstacle to a well-ordered science is one we have noted briefly above, the *Tyranny of the Ignorant*. This arises when:

> Epistemically significant questions in some sciences may systematically be undervalued because the majority of members of society have no appreciation for the factors that make those questions significant.[14]

This problem may arise from the fact that, in any given society, the preferences of the vast majority of the citizens are likely to be untutored. So far as science is based on a type of vulgar democracy, there is a good chance of this unfortunate outcome occurring. Similarly, when scientific paymasters are corporations primarily seeking profits, external elitism will, I think, tend towards this problem. For my purposes I want to add to Kitcher's definition in a way that makes it more obviously applicable to the case of GM crops but that still captures his important idea. I will say that we are also facing cases of the Tyranny of the Ignorant when people's untutored beliefs about the effects of a particular scientific program are the reason for blocking the application of scientific findings. I believe that Kitcher would accept this addition. Importantly, we must also note that not every case where scientists do not receive the resources they think most needful for their projects is a case of the Tyranny of the Ignorant. Tutored preferences need not be identical to the preferences of scientists, so it is possible to be in a state of well-ordered science and yet have scientists feel disappointment over their funding. Finally it is worth noting that many who have formal scientific expertise are often ignorant about other fields—molecular geneticists may be ignorant about ecology and biologists more generally may be ignorant of matters of social causation. Members of traditional societies often have practical knowledge that is unknown to those who would "help" them by introducing new technology. We should expect, then, the tutoring of preferences to flow in multiple directions.[15]

The third possible road-block on the path to a well ordered science is the problem of *False Consciousness*. Kitcher says:

> A research agenda may conform to the tutored preferences of the majority not because the public reasons for the agenda are those that would figure in an ideal deliberation, but because those reasons misrepresent the agenda in ways that cater to the actual (untutored) preferences of the majority. Because these preferences are not tutored, there may be harmful constraints on the pursuit of inquiry and serious threats to the proper application of its results.[16]

Kitcher's example of a case of False Consciousness is the human genomes project, which, he thinks, is not supported by the public for the same reasons it is supported by scientists and their paymasters. The public has been lead to believe that the genomes project will likely produce wide-spread and far-reaching treatments for diseases, and will do so in the short term. In fact, there is little reason to think this is so, and this is not the reason the genomes project is supported by most scientists and their paymasters. Rather, these figures support the genomes project because they believe it will greatly advance basic knowledge in several fields of biology and also will help the US maintain its lead in biotechnology.[17] These are worthy goals, and may even be goals that the tutored preferences of society would favor, but they are not the actual reasons for supporting the genomes project that most people have. So long as this is the case, science is not well-ordered.

The final obstacle to well-ordered science that Kitcher considers is the case of *Parochial Application*. This results when:

> An actual research agenda and a practice of application may be ideally supported by a principle that would license forms of research not currently undertaken or applications of previous research that are not pursued.[18]

Suppose, after a course of deliberation, we decide that one of the goals of science in our society ought to be "reducing cancer." We might approach this goal by means of the genomes project, developing genetic tests that allows us to screen for genetic defects that make the bearer more likely to develop certain forms of cancer. However, once we decide on this goal, we ought to do what we can to achieve it in other ways, too. For example, we might work to further reduce smoking, to monitor radon, and to cut other cancer-causing pollutions in our environment. A small amount of the money spent on the genomes project could make significant in-roads on these problems, and it seems that if our goal is in fact what we say it is, we ought to take the most efficient means to achieving it. There is, of course, an obvious overlap with the problem of false consciousness here, in that one reason why we do not take the most effective means to pursue our stated goal is that, in many cases, the stated goal is not the actual goal of those who control science.

APPLYING THE IDEA OF A WELL-ORDERED SCIENCE TO GM CROPS

From here I turn to the question of how the idea of well-ordered science might help us make sense of controversies surrounding GM crops. When considering GM crops, there are a number of questions that we must consider if we are to evaluate the legitimacy of their development and application. It is my contention that Kitcher's idea of a well-ordered science may help guide us here. Questions of interest include:

- What type of agricultural policy is appropriate to a particular setting?
- What kinds of research will support an adopted policy, and will research into GM crops be among them?
- Is GM food safe to eat?
- How will the environment be impacted?
- Will consumers and producers of crops be given a reasonable choice based on solid, understandable, and comprehensive information?
- How will traditional ways of life be affected?
- Will too much power be concentrated in private hands?
- Can the quality of our food be improved by these means?

- Will agricultural productivity be increased?
- Can GM crops benefit those who live in poor and/or developing countries, or only the wealthy in the west?
- Whose needs does biotechnology respond to?
- What alternatives are there, and might these alternatives serve the needs of the world's hungry better than GM crops?[19]

Space keeps me from dealing with all of these questions specifically. Rather, in what follows I shall look at each of the four roadblocks to a well-ordered science discussed above, and show how certain aspects of the present situation surrounding GM crops fail to reach the ideal. That the situation surround GM crops is far from the ideal of a well-ordered science, is, I think, clear. This will become clearer yet below. Finally, I shall briefly sketch what I think a well-ordered science should say about GM crops.

Who sets the agenda for the development and use of GM crops? At the present time, almost exclusively, first-world scientists and multi-national corporations have set both the research and the application agenda. As Altieri points out, "most innovations in agricultural biotechnology have been profit-driven rather than need-driven. The real thrust of the genetic engineering industry is not to make agriculture more productive, but rather to generate profits."[20] Even proponents of GM crops, such as Pinstrup-Anderson and Schioler, recognize this problem. At present, the agenda for GM crops, they note, is set on "solving the problems of farmers in the wealthy countries."[21] As of yet the poor have little, if any, say in how and where GM crops are developed.[22]

We are, then, facing a problem of Inadequate Representation. Though proponents and opponents disagree about the possible usefulness of GM crops for those in developing countries, they most all agree that at the present time, the needs of people in developing countries are not given proper consideration. Importantly, this may even be the case when projects which are undertaken with the good of the developing world in mind, such as the development of "golden rice." This project, built on good intentions though it may be, clearly came from the west and is dictated to the poor of the world. While it may suit their needs, it is hard to say without consulting them.[23] As Pinstrup-Andersen and Schioler say, "the agenda should be set by those who have to live with the consequences of the resulting action, not by some misguided belief that people in rich counties know what is best for the poor countries and poor people of the developing world."[24] My only contention with this statement is to point out that it applies to the agenda being set in the west by agri-business corporations as well, and that we have yet been given no reason to think that when the poor are given a real choice, GM crops will be the path chosen. While that may be the case, we cannot know beforehand.

These cases show how our present situation is not one of enlightened democracy but rather external elitism edging towards vulgar democracy in search of profits. While these are not the only cases where we can see a lack of adequate representation in matters concerning GM crops, they do show how we have strayed away from the ideal of a well-ordered science, and give us at least some idea of what must be done to get back onto the path. Both Altieri and Pinstrup-Andersen and Schioler agree, for example, that it is necessary that we move towards more publicly funded and controlled research in agricultural policy and development.[25] While publicly funded research does not guarantee adequate representation, it at least moves in that direction and makes it somewhat more likely, especially if the other aspects of a well ordered science are followed through.

We may now turn to the problem of the Tyranny of the Ignorant. Some of the questions here shade into questions about False Consciousness, but I shall try to give them a distinct reading at this point. Of particular interest are questions about public knowledge about GM crops and what prevents it, the possibility of choice and how this is prevented by a lack of knowledge, and the question of labeling.

One of the major problems that prevents us from reaching a state of well-ordered science in the case of GM crops is a terrible ignorance about the issue on the part of most people. As Pinstrup-Andersen and Schioler point out, "a poor grasp of biology" is a serious block to a proper understanding of the issue. For example, only 45 percent of Americans questioned could give the correct answer to the question, "Do ordinary tomatoes contain genes, or is it only genetically modified tomatoes that do?"[26] People in several European countries failed to do much better. Given such a sorry state, it is clear that common consumers do not, for the most part, have tutored preferences about GM crops. Given that they do not have tutored preferences, they cannot make real choices about the issue. It is unlikely that the situation is much different in the developing world in regard to knowledge of scientific biology, though there is some reason to think that farmers who actually work the land and know traditional methods in the developing world might know what the relative options are to a better degree than do even many Western scientists.[27] Additionally, as noted above, it is likely that those who develop GM crops know little about the lives and needs of the potential end users of such crops, especially those in the developing world. So long as scientists remain ignorant of the needs of those they serve, and the (often social) causes of these needs, they too will not have tutored preferences.

If the public is to have tutored preferences about GM crops, they will clearly need more and better information. Unfortunately, at the present time, while the FDA will declare GM crops to be 'substantially equivalent' to regular crops, how to interpret this is beyond the means of a typical western consumer. Furthermore, the vast majority of scientific information used for the testing of GM crops comes not from independent researchers but is based on, "information provided voluntarily by companies producing GE crops."[28] Since few, if any, independent long-term studies have been done on the effects of GM crops, both on consumer health and on the quality of the crops themselves, we do not yet have the information needed to form tutored preferences.[29] While we may agree with Pinstrup-Andersen and Schioler that both western consumers and people in the developing world "should be given a real choice"[30] about using GM crops, it seems clear that while people are ignorant of the science, scientists are ignorant of the complex etiology of problems faced by farmers and consumers in both the western and developing worlds, and the majority of information comes not from independent research but from interested parties, a real choice cannot be made. Before we can overcome the Tyranny of the Ignorant in regards to GM crops, significant work will have to be done to educate those potentially affected, both in the west and in the developing world, and to provide independent, disinterested testing to determine the safety and effectiveness of GM crops, especially as compared to alternative methods.

Many opponents of GM crops, both moderate and extreme, call for labeling of products that contain GM components. At first sight this might seem like an obvious step towards overcoming some aspects of the Tyranny of the Ignorant. While I agree that labeling is a good thing, and necessary if consumers are to have a choice as to whether to buy and eat GM foods, we should not think that this will solve our problems. Labeling will do little good unless it provides consumers with information that they can understand and make use of. Additionally, there are difficult questions to answer about what, exactly, should be labeled as GM. So, while labeling of products as containing GM foods is perhaps a necessary step towards offering consumer choice and building tutored preferences, it is clearly not sufficient.[31]

From here we may turn to the problem of False Consciousness. This is the problem that arises when the publicly offered reasons for supporting some research program do not in fact fit the actual reasons that do support it. I contend that the situation surrounding GM crops fits this description. Proponents of GM crops offer a number of reasons why the public ought to favor them. They say, for example, that GM crops will increase yield and make crops cheaper for consumers.[32] GM crops will provide foods that are healthier, they say, such as cholesterol-free oil, sweeter and more colorful fruits, more starchy potatoes, and foods that do not cause allergic reactions.[33] With the spread of GM crops, we are told, we will be able to use less fertilizer by engineering wheat to fix nitrogen in the soil. Finally, GM crops can serve as a means to give much-needed nutrients to poor people in the third world, thereby combating malnutrition.[34] The claim is that GM crops will go some distance towards letting people, especially the world's poor, control their own lives

While all of these things may be possible, it is very important to note that they are, in fact, quite a ways off, and not just around the corner, as is often suggested. More damning, perhaps, and more relevant to our present concern, is the fact that while these items may all serve the common good, they are not in fact the reason much GM research has been done, nor are they likely to become so unless there is a large profit to be made. Pinstrup-Andersen and Schioler themselves admit this in their more sober moments when they note that, "the major players in (the GM) field have not geared their research towards yield increases in developing countries but towards solving the problems of farmers in wealthy countries."[35] The reason for this is that the "major players" are largely multi-national corporations who make chemicals (usually pesticides and herbicides) as well, and again, as Pinstrup-Andersen and Schioler note, at the present time, "the seeds and chemicals

WELL-ORDERED SCIENCE: THE CASE OF GM CROPS 135

go hand in hand: there is little sense in one without the other."[36] It seems clear that Altieri is right when he says that the developments in the GM field have been "profit, not need, driven."[37] Given these facts, we see that GM crops are in much the same boat as the genomes project—both represent a case of false consciousness, where the reasons for pursuing the projects offered by their proponents are far from the reasons that the project is actually pursued. This is not to say that these goals cannot be reached, but only that there is little reason to think we shall reach them any time soon, and even less reason to think so if serious changes in the structure of scientific practice are not made.

The final obstacle to well-ordered science that we shall consider is the problem of Parochial Application. Recall that this is when the principles that we use to support one research program or application of some scientific findings would also support, perhaps to a higher degree, another research program or application that is not presently undertaken. The discussion of GM crops is ripe with such cases. I shall focus on two particular cases that I think are quite clear—attempting to use GM crops to provide more nutrients to people in poor countries and attempting to give people in poor countries more control over their lives. I start with the former.

Poor people in many developing countries suffer not only from a lack of calories, but also a shortage of vitamins and minerals. This is a result of a lack of a balanced diet. If these people are to be healthy and develop properly, they need to receive more nutrients in their food. So, we may take as our goal increasing the level of nutrients in the diets of poor people. The reason that many poor people in the developing world have a diet low in vitamins and minerals is that their diet consists primarily of rice, which is low in nutrients. So, one way to improve the diet of these people would be to genetically modify rice so that it contains a greater amount of the necessary vitamins and minerals. So called 'golden rice' is the first step on this program.[38] So far this program has had very limited success, producing only a type of rice that is "not particularly common, and (with a) flavor and appearance that leaves something to be desired."[39] Additionally, this product has not been very practical. The diets of the people most likely to benefit from it do not contain the level of fats necessary for the nutrients from golden rice to be absorbed, and it would be necessary to eat over a kilogram of rice a day to receive the recommended daily allowance of vitamin A.[40]

So far, then, this approach to our problem has not paid off. We should not yet conclude that it is hopeless. It is yet possible that we will improve these techniques and overcome the difficulties. But, even if this is so, it is not clear that this is the path best licensed by the principle that guides this research, 'work to provide more nutrients to poor people in developing counties.' As Altieri points out, "one must . . . realize that (vitamin) deficiency is not best characterized as a problem, but a symptom. . . . People do not exhibit vitamin A deficiency because rice contains too little vitamin A, but rather, because their diet has been reduced to rice and almost nothing else."[41] Pinstrup-Andersen and Schioler respond to this argument by noting that the poor cannot easily find or pay for this varied diet.[42] While this may be true, it seems likely that if the same energy and millions of dollars that had been spent merely on the development of golden rice had instead been spent on methods to improve local agriculture, perhaps by other means, then this problem may already have been solved. That other methods have not been seriously undertaken should make us worry that perhaps here we are facing not just a problem of parochial application, but also false consciousness.

From this example it is easy to move on to our second case. Both sides in the debate agree that we should make it our goal to provide the poor of the world with the means to control their lives, but they disagree widely over how this is to be done.[43] While some see GM crops to be the best method to do this, and suggest that we put significant energy into the project, others propose, rather, that we follow the path of agro-ecology.[44] There is some reason to think that the stated goal would in fact support agro-ecology more fully than GM crops. Consider: at the present time, poor people in the developing world are subject to the whims of weather, insects, and disease in growing their crops. GM crops may offer solutions to some of these problems. However, these solutions are partial at best, and cannot be considered long-term in the cases of disease and insects, where adaptation will certainly take place, perhaps quite quickly. GM crops also do nothing to stop, and perhaps even promote, the dangers associated with monocultures.[45] Additionally, using GM crops replaces the dangers of standing at the whims of weather with the dangers of standing at the whims of multi-national corporations, foreign aid programs, and seed dealers. It is not at all clear that the poor will end up with more control if this path is followed.

But there are alternatives. Agro-ecology, for example offers many of the same benefits of GM crops (such as a reduced use of chemical fertilizer, protection from insects, etc.) but does it in a way that does not depend on the whims of foreign aid or multi-national corporations, and does not promote the spread of monocultures. Additionally, it seems that agro-ecology is more likely to help provide a varied diet. While we should not conclude *a priori* that agro-ecology would be favored by the tutored preferences of the developing world, it does seem that there is some reason to think that this is so, and that it better suits our stated goal of providing the poor of the world a way to control their own lives. We should not assume, then, that GM crops will be the best way to meet these problems.

STEPS TOWARDS A MORE WELL ORDERED AGRICULTURAL POLICY

I shall now, very briefly, sketch what I take a well ordered science should say in regards to GM crops. First, it is important that all who are affected by the use of GM crops be represented in the deliberation over them. This should include not only scientists and the heads of chemical and bio-tech corporations, but also farmers, consumers, and other interest groups. Of particular need of representation are poor people from the developing world, whose voices are often ignored, and those who face risks from the use of such products such as Bt[46] corn without benefit, such as organic farmers. For this deliberation to be effective, it will be necessary that it be public, and result in a publicly controlled research program and standards. While we probably do not want to forbid private work in this area, (allowing multiple approaches can, in the right circumstances, give us more options, so long as one approach is not allowed to become artificially dominant) strong and clear safety and ethics standards decided on by all interested parties can and should be placed on all private sector research. Additionally, a serious educational campaign will have to be undertaken to help the deliberators develop tutored preferences, ones that do not reflect baseless fears or prejudices, but that also acknowledge the real dangers and possible alternatives such as agro-ecology. Once we are clear what our goals are, it seems likely that a well ordered science would allow significant research to be done on GM crops, but that they would be given a much smaller role than some of their proponents believe they ought to have. Rather, it seems that many of our goals, when we see them clearly and avoid false consciousness, can be better met by other means, such as agro-ecology and a more just distribution of wealth. I cannot hope to develop these ideas more fully here.

We have seen, then, several ways how the present situation surrounding GM crops falls far short of the ideal proposed by a well ordered science. While I cannot hope to say what, exactly, a well ordered science should say about GM crops, I have tried to at least sketch some ideas. Though there seems to be little hope of such a scheme being put in to place in the near future, I still take this to be an important step.

NOTES

My thanks to Michael Weisberg, Philip Kitcher, and especially Hugh Lacey for their helpful comments, discussion, and encouragement. They are not, of course, responsible for any mistakes in the paper. My thanks also to Ekaterina Dyachuk for her constant support and encouragement.

1. Philip Kitcher, *Science, Truth, and Democracy* p. 133.

2. Ibid., p. 133.

3. I would like to note that I have nothing in particular against the corporate form of business and think that it is likely to be a necessary part of any large-scale industrial society, at least in some form. To my mind the problems that exist with corporate behavior are best dealt with through laws governing corporate set-up and liability rather than largely empty railing at the evil of corporations as such.

4. Kitcher, "Reply to Helen Longino," *Philosophy of Science*, 69, (Dec. 2002) p. 569.

5. Kitcher, *Science, Truth, and Democracy* p. 118. For a helpful overview of the idea of deliberative democracy see Samuel Freeman, "Deliberative Democracy: A Sympathetic Comment," *Philosophy and Public Affairs*, 29 No. 4, 2000, pp. 371–418.

6. Ibid., p. 117.

7. Helen Longino, "Science and the Common Good: Thoughts on Philip Kitcher's *Science, Truth, and Democracy*," *Philosophy of Science*, 69 (Dec. 2002) p. 565.

8. Kitcher, *Science, Truth, and Democracy*, p. 118.
9. Ibid., p. 118.
10. Ibid., pp. 118–121.
11. Ibid., p. 123.
12. Ibid., p. 129. The "Nonrepresentational Ratchet" is a mechanism where an early lack of representation can lead to a preference for a path that continues to favor one group even when adequate representation is established. This is due to the ease of continuing down an established path of inquiry, even if another may be more optimal in some important ways.
13. Ibid., p. 129.
14. Ibid., p. 130.
15. Thanks to Hugh Lacey for pointing out the need to make this point clear.
16. Kitcher, *Science, Truth, and Democracy*, p. 131.
17. Ibid., pp. 130–131.
18. Ibid., p. 132.
19. This list of questions is largely derived from those in Pinstrup-Andersen and Schioler, p. 2, and Miguel A. Altieri, *Genetic Engineering in Agriculture*, p. vii.
20. Altieri, p. 4. Again I would like to note that I have no general objection to profit-making by corporations and think that it is almost certainly a necessary part of a large-scale industrial society. My interest here is only in seeing to what degree the natural and appropriate profit-seeking behavior of corporations might conflict with science being well ordered.
21. Pinstrup-Andersen and Schioler, p. 92.
22. Ibid., p. 5.
23. Altieri, pp. 5–7.
24. Pinstrup-Andersen and Schioler, p. 109.
25. Ibid., p. ix, 4, Altieri, pp. 46–47.
26. Pinstrup-Andersen and Schioler, p. 111.
27. Thanks to Hugh Lacey for bringing out the need to be clear on this point.
28. Altieri, p. 18.
29. Ibid., p. 19.
30. Pinstrup-Andersen and Schioler, p. 85.
31. Oddly enough, Pinstrup-Andersen and Schioler seem to hold that while labeling is a good idea in the west, we cannot expect companies to meet such standards in the developing world. This seems to be clearly at odds with their claim that the poor should be given the choice whether to use GM crops or not. If one does not know that one is using GM crops, it is hard to see how one has a choice in the matter. I do not know how to reconcile this apparent contradiction. It seems to me to be only one of several tensions in their book. See Pinstrup-Andersen and Schioler, p. 132.
32. Ibid., p. 51.
33. Ibid., p. 94.

WELL-ORDERED SCIENCE: THE CASE OF GM CROPS

34. Ibid., pp. 52–55.
35. Ibid., p. 92.
36. Ibid., p. 44.
37. Altieri, p. 4.
38. Pinstrup-Andersen and Schioler, p. 53.
39. Ibid., p. 53.
40. Altieri, p. 7.
41. Ibid., pp. 5–6.
42. Pinstrup-Andersen and Schioler, p. 138.
43. ibid., p. 5, and Altieri, p. 3.
44. Altieri, p. 33.
45. Altieri, p. ix.
46. Bt Corn is corn that has been genetically modified to contain the Bacillus Thuringiensis toxin, which is toxic to many insects that feed on crops. Bt is naturally occurring and has long been used as a natural pesticide by organic farmers. There is significant worry that wide-spread use of Bt corn will result in the quick development of Bt-resistant insects thereby depriving organic farmers of one of the few pesticides which they can use.

ETHICS AND THE LIFE SCIENCES

FRANKENFOOD, OR,
FEAR AND LOATHING AT THE GROCERY STORE

JENNIFER WELCHMAN
UNIVERSITY OF ALBERTA

ABSTRACT: Genetically modified food crops have been called 'frankenfoods' since 1992. Although some might dismiss the phenomena as clever marketing by anti-GM groups, of no philosophic interest, its resonance with the general public suggests otherwise. I argue that examination of the intersection of popular conceptions of monsters, nature, and food at which 'frankenfood' stands reveals significant and disturbing trends in our relationship to organic nature of interest to moral and social philosophy and to environmental ethics.

Though now ubiquitous, the term 'frankenfood' has not been with us long. It first appeared in print in 1992[1] and caught on almost instantly, to the delight of critics of genetically modified (GM) crops and the dismay of their creators and proponents. By 1999, the term was so well established in general usage that the Greenpeace organization, as part of its 'True Food' campaign, introduced one of the first of many spins-offs;[2] 'FrankenTony,' a caricature of the Kellogg's corporation's famous cereal-touting tiger, made up to resemble the Frankenstein monster of the old Universal Pictures films.[3] The purpose was clear: to make the potential threat of frankenfoods more alarming than they already were by calling attention to their possible presence in products marketed to children. Defenders of the genetic modification of organisms for human benefit have fought both the acceptance of frankenfood as a characterization for GM crops generally, and Greenpeace's employment of 'FrankenTony' specifically, arguing that it is (1) misleading in suggesting that GM crops are unnatural monsters analogous to Frankenstein's monster and (2) inflammatory—a tactic employed to forestall reasoned debate by appeal to irrational emotions, especially fear.

But despite these efforts, 'frankenfood' is here to stay. Now one might wonder whether this phenomena merits any particular philosophic attention—is it anything

© 2007 Philosophy Documentation Center   pp. 141–150

more than a case of clever marketing by opponents of GM crops? Well, it is this certainly. But it may also be something more. If, as marketing professionals often claim, the most successful marketing campaigns are those that tap into the prevailing Zeitgeist in some way, then the ready acceptance of frankenfood—the concept not the crops—points to a convergence of social attitudes to which philosophers, especially philosophers interested in ethics, technology, and the environment, might do well to investigate. There is, I think, much of philosophical interest in this intersection between contemporary attitudes towards food, monsters, and nature.

Before I do, however, there is an important preliminary issue to be addressed. Critics of the employment of the term 'frankenfood' sometimes base their criticisms on a common conception of human emotions as pure affective responses to external stimuli that lack any cognitive content. Actions or attitudes arising from emotions, on this view, are, like the emotions themselves, nonrational and on occasion, even contrary to reason. Now if one holds such a view of human emotions, labelling GM crops 'frankenfoods' will seem an illegitimate tactic in public debates. The identification of GM crops with the Frankenstein monster might be legitimate, on this view, if the effect were to persuade our reasons by evidence and argument. But the actual and intended effect is to induce fear. If successful, we will fear and reject GM crops. But we will be induced rather than persuaded to do so. Fear does not simply cloud our judgment, it by-passes judgment altogether.

But this view of emotions as contentless affective responses operating independent of and/or contrary to reason misrepresents the true state of affairs. The relation between emotion and reason is much more complex. A thorough investigation of the nature of our emotions would be out of place here, so, for present purposes, I will simply sketch and defend a view nearer to the working-conceptions most of us actually hold—that many (if not all) emotions involve cognitive and evaluative elements.

There is, as has often been observed, a certain logic to our emotions, a logic to which we regularly make appeal when assessing our own and others' emotional responses. Only when emotions clearly violate this logic do we normally judge them irrational. Say, for example, that as I walk down a darkened street late at night, a shadowy figure suddenly springs out of the bushes. Here, we can all agree, fear is a reasonable response to the appearance of a potential threat. But if it turns out that shadowy figure is really a friend playing a practical joke, and I recognize that no threat exists, then we would (again) all agree that fear has become inappropriate and irrational. Clearly we all believe that in fact beliefs do enter into the formation of emotional responses. So also do evaluations. For example, which emotion replaces the fear I felt prior to realizing my friend was playing a joke on me depends in part on how I evaluate it. If it seems to me cruel or humiliating, I may be indignant, even furious, with my friend. But, if instead, my friend's performance is a continuation of a private game of one-upmanship, where each tries to top the other, then envy or hilarity would seem more appropriate, more reasonable affective responses.

Now, as part of a thorough investigation of emotion and its significance in our intellectual and practical lives, we would have to go further into these cases to determine to what extent these cognitive and evaluative elements constitute our emotions, whether these elements are in fact distinct or reducible to any single phenomena (e.g., perception, reception, or appreciation), and whether in addition, emotions also include forms of affection non-reducible to cognitive or evaluative content, but that is a project for another day. It is enough for the moment, if we can simply agree that in normal cases, emotions have cognitive and evaluative elements that play an important role in their formation. For if we can, then it becomes worth asking what beliefs and values symbolized or represented by 'frankenfood' are revealed by the distinctive emotional responses to which it gives rise.

Evidently 'frankenfood' alludes to Mary Shelley's novel *Frankenstein*,[4] one of the seminal works of the gothic-horror genre that dates from the late eighteenth century and remains a popular staple of literary fiction, theatre, and films. In the novel, the name, 'Frankenstein,' applies only to the scientist who is the story's protagonist, but it is now used to refer primarily to his monster—the composite human being Frankenstein creates by 'unnatural,' technological means; specifically the grafting together of body parts that would never naturally exist together, by processes that do not occur in nature. Frankenstein's is perhaps the best known gothic horror monster, but there are of course many others: Dracula, the Wolfman, the Mummy, the Fly, the Blob, the Thing, X: the Unknown, the Creeping Unknown, Alien, and the Terminators 1 and 2, to name just a very few.

As Noel Carroll points out in his study, *The Philosophy of Horror*, monsters in the gothic horror genre differ from the threatening entities of other genres in the peculiar affective responses that horror monsters engender in their fictional victims and audiences: fear combined with loathing or revulsion. As Carroll notes, the victims of horror monsters display a regular pattern of behavioural responses to the physical presence of the monster, which function as external 'cues' for similar internal and external responses by the audience. In films, for example,

> We often see the character shudder in disbelief. . . . Their faces contort; often their noses wrinkle and their upper lip curls as if confronted by something noxious. They freeze in a moment of recoil, transfixed, sometimes paralysed. They start backwards in a reflex of avoidance. Their hands may be drawn toward their bodies in an act of protection but also of revulsion and disgust. Along with fear of physical harm, there is an evident aversion to making physical contact.[5]

Viktor Frankenstein's reaction to his own newly-live creation fits the formula exactly. He reports

> I had worked hard for nearly 2 years, for the sole purpose of infusing life into an inanimate body. . . . I had desired it with an ardour that far exceeded moderation; but now that I had finished, the beauty of the dream vanished, and breathless horror and disgust filled my heart. Unable to endure the aspect of the being I had created, I rushed from the room.

Victor's response to his creation[6] points to something distinctive about how these monsters are perceived—the feature that makes them horror monsters proper. As Carroll notes "within the context of horror narratives, the monsters are identified as impure and unclean.... They are not only quite dangerous, but they make one's skin creep."[7] Carroll opts for the sociological view that in human cultural and religious traditions, things come to be classified as impure when they pose category confusion for the tradition in question, i.e., they resist classification within the categories the culture tries to impose upon them. As Carroll puts it, "an object or being is impure if it is categorically interstitial, categorically contradictory, incomplete or formless."[8] So, for example, snakes are often viewed as impure in part because they crawl like worms when other vertebrates go about on foot. Faeces, especially those of our own kind, are often impure because they are neither living nor dead tissue, internal nor external, parts of ourselves or distinct from us. And so forth.

Horror monsters are categorically interstitial in a variety of ways. Some, like Frankenstein's monster, Dracula, the Mummy, and the original Terminator, are categorically contradictory in being neither dead nor alive. Others, e.g., the Wolfman, the Thing, the Fly, and the Creeping Unknown, violate species boundaries. Some are incomplete realizations of accepted categories of things, for example, Donovan's Brain, while others, e.g., the Blob, X: the Unknown, and the new and improved Terminator of *Terminator 2*, lack any consistently categorizable form. This sets them apart from the fear-inducing entities of other genres, such as disaster films, thrillers, and fantasies.

In disaster films, the subjects: tornados, tidal waves, earthquakes, and giant asteroids, are all threatening. But they are not unnatural nor does their physical presence characteristically induce loathing or revulsion in their fictional victims or the audience. On the contrary, could they or we approach these entities without risk, many would willingly choose to do so. In thrillers, the human villains evoke fear, but their physical presence is often attractive; so attractive they are actually desirable to both the narratives' heroes and the audience identifying with them. In works of "fantasy," we are sometimes introduced to frightening creatures that would be interstitial in our own world. But in the fantastic realms they inhabit, these creatures are normal, even commonplace. Thus reading or viewing films of Tolkien's *Lord of the Rings*, we are not horrified by hobbits, dwarfs, wizards, elves, or ents because we understand them to be perfectly natural within their own universe. They do not revolt us even when they threaten the narrative's heros.[9]

So through the allusion to *Frankenstein*, 'frankenfood,' (and its various spin offs, FrankenTony, Frankenfish, Frankencorn), suggests that fear and loathing, horror and disgust, are proper responses to foods containing GM organisms and at the same time provides a reason for thinking these responses appropriate. GM laden foods are categorically interstitial in at least one of the ways that Frankenstein's monster is categorically interstitial—the crops they are made from are composite creatures whose constituent elements would not naturally occur together in one entity. Certainly some anti-GM organizations overtly make just this sort of analogy,

FEAR AND LOATHING AT THE GROCERY STORE 145

some via their titles: e.g., Greenpeace's 'True Foods' campaign, Friends of Earth's 'Real Foods' counterpart, and groups such as the Alliance for Biointegrity. Others, such as the US-based Campaign to Label Genetically Modified Food and the Center for Food Safety, choose instead to highlight in their publications particularly striking violations of species boundaries in some GM organisms, for example, the introduction of human growth hormone into pigs, flounder genes into tomatoes, and shark genes into strawberries. In other words, they are going for the 'yuck' factor distinctive to the gothic horror genre.[10]

But most anti-GM groups devote only a small fraction of their publications and press releases to making this connection. And it's not hard to understand why. There is nothing visibly yucky about a field of GM canola in bloom, or of amber waves of either Round-up ready grains or pesticide-producing corn. These plants are either indistinguishable from non-modified varieties or are distinguishable only by their greater strength, beauty, and fruitfulness. They taste, smell, and feel as good, or better, to their intended human recipients. Their physical presence is not revolting or loathsome, even when it invokes fear—as it certainly does in their opponents. For example, when activists with the British organization, GenetiX Snowball, protested GM crops by pulling up plants at test sites (a strategy Greenpeace has since imitated), the members would routinely wear what they called 'protective' clothing, when 'decontaminating' sites.[11] But the purpose of the protective clothing was not to protect the protestors from the test crops. It was only to guard against inadvertent transportation of GM pollens outside the test sites. Although the protestors testified to their fear of the potential dangers they thought these crops posed, none that I am aware of ever complained of being nauseated or disgusted by the appearance, smell, or physical presence of the plants they were destroying.[12]

The affective responses to GM organisms both of activists and of the wider public supporting food labelling campaigns and calls for moratoriums on GM crops seems to have less in common with the fear and loathing characteristic of horror monsters and more in common with another sort of fear: classical 'panic' fear, i.e., fear of the god Pan.[13] The fear of Pan is the peculiar fear that can come over one in wild places when one has gone beyond the boundaries of cultivated, tame, predictable environments. It often begins with a feeling of being watched, but from where or by what one cannot tell. In a sense, it doesn't matter, for what one fears isn't any one particular threat; a bear, cougar, or wolf. Bears, cougars and wolves are just different modes by which the capricious and unpredictable Pan may choose to manifest himself. Thus panic fear is a generalized fear of particular kinds of spaces, i.e., wild, unpredictable spaces whose dangers are such that they cannot be recognized or even anticipated until it is too late.[14]

The information presented in the publications of groups such as GenetiX Snowball, the Sierra Club, Friends of the Earth, and others, seems predominately tailored to produce a kind of panic fear. For they particularly stress the possibility of unknown, unanticipated, and unrecognizable threats that GM processes may introduce into three key environments: undeveloped green spaces, farms, and the interior of one's local grocery store. Genetic modification is routinely likened to pollution that contaminates environmental systems. For example, GenetiX Snowball tells us that "Radioactivity has a "half life," it gradually becomes safer over thousands of years; but gene technology creates a pollution with "multiple life"—it keeps on replicating and cannot be recalled."[15] Mothers for Natural Law warn that "Gene Pollution cannot be cleaned up, " and worry that scientists are: "experimenting with very delicate, yet powerful forces of nature, without full knowledge of the repercussions. Unlike chemical or nuclear contamination, negative effects are irreversible."[16] The Sierra Club uses similar language to call for regulation of the biotechnology industry, arguing "The genetic manipulation resulting from genes inserted by genetic engineering cannot be recalled; the altered characteristics will be passed onto future generations and continue to be reproduced in the environment."[17] The Center for Food Safety argues that:

> Currently, up to 40% of US corn is genetically engineered as is 80% of soybeans. It has been estimated that upwards of 60% of processed foods on supermarket shelves—from soda to soup, crackers to condiments—contain genetically engineered ingredients. A number of studies . . . have revealed that genetically engineered foods can pose serious risk to humans, domesticated animals, wildlife and the environment. . . . [T]he use of genetic engineering in agriculture will lead to uncontrolled biological pollution, threatening numerous microbial, plant and animal species with extinction, and the potential contamination of all non-genetically engineered life forms with novel and possibly hazardous genetic material.[18]

In other words, "be afraid, be very afraid."

Genetic pollution, it is argued, will spread via a range of causal processes, cross-breeding, the food chain, etc., potentially contaminating every environment on the planet: farms, remote and inaccessible mountain peaks, even the aisles of one's local supermarket. Regarding the later, the Campaign to Label GM Foods comments: "if you want to avoid eating genetically engineered food all we can say is *good luck*. In just a few short years, GE foods have swept into the market place, affecting almost all of the foods we eat. In fact, the only way you can be sure to avoid eating genetically mutated food is to buy organic or to grow your own."[19] But only for as long as organic seed producers can avoid cross-pollination by related GM species.

The message being reiterated by these various groups is that the introduction of GM organisms into the environment will collapse the traditional boundary between the tame cultivated lands and the wilds where Pan traditionally holds sway. Farms are becoming spaces inhabited by entities posing unknown threats (and, if GM organisms do escape into the natural environment and cross-breed with other species, our remaining wild spaces may become wilder and more dangerous yet). Thus, cultivated spaces are being transformed into regions where classical panic fear is an *appropriate* response, on both cognitive and evaluative grounds. Nor

FEAR AND LOATHING AT THE GROCERY STORE 147

has the process of transformation stopped there—because Pan has now taken up residence at Safeway, Food Lion, and Stop/VShop. But where exactly? Among the strawberries perhaps, or the cans of baby formula? In the *Frosted Flakes*, or that tofu in the 'healthy choices' section just behind you? Nowhere is safe.

If I am correct, all this would appear to imply that the allusion frankenfood makes to Frankenstein is false or misleading, since the affective responses associated with frankenfood are *not* those associated with gothic horror monsters. In fact, if we look more closely at the novel, we find this is not actually the case. While fear-and-loathing is the first affective response the monster evokes in Viktor, later another reaction follows, closer to the panic fear of dangerous spaces that I have argued GM crops evoke. In the novel, after the monster has recounted its woes to his creator, it begs Viktor to make it a female companion, promising that both shall leave Europe forever to settle in the wilds of South America. Viktor agrees and begins work but then hesitates. But its not horror of the female creature he is creating that stops him—it is his fear of the consequences of releasing a new and reproductively viable *species* into the world.

> One of the first results of those sympathies for which the daemon thirsted would be children, and a race of devils would be propagated upon the earth, who might make the very existence of the species of man a condition precarious and full of terror. Had I a right, for my own benefit, to inflict this curse upon everlasting generations?[20]

Viktor refuses to continue.

So after all Frankenstein's monster *does* pose a threat to which panic fear is an appropriate response and it is Viktor's appreciation of the fearfulness of the environment in which future generations would find themselves that causes him to refuse the monster's request (the event that precipitates the novel's concluding series of tragedies). So if we interpret the analogy implicit in the notion of frankenfood as an analogy to this aspect of the Frankenstein monster, it is neither false nor misleading in regard to GM crops or other organisms.

On the other hand, critics do have some legitimate grounds for complaint, if the effect of labelling GM crops 'frankenfoods' is to generate a kind of panic fear. Although fear per se may not be irrational or bypass judgment, panic fear is a special case. Panic fear is a response to a perception of uncertainty of or about an environment. As it has no specific object, being a general response to a surrounding space, it is much less amenable than other sorts of fears to argument or rational persuasion. How *can* anyone prove that there is nothing important one doesn't know about any thing or event, however familiar, let alone a novel entity such as a GM organism? So critics are perhaps not wholly incorrect in suggesting that campaigns characterizing GM crops as frankenfoods may hinder rather than promote rational debate and policy formation.

These are conclusions of some interest to those debating the ethics of the introduction of GM crops and of activists' efforts in opposition. But more interesting and important, to my mind, is what the phenomena of frankenfood, and the campaigns it has spawned, reveal about the relations of human beings in highly developed cultures like our own to organic nature. It suggests that the only significant points of conscious contact most people have with organic nature on a daily basis are at mealtimes. We *eat* it.

As things stand in the highly artefactual environments so many of us inhabit, supermarkets offer the single greatest direct exposure to bio-diversity most of us will experience in an average week. And as few of us now shop daily, this means that our kitchens and dining rooms provide our richest (and perhaps our only) experience of biodiversity in the course of a typical day,[21] of which we are consciously aware. Outside supermarkets and kitchens, what direct contact do urbanites and suburbanites have with organic nature? In summer, homeowners may mow their lawns a couple of times a month, golfers curse the rough at their local course, perhaps once a week, while urbanites feed pigeons in the park as weather permits. Gardeners see a good deal more of the non-human biotic world during the summer months of course. But for those of us who neither mow, golf, garden, or feed pigeons, non-gastronomic experiences of organic nature may be limited to swatting mosquitos or hosing bird-droppings off our cars. Come Autumn, after we have cleared the dead leaves from our lawns, gutters, or balconies and hung up our bird feeders, our interactions with nature become increasingly limited, and by winter, many a Northerner's non-gastronomic interactions with organic nature shrink to nothing at all. Even those of you who actively seek direct interaction with non-human organic nature will concede that for days and even weeks at a time, the same is true of you: the only place where you have direct *daily* contact with non-human organic nature is in your kitchen.[22]

This seems to me to be a profound transformation in the conscious relationships of humans to the natural world, an almost incredible narrowing and impoverishing of those relationships, one that poses a serious obstacle to general recognition of the deeper array of actual interactions humans have with nature and the affective responses appropriate to them. The phenomena of frankenfood, with its explicit focus on food, is thus a sign of something deeply wrong with our evolving relationship to the natural world; something we need to comprehend if we are to cope both with the ethical and technical challenges that GM organisms present, but also the wider problems of how we should deal with the biosphere to which we belong. To care adequately for the natural environment, people must be able to appreciate it as something more than simply extensions of their larders. To care adequately about bio-diversity, ecological stability, and endangered species and threatened environmental systems, one must relate to the natural world as something directly *constitutive* of a good, a beautiful, and/or a meaningful human life. The present trend towards seeing the organic natural world as merely *instrumentally* valuable for varied and healthy menus is not going to be enough to produce the affective, emotional responses that will be necessary if people are to be motivated to make the sacrifices necessary to preserve it.

NOTES

1. Paul Lewis's letter to the Editor of the *New York Times* is credited with the first use of the term in print: "To the Editor: 'Tomatoes May Be Dangerous to Your Health' (Op-Ed, June 1) by Sheldon Krimsky is right to question the decision of the Food and Drug Administration to exempt genetically engineered crops from case-by-case review. Ever since Mary Shelley's baron rolled his improved human out of the lab, scientists have been bringing just such good things to life. If they want to sell us Frankenfood, perhaps it's time to gather the villagers, light some torches and head to the castle.' Paul Lewis, Newton Center, Mass., June 2, 1992" ("Mutant Foods Create Risks We Can't Yet Guess," *The New York Times*, June 16, 1992). Source: Wordspy Website at www.wordspy.com. Retrieved July 6, 2004.

2. Other spin-offs include 'franken fish,' (although this term is also used in parts of the USA to refer to an exotic, introduced species of snakehead which is not transgenic), 'franken corn,' 'franken salmon', and more recently, 'franken flower' (a rhododendron with a gene from a species of frog.)

3. The original poster of Franken Tony gracing the front of a box of "Genetically Modified Frosted Flakes of 'Corn'" is no longer generally available from Greenpeace Websites, ever since Greenpeace obtained assurances from Kellogg regarding their future use of GM products. It can however, still be found here and there, on the Internet if one is persistent, as can many images of Greenpeace demonstrators wearing Franken Tony costumes at protests around the US.

4. Mary Shelley, *Frankenstein*, ed. Joanna M. Smith (Boston: Bedford Books/St. Martin's Press, 1992).

5. Noël Carroll, *The Philosophy of Horror, or, Paradoxes of the Heart* (New York: Routledge, 1990), p. 22.

6. And that of the thousands of other fictional victims of horror monsters since, as anyone who enjoys stories, books, plays, and films in this genre will readily attest.

7. Carroll, *Philosophy of Horror*, p. 23.

8. Ibid., p. 32.

9. But even Middle Earth has its gothic horrors: the orcs, who we learn are a deliberately created degenerate form of elvish life, and in the case of Saruman's orcs also products of miscegenation with human beings. In keeping with gothic tradition, orcs revolt other more natural inhabitants of Middle Earth by their noxious small and habit of consuming impure food and fluids, and so forth.

10. Images utilized by such groups routinely go for the same effect often by superimposing the facial features of the Universal Pictures' Frankenstein on corn, bananas, fish, etc. One of the most striking in this vein is an image featured on the Democratic Underground Website, in which the Frankenstein Monster eyes a sandwich which it turns out is eyeing him back; see www.democraticunderground.com. Retrieved July 6, 2004.

11. These were usually limited to overalls with sleeves, hats, and or gloves, etc.. Here again Greenpeace has followed suit in later demonstrations at GM test sites.

12. See the GenetiX Snowball Website, at www.fraw.org.uk, for information about the group. Personal statements by activists involved in early 'decontamination efforts' may be found at the site: www.fraw.org.uk/gs/campaign.htm#state. Last retrieved, July 6, 2004.).

13. In Greek, *panikon deima*, fear of that pertaining to Pan.

14. In current psychological literature, 'panic' is defined by the symptoms that sufferers' experience rather than any particular source or stimulus (sources and/or stimuli generally being unidentifiable). Interestingly, however, sufferers of panic attacks apparently associate the attacks or a greater severity of symptoms (e.g., feelings of being out of control, etc.) with particular environments and so develop an aversion to exiting whatever zones they feel are "safe" unless accompanied by trusted individuals. At the experiential level then, there continues to be a felt connection between contemporary conceptions of panic fear and uncontrolled or controllable spaces. See *The Diagnostic and Statistical Manual of Mental Disorders: DSM-IV*, 4th ed., (Washington, D.C.: American Psychiatric Association, 1994).

15. See "Introduction, " GenetiX Snowball, *Handbook for Action: A Guide to Safely Removing Genetically Modified Plants From Release Sites in Britain*, online at www.fraw.org.uk/gs/handbook/handbook.htm. Last retrieved July 6, 2004.

16. See the website of Mothers for Natural law at www.safe-food.org/-campagn/about.html. Retrieved June 28, 2004.

17. See, Sierra Club Policy on Biotechnology, Sierra Club web-site, www.sierraclub.org/policy/conservation/biotech.asp. Retrieved June 24, 2004.

18. Center for Food Safety Website, www.centerforfoodsftey.org/genetical12.cfm, retrieved June 24, 2004.

19. See the website (esp., the "Tutorials" section) of the Campaign to Label Genetically Modified Food at: www.thecampaign.org. Retrieved June 28, 2004.

20. Shelley, *Frankenstein*, p. 140. A similar suggestion is made again later in the novel when the Frankenstein is pressed by a companion for details about the monster's creation. "Are you mad, my friend, said he; or whither does your senseless curiosity lead you? Would you also create for yourself and the world a demonical enemy?" (p. 175). Viktor clearly anticipates that further experiments like his own would result in a self-perpetuating race of creatures, since single individuals, though dangerous and horrific, could not be said to threaten the human world as a whole.

21. Of which we are consciously aware. Other interactions are going on all the time, internally and externally, esp. at the microscopic level, but of these we are not normally conscious.

22. I think this is true even of the significant percentage of people who own houseplants or pets. Many small pets do not require or receive daily attention (fish, snakes, etc.). Many houseplants will not need watering or other care more than once or twice a week for a matter of moments (in between these times they may be hardly noticed). Even people with pets like cats and dogs may often spent far more time purchasing, preparing, and/or consuming food than they do interacting with their animals. So the grocery store and the kitchen will remain the main and perhaps sole points of conscious contact with non-human organic nature that many such people will have for prolonged periods.

ETHICS AND THE LIFE SCIENCES

THE OTHER VALUE IN THE DEBATE OVER GENETICALLY MODIFIED ORGANISMS

J. ROBERT LOFTIS
ST. LAWRENCE UNIVERSITY

ABSTRACT: I claim that differences in the importance attached to economic liberty are more important in debates over the use of genetically modified organisms (GMOs) in agriculture than disagreements about the precautionary principle. I will argue this point by considering a case study: the decision by the U.S. Animal and Plant Health Inspection Service (APHIS) to grant nonregulated status to Roundup Ready soy. I will show that the unregulated release of this herbicide-resistant crop would not be acceptable morally unless one places a very high premium on economic liberty. This is true even if one takes a sound science attitude to unknown risks, rather than a precautionary attitude. I concede that it may not have been within APHIS's legislative mandate to regulate Roundup Ready soy further, but for those of us who do not put a high premium on economic liberty, this only calls for extending regulatory oversight of GMOs.

I. INTRODUCTION

According to Michael Ruse and David Castle, the 'precautionary principle' is "a cornerstone of biotechnology policy" (Ruse and Castle 2002, 250). The precautionary principle is a rule of prudential reasoning designed to compensate for the perceived recklessness of current methods for making decisions when risks are poorly understood, including cost-benefit analysis. It is explicitly written into

European law but has been kept out of U.S. regulation by lawmakers on the right, who prefer the so-called 'sound science' principle. The sound science principle requires that no safety risk be considered in regulation until the causal mechanism that underlies it is thoroughly understood. Because U.S. lawmakers cannot agree on an approach to precautionary issues, regulatory agencies have simply judged genetically modified organisms (GMOs) based on analogies and resemblances to previously known and understood organisms.

The differing approaches to precaution in Europe and the United States have clearly affected the GMO debate. However, I want to highlight the importance of another value at play in this debate, economic liberty. I claim that differences in the importance attached to economic liberty are decisive in deliberations about GMOs. I will argue this point by considering a case study: the decision by the U.S. Animal and Plant Health Inspection Service (APHIS) to grant nonregulated status to Roundup Ready soy. I will show that the unregulated release of this herbicide-resistant crop would not be acceptable morally unless one places a very high premium on economic liberty. This is true even if one takes a sound science attitude to unknown risks, rather than a precautionary attitude. I concede that it may not have been within APHIS's legislative mandate to regulate Roundup Ready soy further, but for those of us who do not put a high premium on economic liberty, this only calls for extending regulatory oversight of GMOs.

Two caveats: First, this is essentially an exercise in rational reconstruction. I am identifying a premise that must be in place to justify a decision. More empirical sociological methods might yield different conclusions about the values in play in the GMO debate. However, the principle of charity in interpretation—the rule that says we should always be kind to our opponents in reconstructing their arguments—guarantees that this sort of analysis must play at least some role in understanding the debate. Second: I am not opposed to all use of GMOs in agriculture. I am only opposed to using the GMOs that worsen the current problems with the global agricultural system. I actually hope this essay will be a contribution to the discussion of the question "What kind of GMOs should there be?"

II. BACKGROUND

The vast majority—81 percent in 2004—of the genetically modified (GM) crops in the environment right now have been modified to tolerate an herbicide (James 2004). Generally the same company that sells the GM seeds makes the herbicide, and the two are sold as a package. The farmer can thus blanket her crops with the herbicide, knowing that it is likely to only affect the weeds. Although many benefits have been cited for herbicide-resistant crops, their only direct benefit is to increase yields relative to cost. They do this by allowing the farmer to kill more weeds with fewer applications of herbicide. Previously farmers would blanket their fields with a wide-spectrum herbicide before the emergence of their crops, followed by many sprayings using targeted herbicides or delivery methods. With herbicide-resistant crops, farmers can simply use a small number of sprayings of a wide-spectrum

THE OTHER VALUE IN THE DEBATE OVER GMOS 153

herbicide at any point in crop development. It is worth noting, however, that using fewer applications of herbicide is not the same as reducing the overall amount of herbicide pumped into the environment.

Since 1996, APHIS has handled most of the regulation of GMOs.[2] APHIS claims jurisdiction over GMOs because they typically contain genes from *Agrobacterium tumefaciens*, the cauliflower mosaic virus, or other known plant pests (APHIS 1987). This policy leads to a couple of oddities. First, ever since the establishment of the "Coordinated Framework for the Regulation of Biotechnology" (Office of Science and Technology Policy 1986), the major complaint against U.S. biotechnology regulation is that it refused to acknowledge any differences between current genetic technology and traditional selective breeding. Yet APHIS effectively goes back on that refusal by using genetic modification to trigger regulatory review. Second, APHIS's claim of jurisdiction contains a curious piece of genetic essentialism. (Genetic essentialism is the almost superstitious belief that the "true nature" of a thing can be found only in its genes.) Often the genetic material taken from the known pest consists only of promoter or stop sequences, short statements of genetic code that say "start reading here" or "stop reading here." The meaning of such statements, and hence their danger, will have much more to do with the context they are placed in than the context they came from.

In any case, once a GMO falls under APHIS's jurisdiction, the seed company generally asks that APHIS grant the product "nonregulated status," which relieves it of all further oversight. Essentially, APHIS declares that it didn't really have jurisdiction after all. Among other things, this absolves the GMO of all postcommercialization monitoring to see what an organism actually does when it is released into the wild. One of the most pervasive unmonitored GMOs is Monsanto's Roundup Ready soy, which was granted nonregulated status in 1994 (APHIS 1994a, 1994b, 1994c). Roundup Ready soy is the herbicide resistant counterpart to Monsanto's flagship herbicide, Roundup. The farmer buys Roundup and Roundup Ready soy together, knowing that the Roundup will kill all the plants in her field besides the Roundup Ready soy. Roundup is a common weedkiller, available to ordinary consumers in hardware stores. Its active ingredient is glyphosate, which blocks an enzyme used in photosynthesis. Glyphosate is benign by herbicidal standards. It is water soluble, so that it does not lodge itself in animal tissues and accumulate as it works its way up the food chain, the way DDT does. It also disperses quickly, so that no traces can be found in the soil a week after spraying. Nevertheless, there are good reasons why the Roundup in the hardware store carries warning labels. Glyphosate itself can damage the liver of mammals (Chan and Mahler 1992). More important, Roundup contains the surfactant polyoxyethyleneamine (POEA), which helps the herbicide spread more evenly. It also can kill you. The twenty people known to have died from directly ingesting Roundup (all probable suicides) were killed by the POEA (Sawanda et al. 1988; Tominack et al. 1991).

When Monsanto petitioned to have Roundup Ready soy deregulated, they submitted results from nine field trials. Thirty-three letters of public comment were also solicited by APHIS in the *Federal Register*. In their response to Monsanto's petition (APHIS 1994c), APHIS made five findings: (1) neither the Roundup Ready gene construct nor its products pose a plant pest risk, (2) Roundup Ready soy has "no significant potential to become a weed," (3) Roundup Ready soy will not increase the weediness of plants it can breed with, (4) Roundup Ready soy will not damage processed agricultural products, and (5) Roundup Ready soy will not harm beneficial organisms. Given these five findings, APHIS determined that Roundup Ready soy was not a plant pest, so it did not fall under their jurisdiction and would not be subject to any further regulation.

III. THE COST-BENEFIT ANALYSIS: WHAT BENEFIT?

In their deliberations, APHIS failed to consider many of the environmental risks posed by Roundup Ready soy at all and treated other risks inadequately. All of these risks are compounded by the lack of postcommercialization monitoring. Furthermore, unless you put a premium on economic liberty, the widespread use of Roundup Ready soy has no direct redeeming benefits.

APHIS did not consider any possible risks from the changing patterns in the use of glyphosate, seeming to take for granted the assertion by the petitioners that Roundup Ready soy would decrease herbicide use and that this would be a guaranteed environmental gain. However, as Brian Johnson and Anna Hope point out (Johnson and Hope 2000), the net effect of herbicide use has as much to do with timing and application methods as it does volume of herbicide used. In this regard, Roundup Ready soy looks dangerous. Farmers who use Roundup Ready soy are more likely to set spray nozzles high or even use aerial spraying, increasing pesticide drift (Johnson and Hope 2000; Lappé and Bailey 1998). The environmental impacts of glyphosate itself are still unknown. It is known to disrupt the soil's microflora, but the long-term impact is unknown (Lappé and Bailey 1998, 80). Overall effects on biodiversity in farmed areas are also unknown (Johnson and Hope 2000). And because soy products are used in animal feed, glyphosate can wind up in the human food supply (Lappé and Bailey 1998).

Two other risks not considered at all are the pleiotropic and position effects of gene insertion. It is well known that genes have multiple effects (pleiotropy) and that these effects are determined by the position in the genome (position effects). But when Monsanto asked to have Roundup Ready soy deregulated, they provided no information about where the Roundup Ready gene construct landed. They could show which portions of the construct were incorporated into the soy genome, and that these portions were inherited in a Mendelian fashion, but the information necessary to evaluate pleiotropic and position effects was not available (APHIS 1994c). Thus there was no way to know what else the Roundup Ready construct did to the soybean besides confer Roundup resistance, again entailing unknown risks.

APHIS also did not adequately consider the risk that Roundup Ready genes might find their way into the soybean's wild and weedy relatives, *glycine soya* and *glycine gracilis* (APHIS 1994b, 6). These plants only grow wild in Asia, but APHIS

THE OTHER VALUE IN THE DEBATE OVER GMOS 155

is required by law to consider the global impact of their decisions. Since many other countries base their regulation in part on U.S. regulation, and the existence of one deregulated market can spur the creation of other black markets, this mandate is well conceived. APHIS made a token effort to consider global effects of their decision in their environmental impact statement by mentioning the existence of international and Asian regulatory agencies and asserting without justification that these agencies could handle any problems that arise (APHIS 1994b). Unfortunately, many Asian governments, especially China, ignore or fail to enforce international intellectual property laws. Pirated seeds could easily become as common as pirated CDs and DVDs and Rolex knockoffs.

Postcommercialization monitoring would help with all of these issues. While many of these risks depend on mechanisms that are well understood—for instance, pollenization—we need large-scale monitoring to measure the effect in this instance. For instance, while there have been plenty of reports of genes from GMOs appearing in wild organisms, there is no general consensus on how likely this is to occur. In 2002 the National Research Council recommended a system for postcommercialization monitoring for GMOs, which have not been implemented (National Research Council 2002). A 2003 report commissioned by the Pew Initiative on Food and Biotechnology argued that none of the agencies involved in biotech regulation were prepared to perform the kind of postcommericalization monitoring needed to achieve the "traditional objectives" of those agencies (Taylor and Tick 2003). Unless we examine the outcome of our actions, we risk repeating mistakes indefinitely.

So there are real environmental risks here; how do they stack up against the benefits? The only *intended* benefit of Roundup Ready soy is to increase yields relative to costs. Other benefits are frequently mentioned by GMO advocates. Half of the letters sent to APHIS during the public comment period suggested that farmers using Roundup Ready could move to no-till agriculture, and several others emphasized the possible decrease in the total amount of pesticides put into the environment (APHIS 1994c). However all of these benefits are speculative at best. The product will not succeed or fail depending on whether it increases no-till agriculture, no efforts have been made to tie the use of this product to no-till agriculture, and indeed we may never know if it increases no-till agriculture. Thus, the focus of our cost-benefit analysis must be on the benefit of increasing yield relative to cost. But here is where the real head scratching begins: Does the world really need cheaper soybeans? While some farmers may try to use the decreased costs to increase their profit margins, competition will quickly force them to drop prices. This effect is positively pernicious in a market where prices are already depressed due to overproduction. According to the Food and Agriculture Organization of the United Nations (FAO), in 1961 the United States produced 18,468,000 metric tons (Mt) of soy. By 2002, that number had more than quadrupled to 85,483,904 Mt (FAO 2005). This is actually less than the total world increase, which is more than sevenfold (FAO 2005). Population growth only puts a dent in the force of this number, since the world population has merely doubled since the 1960s. There has also been a great deal of increased demand due to increased consumption of heavily processed junk food. Nevertheless, the price of soy has been plummeting: In 2000, the price was about 40 percent of what it was in 1972 (World Bank 2000, 56). As a result of this, soy farmers are now heavily dependent on subsidies. Between 1995 and 2004, the U.S. federal government paid out $13,017,619,420 in soybean subsides (EWG 2005). As Kerschenmann (2003) has pointed out, the economic effects of Roundup Ready soy present the same conflict between individual and group rationality seen in arms races. It is rational for an individual farmer to use Roundup Ready soy, because she will be able to underprice her competitors. However it is not rational for every farmer to adopt Roundup Ready soy, because they will only further reduce prices for a product that already has weak demand. Widespread use of Roundup Ready soy will likely simply increase dependence on subsidies.

What about Third World starvation? Supporters of GMOs love to say that they are necessary to feed the 800 million people who are chronically malnourished worldwide. Superficially, it seems like all these soybeans would help, since each year between 30 and 40 percent of them are exported (EWG 2003). The problem is that starvation is not correlated with the underproduction of food, and is rarely caused by it (Sen 1981, 1999). This is shown most clearly in Amartya Sen's work on famines. Sen has shown that famines occur when food production is at its peak, and food production can drop as much as 70 percent in a poor region without triggering a famine (Sen 1999). Famine is caused not by an absence of food in a region but by difficulty accessing that food, often by a particular economic class. In many of the most notorious famines, a particular group went hungry because of a drop in the value of their product relative to the price of staple grains. For instance, in the Bengali famine of 1943, fishermen starved because of a drop in the price of fish relative to rice (Sen 1981, 1999). Something similar can happen if the price of soy drops precipitously. So, as Nottingham (1998) points out, the use of GMOs by First World farmers is likely to increase starvation by undercutting the incomes of Third World farmers.

The main people who stand to benefit from Roundup Ready soy are the employees, executives, and shareholders of Monsanto. There is one other group that benefits a little, though. Farmers get to exercise their economic liberty by purchasing a product of their own free will, which they will need to keep up with the increased production of their neighbors. Let's look at this value in more depth.

IV. THE ROLE OF ETHICAL PRINCIPLES IN THIS ANALYSIS

People who write about the role of values in the GMO debate tend to focus on the precautionary principle, which is written into law in various forms in Europe, and the alternate sound science principle, which has been adopted by American policymakers. Neither of these principles, however, can make sense of APHIS's decision regarding Roundup Ready soy. I claim that this decision only makes sense if it was motivated by a strong concern for economic liberty. An important factor

THE OTHER VALUE IN THE DEBATE OVER GMOS

here is that the precautionary principle and the sound science principle have been given so many different formulations that it is hard to tell what is really being argued over anymore. In fact, it is hard to even distinguish the principles from one another unless you assume that the partisans are making different assumptions about economic liberty.

The precautionary principle is supposed to provide guidance for decision making under scientific uncertainty and is supposed to mandate more caution than ordinary cost-benefit analysis would require. Beyond this general goal, however, there is no agreement about what the precautionary principle says. Neil Manson, in his analysis of various formulations of the precautionary principle, suggests a general logical structure that they all share (Manson 2002). Every formulation specifies a possible negative outcome, a degree of certainty about that negative outcome occurring, and an action that should be taken to avoid the negative outcome. For instance, one popular version of the precautionary principle is the catastrophe principle, which says that when the negative outcome is catastrophic, and the chance of it occurring is small but cannot be ruled out, then any activity that might lead to the outcome should be stopped. The first test of the atomic bomb would have been a nice place to employ this principle: there was a small risk, which could not be ruled out, that the bomb would ignite the atmosphere and incinerate the Earth. The catastrophe principle would bar the atomic test in these circumstances. Not all versions of the precautionary principle are concerned with catastrophe, however. The version of the precautionary principle in the Rio declaration, for instance, merely talks about damages that are "serious or irreversible."

Because the formulations of the precautionary principle have little in common besides a logical structure, the alternatives to the precautionary principle are hard to specify. While the precautionary principle has been contrasted with the sound science principle and with standard cost-benefit analysis, the logical structure is actually compatible with both of them. For instance, the precautionary principle could say: "If the possible damages are worth x (in dollars), and the probability of those damages is y (on a scale of 0 to 1), subtract $x(y)$ from the benefit of the project." Indeed, many of the more reasonable formulations of the precautionary principle say little more than this. This option is open in part because, although the focus of debate about the precautionary principle has been scientific uncertainty, there is no reason that the probabilities involved in the second condition be epistemic. Even the sound science principle promoted by industry advocates can also be put in the logical form of the precautionary principle. The sound science principle is generally taken to say, "Only act to avoid a risk when the causal mechanism underlying the risk is understood." This is a stricture on the probability portion of the precautionary principle, saying that the chance has to be well characterized.

The sound science principle suffers from the same vagueness as the precautionary principle. Chris Mooney, an activist journalist, traces popularization of the sound science approach to the formation of The Advancement of Sound Science Coalition (TASSC) in 1993 (Mooney 2005). Although TASSC claimed to be a grassroots organization interested in science policy in general, internal documents from Phillip Morris reveal that TASSC was created by the tobacco company with the help of the public relations firm APCO with the specific goal of discrediting reports of the dangers of secondhand smoke. In the hands of the tobacco industry, sound science was not so much a principle as a strategy. Mooney suggests that the strategy is best summarized in the much earlier notes for an internal presentation at Brown and Williamson, which were made public as a part of tobacco litigation: "Doubt is our product, since it is the best means of competing with the body of fact that exists in the minds of the general public. It is also the means of establishing a controversy." (Brown & Williamson 1969, quoted in Mooney 2005, p. 67)

It would be unfair to leave the rhetoric of sound science as it stood in the hands of the tobacco industry. As I have said, it can be rendered in the same logical structure as the precautionary principle. Phrased this way, it is essentially an attempt to loosen the restrictions of caution by saying that a high level of confidence in the negative outcome must be established before the preventative action may occur. One can already see the value of economic liberty at work in the justification of this principle. A background assumption in this debate is that the "preventative action" is an action by a government to restrict some form of industry. That is certainly the form that the action takes in this debate, since we are considering whether the U.S. government should allow Monsanto to pursue its business plans. But why raise the standard of evidence, across the board, for any government action? The obvious justification, close to the lips of all promoting sound science, is that companies like Monsanto have a strong *prima facie* right to do business as they please. Conversely, those who want to tighten the restrictions of caution assume that Monsanto's economic rights are quite weak.

The problem is that simply adjusting the probability portion of the precautionary principle is not enough to justify APHIS's action in the case of Roundup Ready soy. There are negative outcomes with probabilities greater than zero involving mechanisms like crossbreeding whose workings are well understood. There is no net benefit to the use of these crops. On any formulation of any of the above principles, the use of Roundup Ready soy is an unjustified risk.

To really justify APHIS's decision, you must appeal directly to the principle behind the sound science principle, the principle of economic liberty. A libertarian understanding of economic liberty supports APHIS's decision three ways. First, it implies that deregulation of Roundup Ready soy automatically brings about at least one good result, since economic liberty is itself a good. Second, it blocks my claim that the market for soy is so glutted that further production of soy would not be a good, because the free market is the only legitimate mechanism for determining when too much of a product is being produced. Finally, it blocks considerations of many of the long term potential harms of Roundup Ready soy as illegitimate attempts at social engineering.

The first piece of support for APHIS's decision comes because the economic freedom is now an intrinsic good. The exchange between Monsanto and individual farmers is, as Robert Nozick would put it, a free act of capitalism between consenting adults (Nozick 1974). Moreover, this free act is no less important to our well being than our freedom of speech or our freedom to choose our romantic partners. Indeed, for some libertarians, economic liberty becomes central to all other liberties: "Economic control is not merely control of a sector of human life which can be separated from the rest; it is the control of the means to all our ends" (Hayek 1944, 92). In the spirit of Mill's *On Liberty* we can say that the state should only interfere with such acts to prevent direct harm to others or the significant risk of such harm. This argument may not be enough to justify APHIS's decision, though, because there Roundup Ready soy does pose potential harm to others. Fortunately for the economic libertarian, there are other factors bolstering APHIS's decision.

The economic libertarian can also claim that a further lowering of prices is also a positive outcome, even though the market for soy seems to be glutted. She can claim this because she believes the only legitimate method for determining how much of a product should be produced is whether sellers can find a market for it. We will know when there is too much soy on the market because farmers won't be able to stay in business selling it. The gap between the individual and collective self-interest of farmers which Kerschenmann described should really be lauded as the source of our affluence, as competition to increase production and lower prices is a part of the genius of modern society. If farmers acted in their collective self-interest to limit production, they would be forming an anticompetitive cartel. A group decision to avoid Roundup Ready soy because increasing production would have no benefit would be similarly anticompetitive. The libertarian would also say that my dismissive description of much of the increased demand as coming from the rise of "junk food" amounts to an elitist sneer at other people's preferences. If the world wants more junk food, then providing it for the world would be a good thing. Concerns that further production of soy would increase famine by undercutting the ability of Third World farmers to sell their product are similarly misplaced. The decline of Third World farming is simply the transfer of production to the regions that can do it most efficiently. There is one problem with the current global soy market the libertarian would acknowledge: the existence of huge subsidies. If there is a glut of soy, it is because subsidies prevent the pricing mechanism from doing its work. But the solution then would be to remove the subsidies, not to block new technology.

Finally, the economic libertarian can dismiss many of the risks I described as illegitimate attempts at social engineering. Many of the risks discussed, such as the risks involved with increased use of Roundup, assume large-scale adoption of Roundup Ready soy. But in considering limiting freedom on the basis of potential harms, one should only look at immediate harms to identifiable individuals. The long-term and large-scale harms and benefits of an action are too complicated for an individual planning agency to predict. It thus must be left to the free market, with its ability to aggregate the values and opinions of the whole society, to decide how to deal with such big picture issues.

Although APHIS did not make an explicit appeal to the value of economic liberty, much of this libertarian style argument is implicit in the APHIS rulings (1994b, 1994c). APHIS made its decision by looking at the immediate circumstances. The benefits considered were all benefits to the individual farmer using Roundup Ready soy. Whether there was a pressing need for cheaper soy was apparently not something they were authorized to consider. Similarly, the only concern considered was the possibility that Roundup Ready soy might be a plant pest. In response to a public comment about the need to change patterns of pesticide use, APHIS claimed that such goals are beyond their jurisdiction. This last point may actually be true. Indeed, the libertarian premises behind APHIS's reasoning may in general be a feature of their legislative mandate, and not ideological biases. But for those of us opposed to economic libertarianism, this merely points to the need to expand the mandate of regulators.

NOTES

This paper was presented to the Fourteenth North American Interdisciplinary Conference on Environment and Community, Saratoga Springs, NY, February 19–21, 2004, in addition to the Ethics and the Life Sciences conference that this volume represents. I thank audiences at both conferences. Some of the arguments and explication of background facts in this paper are expanded and adapted from Loftis (2005).

1. The Environmental Protection Agency (EPA) does have jurisdiction over plants that produce their own pesticides and has enacted some restrictions. Unfortunately, EPA turns over all enforcement of its regulations to the Food and Drug Administration, which effectively leaves the regulations unenforced (Taylor and Tick 2003).

BIBLIOGRAPHY

Animal and Plant Health Inspection Service (APHIS). 1987. 7 *CFR* parts 330 and 340, plant pests; introduction of genetically engineered organisms or products; final rule. *Federal Register.* 52: 22891–22915.

———. 1994a. "Availability of Determination of Nonregulated Status of Monsanto Co. Genetically Engineered Soybean Line." *Federal Register.* 59: 26781–26782.

———. 1994b. APHIS-USDA Petition 93-258-01, Environmental assessment and finding of no significant impact. Washington DC. Available at http://199.89.233.43/biotech/bbasics.nsf/product_information_rrsoy_usda_environ.html?OpenPage. Accessed November 2, 2005.

———. 1994c. "Response to Monsanto Petition P93-258-01 for Determination of Nonregulated Status for Glyphosate Tolerant Soybean Line 40-3-2." Washington, DC. Permit 93-258-01P Available at http://www.biotechknowledge.com/biotech/bbasics.nsf/product_information_rrsoy_usda_decision.html. Accessed November 2, 2005.

Brown & Williamson. 1969. "Smoking and Health Proposal." Document number 332506. Available at http://tobaccodocuments.org/bw. Accessed November 25, 2005.

Chan, P. C., and J. F. Mahler. 1992. "NTP Technical Report on Toxicity Studies of Glyphosate" (CAS 1071-83-6). *National Toxicology Program: Toxicity Reports Series* 16. Available at http://ntp-server.niehs.nih.gov/ntp/htdocs/ST_rpts/tox016.pdf. Accessed November 2, 2005.

EWG (Environmental Working Group). 2003. Farm subsidy database, 2003 edition. http://ewg.org/farm. Accessed October 2004, no longer online. Copy available in the Google cache at http://64.233.161.104/custom?q=cache:P-me5WwIEvQJ:www.ewg.org/farm/findings.php+soybean+export&hl=en&start=3&ie=UTF-8. Accessed November 2, 2005.

———. 2005. Farm subsidy database, 2005 edition: "Soybean Subsides in the United States." Available at http://ewg.org/farm/progdetail.php?fips=00000&progcode=soybean. Accessed November 2, 2005.

FAO (Food and Agriculture Organization of the United Nations). 2005. "FAOStat." Available at http://faostat.fao.org/. Accessed November 2, 2005.

Hayek, F. A. 1944. *The Road to Serfdom*. Chicago: University of Chicago Press.

James, Clive. 2004. "Global Status of Commercialized Biotech/GM Crops: 2004." Ithaca, NY: International Service for Acquisition of Agri-biotech Applications (ISAAA). Brief 32. Executive summary at http://www.isaaa.org/kc/CBTNews/press_release/briefs32/ESummary/Executive%20Summary%20(English).pdf. Accessed November 2, 2005.

Johnson, Brian, and Anna Hope. 2000. "GM crops and equivocal environmental benefits." *Nature Biotechnology*. 18 (242).

Kerschenmann, F. 2003. "Designing a Food and Farming System that Works for People and the Land." Unpublished talk, Auburn University, May 7, 2003.

Lappé, M., and B. Bailey. 1998. *Against the Grain: Biotechnology and the Corporate Takeover of Your Food*. Monroe, Maine: Common Courage Press.

Loftis, J. Robert. 2005. "Germ-line Enhancement of Humans and Nonhumans." *Kennedy Institute of Ethics Journal*. 15 (1):57–76.

Manson, N. 2002. "Formulating the Precautionary Principle." *Environmental Ethics*. 24: 263–274.

Mooney, Chris., 2005. *The Republican War on Science*. New York: Basic Books.

National Research Council (NRC). 2002. *Environmental Effects of Transgenic Plants: The Scope and Adequacy of Regulation*. Washington, D.C.: National Academy Press.

Nottingham, S., 1998. *Eat Your Genes: How Genetically Modified Food is Entering our Diet*. New York: Zed Books.

Nozick, Robert. 1974. *Anarchy, State, and Utopia*. New York: Basic Books.

Office of Science and Technology Policy. 1986. "Coordinated Framework for Regulation of Biotechnology." *Federal Register* 51: 23302–23393.

Ruse, Michael, and David Castle. 2002. *Genetically Modified Foods: Debating Biotechnology*. Amherst, NY: Prometheus Books.

Sawanda, Y., Y. Nagai, M. Ueyama, and I. Tamamoto. 1988. "Probable Toxicity of Surface Active Agent in Commercial Herbicide Containing Glyphosate." *Lancet*. 1: 299.

Sen, A. 1981. *Poverty and Famines: An Essay on Entitlement and Deprivation*. Oxford: Oxford University Press.

———. 1999. *Development as Freedom.* New York: Alfred A. Knopf.
Taylor, Michael, and Jody Tick. 2003. "Post-market Oversight of Biotech Foods: Is the System Prepared?" A report commissioned by the Pew Initiative on Food and Biotechnology and prepared by Resources for the Future. Washington, DC.
Tominack, R. L., G. Y. Yang, H. M. Chung, and J. F. Deng. 1991. "Taiwan National Poison Center Survey of Glyphosate Surfactant Herbicide Ingestions." *Journal of Clinical Toxicology.* 29: 91–109.
World Bank. 2000. "Fats, Oils and Oilseeds." *Global Commodity Markets.* January: 46–56. Report 20306. Available at http://econ.worldbank.org/external/default/main?pagePK=64165259&theSitePK=469372&piPK=64165421&menuPK=64166093&entityID=000094946_00080105305346. Accessed November 2, 2005.

ETHICS AND THE LIFE SCIENCES

BIOMARKETING ETHICS, FUNCTIONAL FOODS, HEALTH, AND MINORS

WHITON S. PAINE AND MARY LOU GALANTINO
RICHARD STOCKTON COLLEGE OF NEW JERSEY
AND UNIVERSITY OF PENNSYLVANIA

ABSTRACT: In the next few years, biotechnology will continue to develop a wide variety of functional foods, foods whose benefits go well beyond basic nutrition. Minors are a major potential market for bioengineered foods that are promoted not as sustaining health but rather as supporting desired lifestyles through the enhancement of physical, athletic, intellectual, or social performance. The experience of other industries suggests that such biomarketing is likely to create a variety of highly public ethical controversies. After a discussion of some of these potential issues, suggestions on how companies and industries can work with marketing ethicists and child advocates to limit negative impacts on children and youth are presented. That discussion includes a preliminarily analysis of some of the considerations that should be involved in the initial development of a model of biomarketing ethics and in the use of that model to prevent ethical abuses.

In a very few years, biotechnology has moved from a focus on benefiting producers through the genetic engineering of plants (and increasingly animals) to the development of products that will be promoted directly to consumers. The general ethical implications of biotechnology have already evoked considerable ethical analysis (for example see Beauchamp and Childress, 2001). To date, that discussion has involved little consideration of the particularly complex issues involved in the commercialization of new biotechnology products (Newton, 2002). Functional foods are one of the first classes of products likely to be broadly commercialized. These foodstuffs are presently defined and used differently in different nations. Health and lifestyle claims for such products will influence consumer behavior and potentially affect public health, particularly when they are consumed by minors.

In an increasingly global economy, health claims for functional foods should meet internationally agreed upon scientific criteria (Clydesdale,1997).

In particular, biotechnology will create almost unlimited possibilities for contentious new functional food products, which, like presently available products, will be marketed as providing desirable benefits that go well beyond basic nutrition (American Dietetic Association, 1999). The technology is already available as illustrated by pharmrice wherein human genes were transplanted into food plants to drive the production of lactoferrin proteins commonly found in tears, saliva, bronchial fluids, and mother's milk. Despite serious concerns by a variety of environmental and consumer organizations (Freese, Hansen, and Gurian-Sherman, 2004) this effort has already passed a USDA environmental impact analysis (US Department of Agriculture, n.d.). Some of these foodstuffs will be of particular interest to minors.

Recent controversies with respect to marketing to this group (Linn, 2004; Acuff, and Reiher, 2005) suggest that such promotion is likely to become a particularly contentious and public ethical arena. As a new field, biobusiness has an opportunity to proactively define and address areas of potential ethical uncertainty with respect to minors and their parents.

This assertion is consistent with Dhanda's recent (2002) conclusion that:

it is not enough for companies to have faith in its technology anymore; the company must also be responsible for it, and responsible to the many stakeholders who are affected by the progress of biotechnology. Furthermore, the definition of stakeholders has to be broadened. Biotechnology is too pervasive for industry to err on the side of exclusion when identifying stakeholders. (p. 5)

Ethically and economically, minors and their parents are key stakeholders. Because of their growing size, increasing economic power, and ability to influence adults with respect to new technology, children and youth are likely to play an increasing role in biomarketing. However, they will have varying abilities to understand this new category. For example, studies have highlighted differences in lay knowledge about food and health across social class. They emphasize the need for public health nutrition policy-makers and practitioners to pay attention to lay knowledge on its own terms, rather than attempting to educate from predetermined assumptions, principles, and standards (Coveney, 2005). Ignoring this reality raises the risk, as Davidson (2002) noted in a recent marketing ethics book published by the American Marketing Association, that companies will: "Push too hard or manipulate too aggressively and society will turn on the marketer with a vengeance (p. 19)."

Bioengineered functional foods have the potential to be promoted as physiologically active natural products that reduce illness or support optimal health and an enhanced lifestyle beyond the basic nutritional value of the original food. The promise of biotechnology here is the development of foodstuffs that provide a wide range of potential benefits (Shah, 2002). For example, it is theoretically possible to produce potatoes that change moods, reduce stress, increase libido, support memory or add energy. Such nutritionally enhanced products, known as "phoods" in the industry, are likely to play a major role in the marketing efforts of companies like Nestle (Ball, 2004). They also offer the potential to counter the global resistance to genetically modified crops, which has limited the growth of this area (Corporate Watch, 2000). However, with minors this may have exactly the opposite effect, if children suffer damage. The knowledge base underpinning the setting of nutrient requirements for children and adolescents is not very secure. Target functions have been identified for growth development and differentiation and for behavioral and cognitive development. However, ideal markers or effects for these are not generally available. It is suggested that functional effects should include markers of reduction of risks of disease (nutritional safety) as well of benefits for health and well-being. Such markers of functional effects should be expected to arise from fundamental studies of nutrient-gene interactions and post-genomic metabolism (Aggett, 2004). Some of the general reasons for this possible negative consequence and a number of preliminary suggestions on how it can be avoided or limited through the development of biobusiness ethics are presented below after a brief discussion of functional foods and minors as a vulnerable target market.

CHILDREN AND YOUTH AS A POTENTIAL AND VULNERABLE MARKET

Minors are likely to be a particularly important set of marketing targets when these new foods are launched. More often than adults, they tend to be the innovators and early adopters of products that meet their needs through new technology (Lanctot, 1997). DVD and CD players and the Internet are all examples of areas where the interest of children has facilitated the adoption of new technologies. In these situations, kids also can serve as experts within the family who will support adults in learning about, buying, and using these new product categories. They easily adopt new and innovative food products like green catsup and are less likely to be encumbered with adult fears about the possible differences between "Frankenfoods" and "Superfoods" noted in the media (Byfield, 2000).

The Vulnerability of Children

In general, minors are a group that has received little attention by bioethicists outside such broad areas as pediatric medical practice, participation in research, and reprogenetics (Miller, 2003; Committee on Clinical Research Involving Children, 2004; Knowles, Murray, and Parens, 2003). A leading business ethicist has validly asserted that in general "vulnerability implies a special responsibility of protection and avoidance (Brenkert, 1998, p. 517)." This is good advice for biobusiness particularly since as Davidson (2002) has indicated: "Children are the quintessential 'vulnerable' target market" (p. 19).

Minors in particular exhibit some unique elements of vulnerability. Research suggests that younger children are inherently less able to accurately assess persuasive marketing messages (American Psychological Association, 2004). Their older siblings are predisposed to experiment with novel adult-like experiences despite suboptimal decision-making about attractive risks (Chambers, Taylor, and Potenza, 2003). Biomarketers will be tempted to take potentially unethical advantage of these differences from adults. Any company, including those in biotechnology, that are perceived by consumers as ignoring or taking advantage of this vulnerability is likely to suffer an abrupt erosion of trust and, potentially, of sales as well.

This increased level of risk in the marketplace is likely because of the moral intensity inherent in unethical marketing to minors discussed below and the related possibility of evoking the types of moral outrage adults have shown in some classic business cases involving perceived damage to children (May and Pauli, 2002; Paine, 2004). Some of the intensity of this reaction is due to beneficence being as close to a Kantian Imperative or societal hypernorm as biobusiness decision makers are likely to encounter. Obviously, this norm is on occasion violated as shown by such problems as child labor exploitation, abuse and neglect, sexual slavery and, Linn (2004) would argue, marketing to children. Despite their occurrence, such violations do appear to have a unique ability to evoke moral outrage in many adults.

Health, Lifestyle, and Children

In this context, it is important to remember that children, as opposed to adults, tend to take wellness as a given and feel largely invulnerable to potential threats to their health status. Thus, their main interest will be in how a functional food might positively influence their lifestyle. The most attractive products are likely to be those that claim to enhance their physical (athletic and sexual), intellectual, emotional (as a result of desired mood changes), or social (particularly through appearance) performance. This preference can already be seen in products like sports and energy drinks that are consumed because kids think they will enhance performance.

An important related concern is that the tendency noted above of minors ignoring risks when seeking desired benefits might make them particularly vulnerable with respect to functional foods. This vulnerability is already being exploited in the promotion of dietary supplements like FocusFactor, a product that is no longer allowed to promise children and teenagers that they would "feel more alert, focused, and mentally sharp; improving students' ability to concentrate and their academic performance" (Federal Trade Commission, 2004, paragraph 2). This vulnerability also underlies the escalating controversy regarding perceived links between nutrition, children's obesity, and health. The most intense debate is occurring in the United Kingdom and was sparked by the negative Hasting's Report on junk foods commissioned by the government's Food Standards Agency (Hastings et al., 2003).

However, methodological and conceptual issues make the cited linkages between promotion and damage to children rather problematic (Paliwoda and Crawford, 2003; Young, 2004). This lack of evidence has not prevented *Commercial Alert*, a major advocacy organization, from launching an attempt to incorporate a ban on marketing junk food to children (twelve and under) into the World Health Organization global anti-obesity initiative (http://www.commercialalert.org/junkfoodstatement.pdf). As of early 2004, over 250 influential organizations and individuals from some eighteen countries had signed onto this campaign. Many of the social and political lessons learned that led to the Tobacco Settlement are now being applied in this area. Many of the ethics issues involved in the promotion of junk food and dietary supplements may also arise for functional foods.

Another area of ethical controversy that biobusiness should monitor carefully involves the marketing of pharmaceuticals. This industry is experiencing growing problems with the promotion of both on-label and off-label drugs for the treatment of children and youth. For example, a recent analysis of the ongoing controversy over the marketing of Ritalin and other stimulants to treat ADHD identified eight separate areas of significant ethical issues (Paine, 2003; Radigan, Lannon, Roohan, Gesten, 2005). The etiology of ADHD is acknowledged to be both complex and multifactorial, and stimulant drugs may not solve the problem. Some features of ADHD may reflect an underlying abnormality of fatty acid metabolism. One implication here is that fatty acid treatment may be relatively safe compared to existing pharmacological interventions. Further studies are still needed in order to evaluate its potential efficacy in the management of ADHD symptoms (Richardson and Puri, 2000), but it is quite conceivable that functional foods will also be promoted for the treatment of this condition.

In a related area, the aggressive promotion of selective serotonin reuptake inhibitors (SSRIs) to physicians has led to their use in treating a wide range of emotional problems in children (Rushton, 2002). Emerging data from several clinical trials show that the SSRIs provide moderate benefits for youth with depression. However, SSRI treatment may be associated with increased risk of behavioral activation, self-harm, and suicidal ideation. Appropriate use of the SSRIs in children and adolescents requires careful diagnostic assessment, evaluation of comorbidity, and close monitoring, especially early in treatment (Hamrin and Scahill, 2005). The initially intensive marketing occurred despite unpublished studies by manufacturers that found SSRI to be generally ineffective and even potentially dangerous to suicide-prone teenagers (Whittington et al., 2004; Kondro, 2004; Vedantam, 2004). A whistleblower bringing this situation to public attention and similar actions by biobusiness employees are probably much more likely when kids are the target for questionable marketing.

In this context, it is also important to note that the commercialization of technology has already led to ethical issues with minors. For example, the Internet dramatically increases the accessibility and availability of information and potentially empowers by closing the information gap between the manufacturer, intermediaries, and consumers. Biobusiness is likely to use the Internet to reach both professionals and consumers. This may help bridge the gap between knowledgeable biomarketers and parents and children without the needed scientific information to make more informed decisions about the selection or consumption of functional foods.

However, with respect to health- and illness-related products, there are already abuses in this area including questions about the accuracy of the information, unregulated and possibly dangerous sales through Internet pharmacies, the protection of personal privacy (Shani, 2003), and informed consent (Rowan, 1998). These questions are likely to reoccur with functional foods. On the Internet, the escalating abuses of minors relatively quickly led to the Children's Online Privacy Protection Act to protect children accessing Web sites specifically designed for this market (Austin and Reed, 1999; Davis, 2002). For example, web-based data on psychoactive substances seem to influence a broad range of drug-use behaviors in adolescents. Information on the ways that the Internet is being used by this vulnerable population should be considered in the design of Web sites to prevent the initiation and use of inappropriate or psychoactive substances (Boyer, Shannon, and Hibberd, 2005). Future biomarketing abuses on the Net are thus likely to spark legislation and regulation if they involve the perceived manipulation of kids' vulnerability, invasion of privacy, misstatements of fact, or inappropriate content.

More specific issues arising in connection with genes and nutrition and social policy include both nutrigenomics and nutrigenetics. Nutrigenomics considers the relationship between specific nutrients or diet and gene expression and will facilitate the prevention of diet-related common diseases or of self-administered changes in lifestyle. Nutrigenetics is concerned with the effects of individual genetic variation (single nucleotide polymorphisms) on response to diet, and in the longer term may lead to personalized dietary recommendations (Chadwick, 2004). This raises the possibility of functional foods legally or illegally tailored to the individual child. It is important also to consider the surrounding context of other issues such as novel and functional foods in so far as they are related to potential genetic modification.

Here a variety of ethical issues fall into a number of categories:

(1) Will nutrigenomics have important public health and private lifestyle benefits?
(2) Questions about research, for example, concerning the acquisition of information about individual genetic variation;
(3) Questions about who has access to this information, and its possible misuse;
(4) The applications of this information in terms of public health policy, and the negotiation of the potential tension between the interests of the individual in relation to, for example, prevention of conditions such as obesity and allergy;
(5) The appropriate ethical approach to the issues, for example the moral difference between therapy and enhancement in relation to individualized diets and whether the 'technological fix' is always appropriate. This is important especially in the wider context of the purported lack of public confidence in science, which has special resonance in the sphere of nutrition (Chadwick, 2004).

WHAT MIGHT BE DONE?

Bioengineered foods will ultimately become part of the life of minors since it is impossible to prevent them from hearing about, diverting, and using in creative and unexpected ways foodstuffs that are publicly intended for adults. Tobacco is the obvious example of this process. So there is a need to develop a new area, *Biomarketing Ethics*, which can assist managers, regulators, child advocates, and parents in protecting children from inappropriate marketing. That area would have to integrate the literatures of marketing and marketing ethics, business ethics, bioethics, and philosophical ethics, with the new knowledge emerging from biotechnology and with what is known about ethical issues in related categories of products. In this process, there are some real advantages in managers, industries, and companies learning to work more closely with ethicists, professionals in the health fields, regulators, and child advocates without attempting to co-opt them (Davidson, 1997). Unfortunately, biobusiness already has been drawn into adversarial relationships with outside critics (Stecklow, 1999, Newton, 2002).

In particular, biobusiness companies and industries should take the opportunity to initiate actions now to avoid the often-expensive problems experienced in other areas of marketing. Those actions include participating in the development of a useful model for biomarketing ethics, creating new codes of conduct, and structuring their organizations to limit unethical biomarketing to children and youth. These areas are briefly discussed below.

Biomarketing Ethics and Codes of Conduct

One important joint task would be to develop a useful ethical model and embody it in industry Codes of Conduct and professional codes for key gatekeepers: nutritionists, physicians, and practitioners of alternative and complementary medicine. However, such codes are inherently misleading if they do not embody both an effective mechanism for monitoring compliance and sanctions for unethical marketing to children and youth. This development process would be most effective if it recognizes that outside critics and internal whistle blowers both will play key roles in the identification of unethical corporate behavior particularly with respect to kids.

Protection of Minors through Professional Stakeholder Communities

Physicians, complementary and alternative medicine practitioners, nutritionists, and other health professionals are likely to play a key gatekeeper role in the growth of functional foods for minors. As such, they are uniquely positioned to both protect kids and to alert companies, parents, and regulators to possible problems. However, past marketing practices of pharmaceutical companies, particularly with respect to such issues as the promotion of off-label uses with children, direct-to-consumer marketing, and informed consent, suggest some major potential problems for biobusiness. Biomanagers should carefully study the widespread abuses of promotional practices with professionals that led to the second attempt at a PhRMA Code on Interactions with Healthcare Professionals (Tsai, 2003; Pharmaceutical Research and Manufacturers of America, 2004).

Gate keeping professionals also need to be acutely aware of potential conflicts of interest as they receive income from the development, promotion, and sale of functional foods. This conflict is likely to be particularly troublesome when biobusiness aggressively promotes products despite limited information on their benefits as well as their short and long-term effects on kids. The controversy over direct-to-consumer promotions, which bypass professional gatekeepers, is also relevant here (Federal Drug Administration, 1999). It raises issues similar to the major complaint of advocates that companies are bypassing parents when marketing to kids (Linn, 2004).

Building a Model of Biomarketing Ethics

Any changes in corporate operations should reflect a basic model of biomarketing ethics. In recent years, considerable work has been done on marketing ethics (Murphy, 2002: Murphy et al., 2005); but relatively little of this literature has focused directly on the specific issues involved when the primary or secondary marketing target is a minor. Most of the ethical debates have occurred in the media and have been summarized by Linn (2004). Her adversarial analysis starkly illustrates the growing gap between marketing managers and child marketing opponents. Ultimately, both sides in these intensely emotional debates will have to find a common ground. The model of ethical decision-making developed by Jones (1991) has received some empirical verification (May and Pauli, 2002) and may be relevant here. That model posits four stages: (1) recognition of a moral issue leading to (2) a process of moral evaluation/judgment that fosters a (3) moral intention that is ultimately expressed in (4) moral behavior. Each of these is briefly discussed below with the caveat that this analysis is quite preliminary.

Moral Issue Recognition

Jones (1991) hypothesized that ethical problems are unlikely to be recognized unless the marketing situation inherently involves a high enough level of moral intensity to evoke a perception of possible issues. Intensity is hypothesized to be related to six factors: the magnitude of the consequences of a decision, its relationship to a specific social consensus, the probability that the negative effect will occur, how quickly it happens, how large are the differences between the decision maker and the impacted group, and how concentrated is the effect. These considerations have not been well analyzed for minors and the inherent nature of a developing child complicates all six factors.

With minors, some significant problems arise because it is often difficult to quantify, or even identify, the magnitude, probability, or speed of potential damage resulting from an unethical marketing decision. This is particularly true if the consequences are seen as the result of industry-wide marketing practices such as the promotion of sugary cereals or of violence in the media. As noted above, beneficence and the avoidance of maleficence operate as a Kantian Imperative or hypernorm and so probably reflect a general social consensus. In addition, as the tobacco industry discovered, advocates can effectively shift a social consensus when marketing actions are successfully reframed as the unethical promotion of an attractive and damaging product directly to vulnerable children and youth. A significant additional challenge is that children are vulnerable to harm at different ages, and there are wide variations in vulnerability to physical, emotional, moral, intellectual, or social harm within any age cohort. Also any "damage" can be construed as accelerating, decelerating, stopping, or diverting the typical pattern of development. Finally, a generation is now roughly four years and thus a generational distance been adult decision makers and target markets is always present, particularly if the managers involved do not have children of their own.

In addition, a review of some enduring controversies in business and marketing ethics suggests that a major barrier to moral recognition by marketing executives is the Four Horsemen of unethical marketing to children: *Ignorance, Stupidity, Greed and Hubris* (Paine, 2004). In a new and exciting field like biobusiness, all four are likely to ride roughshod as biomanagers with little experience with minors discover the joy of profitably using the increasingly sophisticated and varied tools of children's marketing to manipulate vulnerabilities without due consideration for possible longer term consequences. In particular, they are likely to follow some common and ethically dubious assumptions including: "Parents will protect their kids in the marketplace" and "If it sells, it must be ethical." This model was recently found useful in the analysis of a potentially ethically dangerous assumption within modern marketing theory—that marketers should drive markets rather than be driven by them (Paine, 2005).

Moral Evaluation

Even if an ethical problem is recognized, the ensuing process of moral evaluation is difficult in this area since the nature of children complicates teleological and deontological considerations. Any utilitarian analysis is confounded by the complexity of possible impacts noted above. In terms of defining the most applicable ethical rules, a recent paper by a philosopher and a marketing professor found that the standard ethics rules used by business people probably should not be applied the way they are in ethical situations involving adults (Jackson and Paine, 2001). Even the Golden Rule becomes problematic since "Doing unto children as you would have them do unto you" almost guarantees unethical corporate behavior. The alternative of "Do unto other's families as you would have them do unto your family" may be insufficient depending on the nature of a marketer's family.

A related issue is confusion as to how the principles of distributive, procedural, and interactive justice should apply in this area. Critics would argue that there can be no distributive justice since marketing to minors almost involves taking advantage of the vulnerabilities of this group. Marketing messages that rely on persuasion rather than on providing information may not be procedurally just given the assumed inexperience of minors with respect to products. Similarly, the assumption that children are often unable to fully understand inaccuracies and manipulations in marketing communications raises questions of interactive justice.

Moral Action

Finally, making the leap from moral intention to moral action is often complicated by conflicts between corporate cultures attempting to support ethical behavior and marketing department cultures that may not be as sensitive to ethical issues involving either adults or minors. For example, an insider in the highly regulated pharmaceutical industry recently decried the chronic battles that occur between those responsible for regulatory compliance and marketers attempting to protect their ability to use their creativity to increase off-label sales (Pines, 1997).

One potentially important contribution of business ethics here is the insight that ethical behavior by managers is strongly influenced by organizational contexts (see for example Travino and Nelson, 2004; Ferrell, Fraedrich, and Ferrell, 2005). This insight suggests that companies can modify their structure and operations to greatly decrease the risk of unethical behavior by managers. Apparently, the first attempt to utilize this knowledge with respect to the children's market was recently published in Europe (Paine, Stewart, and Kruger, 2002). It suggests specific ways of modifying policies and procedures, training, company mission statements and codes of ethics, to protect children. Those suggestions would also be relevant to biobusiness organizations.

Finally, it is not clear how the increasing interest in Aristotelian Virtue Ethics would apply here (Hartman and Beck-Dudley, 1999; Whetstone, 2001). Some of the problems noted above would be presumably less likely if biobusiness employed and supported moral individuals who are assumed to be more prone to make ethical management decisions. This is an area that requires considerable analysis of the most relevant virtues since it seems unlikely that such exemplars of morality with respect to children as Mr. Rogers or T. Berry Brazleton would operate effectively as the managers of children's marketing within a large corporation. This is particularly true given the dangerous tendency of biotechnology to attract entrepreneurial visionaries who may not see the ethical pitfalls.

Unfortunately, this very preliminary analysis illustrates only some of the challenges that will face biobusiness, biomarketers, and bioethicists interested in the commercialization of biotechnology. If these challenges are not well met with minors, functional foods and other areas of biotechnology are likely to suffer the same types of highly public ethical controversies that have afflicted other areas of marketing. On the other hand, since the area is so new, it has the potential to learn from the past mistakes of others.

BIBLIOGRAPHY

Acuff, Daniel S., and Robert H Reiher. 2005. *Kidnapped.* Chicago: Dearborn Trade Publishing.

Aggett, Peter J. 2004. "Functional Effects of Food: What Do We Know in Children?" *British Journal of Nutrition.* October; 92 Suppl 2: S223–6.

American Dietetic Association. 1999. "Functional Foods." *Journal of the American Dietetic Association* 99: p. 1278. www.eatright.org/Public/Other/index_adap1099.cfm (accessed May 5, 2004).

American Psychological Association. 2004. "Report of the APA Task Force on Advertising and Children." http://www.apa.org/releases/childrenads.pdf (accessed May 11, 2004)

Austin, M. Jill, and Mary Lynn Reed. 1999. "Targeting Children Online: Internet Advertising Ethics Issues." *Journal of Consumer Marketing* 16 (6): pp. 590–602.

Ball, Deborah. 2004. "With Food Sales Flat, Nestle Stakes Future on Healthier Fare." *The Wall Street Journal,* March 18, p. A1.

Beauchamp, Tom L., and James F. Childress. 2001. *Principles of Biomedical Ethics, 5th edition.* Oxford, UK: Oxford University Press.

Boseley, Sarah. 2003. "Company Held Back Data on Drug for Children: Antidepressant Had No Effect, Leak Reveals." *The Guardian,* February.

Boyer, Edward W., Michael Shannon, Patricia.L. Hibberd. 2005. "The Internet and Psychoactive Substance Use Among Innovative Drug Users." *Pediatrics.* Feb; 115(2): pp. 302–305.

Brenkert, George. 1998. "Marketing to the Vulnerable." In *Perspectives in Business Ethics,* ed. Laura P. Hartman. Chicago, Irwin McGraw-Hill, pp. 516–526.

Byfield, Mike. 2000. "Superfoods or Frankenfoods?" *Report/Newsmagazine,* May 22, 27 (2): pp. 36–37.

Chadwick, Robert. 2004. "Nutrigenomics, Individualism and Public Health." *Proceedings of the Nutritionists Society.* 2004 Feb;63(1): pp. 161–166.

Chambers, R. Andrew, Jane R. Taylor, and Marc N. Potenza. 2003. "Developmental Neurocircuitry of Motivation in Adolescence: A Critical Period of Addiction Vulnerability." *American Journal of Psychiatry* June, 160(6): pp. 1041–1052.

Clydesdale F. M.1997. "A Proposal for the Establishment of Scientific Criteria for Health Claims for Functional Foods." *Nutritional Review.* December; 55(12): pp. 413–422.

Committee on Clinical Research Involving Children. 2004. *The Ethical Conduct of Clinical Research Involving Children.* Washington, D.C.: Institute of Medicine of The National Academies. Prepublication draft //www.nap.edu/books/0309091810/html/ (accessed 5/3/04).

Corporate Watch. 2000. "The Industry Strikes Back: Functional Foods Good for Monsanto's Health." *GE Briefing Service,* May. www.corporatewatch.org.uk/publications/GEBriefings/funcfoods.pdf (accessed March 10, 2004).

Coveney, John. 2005. "A Qualitative Study Exploring Socio-economic Differences in Parental Lay Knowledge of Food and Health: Implications for Public Health Nutrition." *Public Health Nutrition.* May; 8(3): pp. 290–297.

Davidson, D. Kirk. 1997. "Lessons for Marketers from Cloning." *Marketing News.* April 14, 31(8): p. 7.

———. 1998. "Corporations, Advocates should Work Together." *Marketing News.* October 26 32 (21): p. 8.

———. 2002. *The Moral Dimension of Marketing: Essays on Business Ethics.* Chicago: American Marketing Association.

Davis, Joel J. 2002. "Marketing to Children Online." *S.A.M. Advanced Management Journal.* 67(4): pp. 11–23.

Dhanda, Rahul K. 2002. *Guiding Icarus: Merging Bioethics with Corporate Interests.* New York: Wiley-Liss.

Federal Drug Administration. 1999. "Guidance for Industry: Consumer-directed Broadcast Advertisements." www.fda.gov/cder/guidance/1804fnl.htm (accessed May 5, 2004).

Federal Trade Commission. 2004. "Marketers of the Supplements 'Focus Factor' and 'V-Factor' Agree to Settle FTC Charges and Pay $1 Million." www.ftc.gov/opa/2004/03/vitalbasics.htm (accessed April 2, 2004).

Ferrell, O. C., John Fraedrich, and Linda Ferrell. 2005. *Business Ethics.* Boston: Houghton Mifflin.

Freese, B., M. Hansen, and D. Gurian-Sherman. 2004. "Pharmaceutical Rice in California." Published jointly by Friends of the Earth, Center for Food Safety, Consumers Union, and Environment California. Retrieved 8/26/05 from http://www.centerforfoodsafety.org/pubs/CARiceReport7.2004.pdf.

Hamrin, Vanya, and Lawrence Scahill. 2005. Issues Mental Health Nurs. May;26(4): pp. 433–450.

Hartman, Cathy L., and Caryn L. Beck-Dudley. 1999. "Marketing Strategies and the Search for Virtue: A Case Analysis of the Body Shop International." *Journal of Business Ethics* 20 (3): pp. 249–263.

Hastings, Gerard et al. 2003. "Review of Research on the Effects of Food Promotion to Children. Final report prepared for the Food Standards Agency, September 22. www.foodstandards.gov.uk/multimedia/pdfs/foodpromotiontochildren1.pdf. (accessed March 28, 2004).

Jackson, Rodger L., and Whiton S. Paine. 2001. "Salvaging Kids and Ethical Principles at the Same Time." *Proceedings of the 16th Annual Conference of the Atlantic Marketing Association,* Portland, Maine, September 26–29.

Jones, Thomas. M. 1991. "Ethical Decision Making by Individuals in Organizations: An Issue-contingent Model." *Academy of Management Review* 16: pp. 366–395.

Knowles, Lori P., Thomas H. Murray, and Erik Parens. 2003. "Reprogenetics and Public Policy: Reflections and Recommendations." Hasting's Center. www.thehastingscenter.org/pdf /reprogenetics_and_public_policy.pdf (accessed may 1, 2004).

Kondro, Wayne. 2004. "Drug Company Experts Advised Staff to Withhold Data about SSRI Use in Children." *Canadian Medical Association Journal.* 2004 March 2 1705: p. 783.

Lanctot, Roger C. 1997. "Kids: The Original Early Adopters." *Computer Retail Week* May 5, 7 (169): p. 16.

Linn, Susan E. 2004. *Consuming Kids: The Hostile Takeover of Childhood.* New York: New Press.

May, Douglas R., and Kevin P Pauli. 2002. "The Role of Moral Intensity in Ethical Decision-making." *Business and Society* 41(1): pp. 84–117.

Miller, Richard B. 2002. *Children, ethics and modern medicine.* Bloomington, Ind.: Indiana University Press.
Murphy, Patrick E. 2002. "Marketing Ethics at the Millennium: Review, Reflections and Recommendations." In *Blackwell Guide to Business Ethics*, ed. Norman Bowie. Malden, Mass.: Blackwell Publishers, pp. 165–185.
Murphy, Patrick E., Gene. R. Laczniak, Norman E. Bowie, and Thomas.A. Klein. 2005. *Ethical Marketing* Upper Saddle River, N.J.: Prentice Hall.
Newton, Lisa H. 2002. "The Ethical Dilemmas of the Biotechnology Industry." In *Blackwell Guide to Business Ethics* ed. Norman Bowie. Malden, Mass.: Blackwell Publishers, pp. 313–333
Paine, Whiton S. 2003. "Some Ethical Considerations in Marketing Psychotropic Medications for Minors." *Proceedings of the 17th Annual Conference of the Atlantic Marketing Association*, Portland, Maine, October 2–5.
——————. 2004. "What Biomarketing could Learn about Kids from Business Ethics." Emanuel and Robert Hart Lecture, Center For Bioethics, University of Pennsylvania, May 4.
——————. 2005. "Some Potential Issues Related to Driving the Kids Market." Submitted for presentation at the 2006 Academy of Marketing Science Annual Conference on Revolution in Marketing: Market Driving Changes. San Antonio, Tex., May 24–May 27.
Paine, Whiton S., Karen L. Stewart, and Evonne Kruger. 2002. "Preventing Ethical Problems when Marketing to Minors." *International Journal of Advertising and Marketing to Children*, January–March, pp. 69–80.
Paliwoda, Stan, and Ian Crawford. 2003. "An Analysis of the Hastings Review: The Effects of Food Promotion on Children." www.adassoc.org.uk/hastings_review_analysis_dec03 .pdf (accessed April 4, 2004).
Pharmaceutical Research and Manufacturers of America. 2004. "PhRMA Code on Interactions with Healthcare Professionals." www.phrma.org/publications/policy//2004-01-19.391.pdf (accessed May 7, 2004).
Pines, Wayne L. 1997. "Major issues in marketing regulation." *Food And Drug Law Journal.* 52. pp. 297–302. www.fdli.org/pubs/Journal%20Online/52_3/5art.pdf (accessed May 7, 2004).
Radigan, Marleen, Peter Lannon, Patrick Roohan, and Foster Gesten. 2005. "Medication Patterns for Attention-deficit/Hyperactivity Disorder and Comorbid Psychiatric Conditions in a Low-income Population." *Journal of Child and Adolescent Psychopharmacology*, February; 15(1): pp. 44–56.
Richardson, A. J., and B. K. Puri. 2000. "The Potential Role of Fatty Acids in Attention-deficit/Hyperactivity Disorder. Prostaglandins Leukot Essent Fatty Acids." *Neuroendocrinology.* July–August; 63(1–2): pp. 79–87.
Rowan, John R. 1998. "Informed Consent as an Ethical Principle for Business." *Business and Professional Ethics Journal*, 17 (1/2): pp. 329–343.
Rushton, Jerry L., Sarah J. Clark, and Gary L. Freed. 2000. "Pediatrician and Family Physician Prescription of Selective Serotonin Reuptake Inhibitors." *Pediatrics*, June 105 (6): pp. E82–88.
Shah, Anup. 2002. "Functional Foods, The Next Wave of GE Foods." Global Issues That Affect Everyone, May 28. www.globalissues.org/EnvIssues/GEFood/Functional.asp (accessed March 11, 2004).

Shani, Seklar. 2003. "E-commerce of Pharmaceuticals." *Harefuah*, May, 1425: pp. 372–376, 397, 396.

Stahlberg, Alicia, Denise Webb, and Marsha Hudnall. 2001. "How to Evaluate the Safety, Efficacy, and Quality of Functional Foods and their Ingredients." *Journal of the American Dietetic Association*, July, accessed March 3, 2004 at www.findarticles.com/cf_0/m0822/7_101/78048904/print.jhtml

Stecklow, Steve. 1999. "Germination: How a U.S. Gadfly and a Green Activist Started a Food Fight." *Wall Street Journal*. Eastern edition, November 30, p. A.1

Steffens, Maryke. 2004. "The Viagra Myth." *The Health Report*, Australian Broadcasting Corporation. www.abc.net.au/rn/talks/8.30/helthrpt/stories/s1051896.htm (accessed April 2, 2004).

Travino, Linda K., and Katherine, A Nelson. 2004. *Managing Business Ethics*. Hoboken, N.J.: Wiley.

Tsai, Alexander C. 2003. "Policies to Regulate Gifts to Physicians From Industry." *Journal of the American Medical Association*, 290: p. 1776.

U. S. Department of Agriculture. 2005. "USDA/APHIS Environmental Assessment In Response to Permit Application (04-302-01r) Received from Ventria Bioscience for Field-testing of Rice, Oryza Sativa, Genetically Engineered to Express Human Lactoferrin." USDA/APHIS Environmental Assessment. Retrieved 9/5/05 from http://www.aphis.usda.gov/brs/aphisdocs/04_30201r_ea.pdf.

Vedantam, Shankar. 2004. "Antidepressant Makers Withhold Data on Children." *Washington Post*, January 29, p. A01.

Whetstone, J. Thomas. 2001. "How Virtue Fits Within Business Ethics." *Journal of Business Ethics*, 33: pp. 101–114.

Whittington Craig J, et al. 2004. "Selective Serotonin Reuptake Inhibitors in Childhood Depression: Systematic Review of Published Versus Unpublished Data." *Lancet*. 363 (9418): pp. 1341–1345.

Young, Brian. 2003. "Advertising and Food Choice in Children: A Review of the Literature." Food Advertising Unit, The Advertising Association, August. www.fau.org.uk/content/pdfs/brian_youngliteraturereview.pdf. (accessed April 4, 2004).

ETHICS AND THE LIFE SCIENCES

WHAT'S WRONG WITH FUNCTIONAL FOODS?

DAVID M. KAPLAN
UNIVERSITY OF NORTH TEXAS

ABSTRACT: A "functional food" is a food-based product that provides a demonstrable physiological benefit beyond its dietary or nutritional value. This class of foods for specific health uses are designed to assist in the prevention or treatment of disease, or to enhance and improve human capacities. They include products like vitamin-fortified grains, energy bars, low-fat or low-sodium foods, and sports drinks. Three sets of concerns about functional foods deserve attention. 1) Their health benefits are greatly exaggerated and, in many cases, non-existent; practical questions remain about their efficacy. 2) Their medicinal properties blur the boundaries between food and drugs; public health questions remain about their appropriate use, distribution, and regulation. 3) Their proliferation is fueled by the food industry, not by the medical profession; political questions remain about the role of market forces that too often benefit producers more than consumers.

FUNCTIONAL FOODS DEFINED

All food is in some sense functional insofar as it contains calories and nutrients that support health. The more narrowly construed sense of functional foods are those that have added ingredients believed to provide additional health benefits. Functional foods are not new. They have existed since the early 1900s when iodine was first added to salt to prevent goiter. Vitamin D has been added to milk since the 1930s, extra vitamins and minerals to breakfast cereals since the

1940s, and water fluoridated shortly thereafter. The difference between these fortified foods and the newer generation of functional foods is that more recent ones are designed to replace medicine with food, or sometimes to eliminate qualities from the food to make them (seem) more healthy. Examples of include *Benecol* (a cholesterol-lowering margarine), *Kitchen Prescription Soup* (with the herbal supplement Echinacea), *EggsPlus* (nutritionally enhanced eggs with extra omega-3 fatty acids), *Viactiv* (calcium chews), *Gatorade* and *Vitamin Water* (supplement beverages), *Wow Potato Chips* (fat free, fewer calories), *Ensemble* food products (with soluble fiber to promote heart health), low-carb food products (from beer to frozen food to fast food), and products geared toward the specific health needs of infants, toddlers, and the aging.

Often genetically-modified foods are engineered to be nutritionally enhanced. The most notable example is the highly publicized, Vitamin-A enriched, *Golden Rice*, which has been touted for its ability to reduce blindness in malnourished children. Other genetically-modified products currently promised are high-protein and vitamin-enriched cassavas, milk and peanuts that are allergen-free, tomatoes with three-times the usual amount of lycopene, a cancer-fighting anti-oxidant, carrots with a hepatitis-B vaccine, and potatoes with a vaccine for cholera.

What counts as a functional food varies from nation to nation. But in each instance the definition is bound up with the kind of health claims a product is allowed by law to make. For example, Japan, where the very concept of contemporary functional foods was invented, is the only nation in which functional foods have their own legal designation and regulatory body. Foods for Specific Health Uses (FOSHU) are defined as those foods and beverages with ingredients added for a determined health effect or to reduce the risk of disease or health-related condition. Applications for FOSHU certification are reviewed by the Japan's Ministry of Health and Welfare and must include scientific documentation established by clinical trials performed by approved research institutions. Only FOSHU-approved products are permitted to make health claims on food labels. They are a separate category in the Japanese food system. Participation in FOSHU is, however, voluntary. Food companies can produce items that make *general* health claims (to promote health) so long as they make no *specific* claims (to treat diseases). Products making general, unregulated health claims make up 90 percent of the health food market in Japan.[1] To encourage greater participation in FOSHU, the government lowered the scientific requirements, allowed private-sector laboratories to make legitimate health claims, and streamlined the application process. Still, non-FOSHU-approved functional foods dominate the Japanese market.

In the United Kingdom, there is no legal definition of functional foods, only a working definition by the Ministry of Agriculture Fisheries and Food (MAFF). It defines functional foods as those foods enhanced to have additional health benefits beyond their nutritive benefits.[2] As in Japan, food products are allowed to make general, but not specific, health claims. If a product claims to be capable of preventing, treating, or curing human disease then the food must be licensed as medicine. Food manufacturers are prohibited from making any medicinal claims. They are, however, allowed to make claims which refer to possible disease factors ("can lower cholesterol"), to nutrient function ("Vitamin A is essential for normal vision"), or to recommended dietary practice ("part of a nutritious breakfast"). Other EU countries have adopted a similar strategy: they allow a wide range of generic health claims and have established procedures to assess the evidence for specific health claims. Common to all definitions of functional foods in the EU are that they be recognizable as food, not pills, capsules, or other drug-like forms.

The case in the United States is somewhat more vague. Functional foods are part of an overlapping family that includes food additives, food supplements, and genetically-modified foods. The Food and Drug Administration (FDA) defines a "food additive" as any substance designed to help prevent spoilage, contamination, or make food look and taste better. Additives are things like flavor enhancers (MSG), artificial colors and flavors, preservatives, stabilizers, sulfites, and nitrates. The FDA defines a "dietary supplement" (somewhat unhelpfully) as additional ingredients with either *nutritional* or *non-nutritional* properties, such as vitamins, minerals, proteins, herbs, enzymes, or extracts. They can either take drug-like forms or they can be added to foods. Finally, the FDA defines function foods as any food product fortified with dietary supplements, food additives, genetically-modified organisms, or vaccines with health benefits beyond that of conventional foods. These categories of modified foods are very rough and vague. It does not help clarify things when food technology industry representatives say things like, "fruits and vegetables, being natural sources of beneficial nutrients like vitamins, antioxidants, and fiber, are in essence the ultimate functional food."[3]

There is no legal definition for functional foods in the United States. Although food additives must receive pre-market FDA approval as "Generally Regarded As Safe" (GRAS), dietary supplements and functional foods do not. Under the Dietary Supplement Health and Education Act of 1994 (DSHEA) no pre-market approval is required for dietary supplements and extra-nutritional ingredients. In fact, the FDA must demonstrate that a product is *unsafe* for a product to be pulled from the market. Functional foods are often marketed as dietary supplements to avoid proving their ingredients are GRAS. Yet, only dietary supplements labels must include the disclaimer: "This statement has not been evaluated by the FDA. This product is not intended to diagnose, treat, cure or prevent any disease." Functional food labels need not include a disclaimer about proven effectiveness. Given that functional foods are most often conventional foods with dietary supplement ingredient added, the lack of consistency in labeling is, if nothing else, puzzling.

Health claims, however, are more carefully regulated than ingredients. The FDA regulates "foods for special dietary use," which includes products used for supplying a special dietary need that exists "by reason of a physical, physiological, pathological, or other condition including but not limited to the conditions of disease, convalescence, pregnancy, lactation, infancy, allergic hypersensitivity to food, underweight, overweight, or the need to control the intake of sodium."[4] Health claims for foods for special dietary use must have pre-market approval. Because compliance is voluntary and more strict than what is required for functional foods, very few food products are identified as "for special dietary use." Incredibly, the FDA does not regulate "medical foods." These foods are prescribed by a physician for a patient with "special nutrient needs" in order to manage a disease or health condition. They are not intended for the general public. Examples of medical foods include *UltraClear* (for liver failure), *Vistrum* (for gastrointestinal balance), and *Nephrovite* (vitamin supplements for dialysis patients). The FDA does not require that medical foods have nutritional information labeled, nor must their health claims meet specified standards. In 1996, the FDA conceded that the lack of regulation is a problem and it has relied too much on the medical profession to regulate itself to prescribe and oversee the safety of medical foods.[5] As of 2006, the FDA's webpage continues to states that it is "exploring ways to more specifically regulate medical foods. This might include safety evaluations, standards for claims, and requiring specific information on the labels."[6]

As in Japan and the EU, food and supplement companies in the United States are permitted to make general health claims ("Structure/Function Claim") without FDA approval, whereas a specific health claim ("Disease Claim") does require approval. Unlike other countries, the U.S. permits health claims to be made for nutrients already contained in conventional food. Nothing has to be added to food to warrant a health claim. Another difference between the U.S. and other countries is in the language used to distinguish between a general and specific health claim: it is parsed exceptionally thin. According to the FDA:

> An example of an acceptable claim is "a good diet promotes good health and prevents the onset of disease" or "better dietary and exercise patterns can contribute to disease prevention and better health."
>
> An example of a disease claim is "Promotes good health and prevents the onset of disease" because the claim infers (sic) that the product itself will achieve the intended effect.[7]

It is hard to imagine that language like this does anything but confuse consumers.

FUNCTIONAL FOOD EFFECTIVENESS

The first concern about functional foods is practical, not philosophical. The fundamental practical problem with functional foods is that they do not work very well, and when they do work their health and nutritive affects are far less significant than their advocates would have us believe. That is because the very reductivist premise of functional foods—that food is the kind of thing that can be understood in terms of its component parts—is mistaken. When food is understood in terms of parts rather than wholes it usually does not deliver its promised effect as well as conventional food. There is increasing evidence that food broken down into its component parts and then reassembling as processed food is less nutritious than conventional food. It has been shown that ingredients isolated in laboratories do

WHAT'S WRONG WITH FUNCTIONAL FOODS? 181

not function in the same way they do in whole foods.[8] The Center for Science in the Public Interest warns that too often manufacturer claims about functional ingredients are "misleading and unsubstantiated by scientific evidence," and until governments establish adequate regulatory controls "functional foods may merely amount to little more than 21st Century quackery."[9] Even the nutritionists and industry experts who contribute to *Food Technology*, the leading industry journal, caution that the "single-nutrient approach is too simplistic."[10] Food, it appears, is more than the sum of its chemical parts, therefore treating it as collections of single nutrients to be mixed and matched, rather than as the complex biological system it is, simply may not work.

It is true, however, that food fortification for some nutrients does work. The fluoridation of drinking water in the U.S. has helped prevent tooth decay, vitamin-D fortified milk has eliminated rickets, iodized salt reduced goiter, and niacin-enriched flour, pellagra.[11] The increased fortification of these nutrients has very effectively prevented deficiencies of the nutrients added and eliminated a number of sources of disease. Yet, in complex matters of public health, it is often difficult to isolate single causal explanations. For example, is impossible to know precisely how effective niacin fortification was in the reduction of pellagra deaths in the 1940s since the decrease corresponds with changes in social and economic mobility, food safety, and food availability. If more people were eating healthier, more nutritious diets anyway, it is difficult to explain the reduction of the disease exclusively by niacin fortification. The situation is similar today with grain products fortified with folic acid to reduce the number of infants born with neural tube defects (anencephaly and spina bifida). On the one hand, higher levels of folate are now present in adults in the U.S. since fortification began in the 1980s, and fewer babies have been born with birth defects. On the other hand, the public is already more informed about the link between diet and fetal health—especially wealthier, more educated members of society. It is difficult to determine the effects of fortification on people who are already concerned about maintaining a healthy diet. Other causal factors may explain the reduction in birth defects.[12]

Some nutritionists worry that the single-nutrient approach drives functional food research and marketing, misleading the public to believe that there are dietary magic bullets in their food that will ensure a healthy diet regardless of what they eat. Enhancing food with dietary supplements is a quick techno-fix for more complicated issues of dietary patterns, lifestyle, and public health.

> Can we really accept that super-fortification will eliminate our need to select widely from conventional foods to balance nutrient intake? Americans are intrigued with the notion that a pill or a portion can settle all nutritional needs. Thus, we regard fortified cupcakes and synthesized orange juice as necessary steps in achieving that goal.... Dumping nutrients into such foods will not neutralize their detrimental effects or make them more healthful. Furthermore, fortification schemes serve primarily to add to the public's confusion about nutrition. By their nature, fortification practices discourage the most desirable modifications in food selection behavior.[13]

Although techno-solutions are often a short-cut, they should not be dismissed out of hand. It is much easier and more effective to supplement food than to address the more persistent underlying causes of malnutrition, such as poverty or insufficient education. But the small number of successful examples of food fortification should not lead us to assume that all food fortification will work as well. The single-nutrient approach to diet works only on rare occasions. The majority of functional foods are market-driven consumer goods that have not been proven to work at all.

FUNCTIONAL FOODS AS MEDICINE

The second concern about functional foods is that they blur the line between food and medicine. The FDA concedes that there is greater need for regulating the health claims made by functional food producers but has been negligent in its obligation to provide consumer protection. Meanwhile, the market in functional foods is booming. In 2004 sales of functional food products reached $22 billion in the U.S and $47 billion worldwide.[14] Millions of people in the U.S., Western Europe, and Japan manage their own health by eating dietary supplements and functional foods instead of using prescription or over-the-counter drugs. A nationwide survey conducted recently by the Centers for Disease Control and Prevention (CDC) found that 36 percent of American adults use complementary and alternative medicines ranging from diet to acupuncture to prayer. That means that a sizeable percentage of the public puts their health into their own hands. In 2003, 158 million Americans used some form of dietary supplements instead of over-the-counter drugs in order, they said, to save money, take control of their own lives, and to live healthier.[15] The trend is toward a public increasingly interested in maintaining better health through diet rather than spending money on health care and prescription medications. Under these conditions, the market for functional foods will only continue to grow.

In many ways, there is nothing new about this do-it-yourself approach to health care. It is a technologically-mediated version of long standing traditions that connect moral conduct with self-mastery of one's body. This connection between a self-imposed dietary regimen and moral conduct can be found in religious traditions throughout the world. For the ancient Greeks and Romans temperance and moderation of all of the appetites were central to moral conduct—especially sexual restraint but also control of diet, exercise, and strong emotions. Asian traditions also emphasized the relationship between diet, regimentation, and health. Taoism, Ayurveda, and Zen Buddhism are just some philosophical-religious systems that specify how bodily health connected to moral conduct leads to spiritual salvation. Although we have retained quite a bit from these traditions, the difference between our contemporary notions of diet and health and ancient and religious dietary practices is not only a greater understanding of physiology and nutrition but also the availability of technologies that extend our capacities in ways non-technological dietary and health practices cannot. Our current dietary practices are much better at reducing risk of disease, treating disorders, and fostering health.

The widespread use of dietary supplements, functional foods, and medical foods are twenty-first century versions of long-standing, tradition-bound, dietary/health/ self-management practices.

Yet the regulatory oversight for these edible technologies is terrible: existing regulations do not provide clear guidance—much less enforceable laws—on products ingredients, safety, and health claims. The most serious problem is the lack of regulation on medical foods. Although they are supposed to be used by patients under medical supervision, there is nothing stopping a food producer from calling any product a medical food and making it available to the public. Even when a medical food is used properly, there are no guarantees that the specific health claims made are supported by adequate scientific evidence. The FDA needs to change its current approach to the regulation of medical and functional foods to ensure safety and truthful labeling. It needs to clearly distinguish between medical and functional foods in unambiguous language, establishing standards and procedures for product composition, manufacturing practice and controls, and labeling requirements. The FDA should require that manufacturers notify the agency before marketing the product, submit evidence it is GRAS, and that the claims made about its health benefits are supported by what it calls "sound science." The quantity and quality of scientific evidence required might be modeled after FOSHU. That would clearly distinguish between medical foods and functional foods, and establish standards for what kind of health claims functional foods can legitimately make.

As the line between food and drugs becomes increasingly blurry, the FDA should require that functional food labels carry the same disclaimer ("This statement has not been evaluated by the FDA. This product is not intended to diagnose, treat, cure or prevent any disease") as dietary supplements. That would remove an arbitrary loophole in the food regulatory system and take a minimal step toward informing consumers of scientific validity of the health claims being made. It should require that all functional ingredients, like food additives, are GRAS before, not after, they are marketed.

The current burden of proof placed on the consumer to demonstrate a product is unsafe is unfair and unreasonable. Individuals lack the resources and know-how to provide scientific evidence for food safety. If "sound science" takes place in laboratories and large-scale research facilities, then it is the obligation of those with access to such places to ensure food safety and to verify health claims, not individuals. It is the obligation of the government to enforce laws and punish offenders for unsafe ingredients and false health claims. Only it has the legitimate power and authority to do so. Food safety is a matter of social justice. A government that fails to protect the safety of its citizens fails in its obligations to protect our rights—for what value do rights have if a citizen is unable to safely exercise those rights? How can we freely choose if the knowledge needed to make informed choices is hidden from us? Even the most minimal conceptions of social justice require the State to protect public safety. The market cannot guarantee food safety, health claims, and credible medical practice. That is the proper role of government.

FUNCTIONAL FOODS AS CONSUMER GOODS

The third concern about functional foods is with the role of the market. The food industry runs up against the troublesome fact, from its perspective, that each person can only eat so much food. On average we eat about 1,500 pounds of food in a year. Yet unlike other consumer goods there is a limit to how much we can consume. Although the epidemic of obesity might seem to suggest that this limit is flexible, the reason Americans are obese has less to do with the total mass of food consumed than with total calories, fats, and lack of exercise. Try as it might, the food industry has to convince us to eat more than we need to. The best way to do this is by adding value to cheap raw materials, usually in the form of convenience or fortification. The food industry has learned that selling unprocessed or minimally processed food is far less profitable than modifying existing food items by enhancing elements they already have in them (like vitamins and minerals) or by adding new elements to them. There is not a lot of money to be made selling oranges, somewhat more money to be made selling orange juice, but even more to be made selling orange juice that claims to provide the recommended daily allowance of calcium.[16]

Functional foods once played a crucial role in public health in eliminating nutritional deficiency disorders. It is conceivable that they may do so again. They may indeed, in some social context, be an intelligent way to support health and treat or prevent disease for people suffering from food restrictions and shortages. When functional foods do provide genuine public health solutions, they contribute immensely to the public welfare. They help to provide the very conditions for life; they help us to increase our capacities, to exercise our rights, and to live well together. The use of functional foods under these circumstances is, of course, a morally permissible policy for a government. In extreme cases, such as malnutrition or famine, a policy of functional food distribution might be required by a government to promote public health or even to protect our food rights.[17] Or, in less extreme circumstances, a government might be required to manage long-standing nutritional needs through the distribution of functional foods, if necessary for the public welfare. In the United States, the greatest challenges to nutritional health are currently obesity, chronic diseases (many of which are associated with obesity), the needs of an increasing aging population, and food safety. If functional foods can treat hypertension, diabetes, heart disease, arthritis, and eliminate the risks of food-borne illness and disease, then it would not only be wise to continue to develop and distribute them, but it is conceivable that it would be the obligation of the federal government to do so. This might take the form of food relief and food commodity distribution, school feeding programs, nutrition education programs, or incentives for private sector research and development.

It is morally defensible to rely on markets to provide functional foods to maintain public health, so long as no greater harms are inflicted, capacities diminished, or rights abused. If these conditions are met then markets and health are perfectly compatible. If individuals choose to support their health or treat disease by purchasing functional foods, and they are safe, effective, and consumed with knowledge, then there again is little reason to oppose them. The current case with folate-fortified grains is instructive: the market might presently be serving a genuine public health need by providing a functional food that reduces the instances of neural tube birth defects (assuming for the moment that little or no government subsidies were involved). If this is the case, then privatized food production and distribution should be encouraged as matter of policy to support public health.

The problem with relying on market mechanisms is that they are fickle. Markets may or may not solve public health problems. That is not what they are designed to do. Consequently, to rely on them is, at best, unwise for a government, at worst, negligent and a failure to protect its citizens. The food industry very aggressively influences and distorts nutrition science, federal regulation, and consumer choice. It functions like any other industry: it seeks to maximize profit and increase market share. The food industry does so by creating a favorable sales environment for its products. This includes lobbying political representatives to eliminate unfavorable regulations and pressure regulatory agencies not to enforce regulations, co-opting nutrition experts by supporting favorable research, and marketing and advertising, often to children who are unable to read ads critically. The food industry is, of course, free to sell people whatever people want, but it also relies heavily on its influence on the political process, marketing, and its version of nutritional advice in order to persuade people that they want what the industry is selling. Sometimes the food industry succeeds in producing and publicizing goods that people actually want and need; other times its means are less honest and serve to deceive people into thinking they want and need things they really do not.[18] Once functional foods are seen as one among many products that are a part of a sprawling food industry, then there is reason to question how vital they truly are. Functional foods should be seen as commodities with exchange-value rather than goods with use-value, as Marx would explain it.

When food and medicine are treated like any other consumer goods there is a real danger that our very dietary and medical practices ultimately serve the interests of others more than our own interests. Commerce in functional foods, then, is a profoundly moral and political matter. The more dietary practice becomes a matter of consumer choice, the less it becomes a matter for mechanisms of distribution other than the market. Yet that is precisely the social context in which functional foods exists. On one hand, they are commodities like any other to be manufactured, sold, and consumed; on the other hand, they are uniquely situated at the nexus of diet, health, and commerce, spanning the worlds of optional consumer goods and vital human needs. This puts us all in a tenuous position: commercial interests have the potential to transform how we eat and how we care for ourselves, yet the very future of food and medicine is in the hands of those who may not have our best interests in mind. That may be the most important thing wrong with functional foods.

NOTES

1. Michael Heasman and Julian Mellentin, *The Functional Foods Revolution: Healthy People, Healthy Profits?* (London: Earthscan Publications, 2001), p. 134.

2. Ministry for Agriculture Fisheries and Food, Food Standards Agency, www.food.gov.uk/regulation_health_claims.

3. Linda Orh, "Nutraceuticals and Functional Foods," *Food Technology*, May 2004, vol. 58, no. 5, p. 64.

4. U.S. Food and Drug Administration, Center for Food Safety and Applied Nutrition, "Food Labeling and Nutrition." www.cfsan.fda.gov/label.html.

5. "The agency believes that there is a need to reevaluate its policy for regulating medical foods because of a number of developments, including enactment of a statutory definition of 'medical food,' the rapid increase in the variety and number of products that are marketed as medical foods, safety problems associated with the manufacture and quality control of these products, and the potential for fraud as claims that are not supported by sound science proliferate for these products." U.S. Food and Drug Administration, "Regulation of Medical Foods," *Federal Register*, November 29, 1996 (vol. 61, no. 231).

6. www.cfsan.fda.gov/~dms/ds-medfd.html.

7. "Structure/Function Claims: Small Entity Compliance Guide." U.S. Food and Drug Administration, Center for Food Safety and Applied Nutrition, January 9, 2002. www.cfsan.fda.gov/~dms/sclmguid.html.

8. Bruce Silverglade and Michael Jacobson, eds. *Functional Foods: Public Health Boon or 21st Century Quackery?* (New York: Center for Science in the Public Interest, 2000).

9. Ibid., p. 19.

10. Robert Ward and Herbert Watseka, "Bioguided Processing: A Paradigm Change in Food Production," *Food Technology*, May 2004, vol. 58, no. 5, pp. 44–48.

11. Centers for Disease Control and Prevention, "Ten Great Public Health Achievements in the 20th Century," *Morbidity and Mortality Weekly Report*, October, 15, 1999: 48(40), pp. 905–913.

12. For an analysis of food fortification, see, Marion Nestle, *Food Politics* (Berkeley: University of California Press, 2002), pp. 298–314.

13. C. Christopher, "Is Fortification Unnecessary Technology?" *Food Product Development*, 1978: 12 (4), pp. 24–25. Quoted in Nestle, *Food Politics*, p. 314.

14. A. Elizabeth Sloan, "Top 10 Functional Food Trends 2004," *Food Technology*, April 2004, vol. 58, no. 4, p. 32.

15. "Complementary and Alternative Health Medicine Use Among Adults, United States, 2002." Centers for Disease Control and Prevention National Center for Health Statistics, *U.S. Department of Health and Human Services Publications*, 2004.

16. For more of this argument, see, Greg Critser, *Fat Land: How Americans Became the Fattest People in the World* (New York: Mariner Books, 2003).

17. "Universal Declaration of Human Rights," http://www.un.org/rights, p. 5. The right to food is recognized directly or indirectly by every country in the world, either written into

their constitutions or by virtue of their membership in the United Nations. Article 25 of the 1948 Universal Declaration of Human Rights states that "everyone has the right to a standard of living adequate for the health and well-being of himself and of his family, including food, clothing, housing, medical care and necessary social services, and the right to security in the event of unemployment, sickness, disability, widowhood, old age or other lack of livelihood in circumstances beyond his control."

18. For evidence of precisely how the food industry creates a favorable sales environment through lobbying, marketing, and co-opted nutrition experts, see, Marion Nestle, *Food Politics*, pp. 95–136, 175–218.

ETHICS AND THE LIFE SCIENCES

THE MAGIC BULLET CRITICISM OF AGRICULTURAL BIOTECHNOLOGY

DANE SCOTT
UNIVERSITY OF MONTANA

ABSTRACT: One common method of criticizing genetically modified organisms (GMOs) is to label them as "magic bullets." However, this criticism, like many in the debate over GMOs, is not very clear. What exactly is the "magic bullet criticism"? What are its origins? What flaw is it pointing out in GM crops and agricultural biotechnology? What is the scope of the criticism? Does it apply to all GMOs, or just some? Does it point to a fatal flaw, or something that can be fixed? The goal of this paper is to answer these questions and clarify the magic bullet criticism of agricultural biotechnology. It is hoped that the results of this exercise will be helpful in advancing deliberation over the role GMOs and agricultural biotechnology should play in twenty-first-century agriculture.

Genetically engineered crops are sometimes criticized as being "magic bullets." For example, in his essay, "The Myths of Agricultural Biotechnology," the UC Berkeley agroecologist, Miguel Alteiri writes:

> By challenging the myths of biotechnology, we expose genetic engineering for what it really is; another "technological fix" or magic bullet aimed at circumventing the environmental problems of agriculture (which themselves are the outcome of an earlier round of technological fixes) without questioning the flawed assumptions that gave rise to the problems in the first place.[1]

It is clear from these comments that Alteiri does not think genetically modified organisms (GMOs) mark a substantial break with the environmentally harmful past of technologically intensive agriculture. According to this position, biotechnological solutions to the environmental problems of industrial agriculture will be ineffectual because they arise from a flawed research paradigm, which focuses on magic bullets and technological fixes.

While it is not clear in the above remarks, the notion of a "magic bullet" is conceptually distinct from that of a "technological fix." In general, the magic bullet criticism aims to expose a conceptual flaw in the dominant research paradigm in agriculture that causes it to generate environmental side effects. The technological fix criticism aims to expose flaws in the research paradigm that cause it to generate social side effects. The goal of this paper is to clarify the magic bullet criticism of agricultural biotechnology.

In this effort to clarify the magic bullet criticism of biotechnology, it will be helpful to look at the origins of the term in modern biomedicine. Paul Ehrlich, one of the founders of the modern biomedical paradigm coined the term "magic bullet." He writes: "antibacterial substances are, so to speak, *charmed bullets* which strike only those objects for whose destruction they have been produced."[2] Ever since Ehrlich's day a central objective of biomedical research has been to discover magic bullets through controlled laboratory experiments. These therapeutic agents are designed to target specific disease-causing agents without affecting the healthy parts of an individual's body. This approach is related to an agent-host-environment epidemiological model, which evolved out of the work of Ehrlich, Pasteur, Koch, and other nineteenth-century researchers. This model sees the "host" and the "environment" as modifying rather than causal factors.[3] In so doing, it reduces *the* cause of the disease to a specific agent. Further, this approach gives rise to the doctrine of specific etiology, which enshrines the search for "magic bullets" as a central puzzle-solving task of normal biomedical science.

Ironically, it is narrowness of this approach that proves to be both its greatest strength and greatest weakness. For example, thirty years or so ago addressing bacterial infection with antibiotic magic bullets was seen as an unqualified success. However, taking the long-view this approach may ultimately undermine its "early" successes. To briefly explain, as noted above, this research paradigm places cultural, ecological, and evolutionary factors in the background; in so doing, the effects of these factors are not sufficiently anticipated. For example, the side effect of antibiotic resistant strains of bacteria arose because the cultural reality of antibiotics use was not adequately modeled in relation to bacterial evolution and ecology.

In sum, the magic bullet approach, as guided by the doctrine of specific etiology, was too narrow to anticipate and prevent the unintended consequence of resistant strains of bacteria. For these reasons, biomedicine is now on an anti-biotic treadmill: as the efficacy of one antibiotic is diminished another generation must be developed, as their efficacy is diminished, yet another must be developed, and so on. But this treadmill is ultimately dangerous, expensive, and unsustainable. Success in getting off it has only been made by widening the focus of research to include the cultural, ecological, and evolutionary factors.

Generalizing from the above discussion, the "magic bullet" criticism aims to expose the narrowness of a research paradigm. The essence of the criticism is that an approach that targets specific problem-causing agents with specific technological solutions, without adequately modeling cultural, ecological, and evolutionary

MAGIC BULLETS AND AGRICULTURAL BIOTECHNOLOGY

factors leads to a technological, treadmill phenomenon. Moreover, the treadmill phenomenon is dangerous, expensive, and unsustainable.

The application of the magic bullet criticism to the current research paradigm in agriculture seems appropriate, at least in places. There are relevant parallels between the puzzle solving activities of normal biomedical science and normal agricultural science. In agriculture, as in biomedicine, the conceptual flaw with this research paradigm is its narrowness: it does not adequately model cultural, ecological, and evolutionary factors. This inadequacy leads to the multiplication of unintended consequences and the technological treadmill phenomenon. Agricultural scientists, like medical scientists, are continually forced to create technologies to address side effect problems created by previous technologies. The clearest example of this is the so-called pesticide treadmill.

The pesticide treadmill roughly parallels the anti-biotic treadmill in biomedicine. To explain, in the resent past, the narrow focus of the research paradigm in pest management did not factor in how synthetic insecticides would actually be used by farmers, nor how their actual use would interact with the ecological and evolutionary dynamics in the field. In any given field there exists a dynamic equilibrium between consumers and producers, predators and prey. Insects become classified as pests when their numbers become great enough to significantly impact profitability. Synthetic insecticides approximate the ideal of a magic bullet in that they kill the pest while leaving the crop unharmed. However, they also kill a broad range of nontarget insects. This disrupts the ecological dynamics in the field, as both pests and beneficial insects (i.e., insects that prey on the pest, keeping their numbers in check) are exterminated. After the spraying, because not all the pests are killed, their population rebounds and surges due to the lag time in the return of beneficial insects. For this reason another round of spraying is required, creating a pattern of dependence on the technological solution of toxic chemicals for pest management. All this spraying creates a strong selective pressure favoring the evolution of strains of pests that are resistant to the insecticide; hence, in time, rendering the insecticide useless. Scientists must then develop new insecticides—another round of magic bullets—to control the pest, thus initiating the technological treadmill phenomenon. As in the case of antibiotic resistance in biomedicine, this treadmill is hazardous, expensive, and ultimately unsustainable.

The sustainable, agroecological response to the pesticide treadmill is to replace the narrow, magic bullet approach with a multi-factorial research paradigm that better models cultural, ecological, and evolutionary factors. Integrated Pest Management (IPM) is the name given to this approach. IPM has now been in use for over thirty years and it is acknowledged as being a scientifically sound approach. The University of California's IPM Web site describes this alternative paradigm in pest management, as an

> ecosystem-based strategy that focuses on the long-term prevention of pests or their damage through a combination of techniques such as biological control, habitat manipulation, modification of cultural practices, and the

use of resistant varieties. Pesticides are used only after monitoring indicates they are needed according to established guidelines, and treatments are made with the goal of removing only the target organisms.[4]

It is appropriate to call IPM an alternative paradigm. If widely adopted, the change to IPM qualifies as the kind of gestalt switch that Thomas Kuhn describes as a paradigm shift. Kuhn writes that, "Paradigm changes do cause scientists to see the world of their research-engagement differently."[5] Hence, looking at the problem of pest management in terms of "long-term prevention" in light of the dynamics of cultural, ecological, and evolutionary factors creates a new set of puzzles for normal agricultural science to solve. Generally speaking, the primary puzzle-solving activity for scientists is to discover ways of manipulating cultural and ecological factors to prevent pest populations from reaching harmful numbers. This is in contrast to the magic bullet approach where the primary puzzle solving activity is to develop toxins to target specific pests. In sum, IPM is an effort to get off the pesticide treadmill by prioritizing the management of cultural and ecological factors over the "magic bullet" solution of insecticides. The goal is not to eradicate the pest, but to keep its numbers in check by controlling the ecological dynamics in the field between pests and beneficial insects. In this management plan insecticides are used sparingly and judiciously.

There are clear parallels between IPM and the strategy proposed by the Center for Disease Control (CDC) to get off the antibiotic treadmill. The CDC's plan calls for "accelerating research that focuses on . . . developing infection control strategies to prevent disease transmission.[6] Also, it calls for educating "physicians to prescribe antibiotic more prudently."[7] In other words, antibiotics must no longer be seen as magic bullets. To preserve the efficacy of antibiotic they can no longer be used liberally. This strategy requires a new research paradigm that better models cultural, ecological, and evolutionary factors, and, further, one that seeks to develop strategies to prevent infection and promote the limited and carefully regulated use of antibiotics. Clearly, the wonderful technological innovation of antibiotics is not driving the treadmill phenomenon; it is the narrow magic bullet model. To make another generalization, it is not new technologies that are driving the technological treadmill; it is the narrowness of the inherent magic bullet approach of much modern medical and agricultural research.

With the magic bullet criticism hopefully clarified, and some possible reactions to it identified, it is time to turn to the magic bullet criticism of agricultural biotechnology.

To begin: In what sense are GMOs open to the magic bullet criticism? The GMOs most obviously open to the magic bullet criticism are those designed to be resistant to pests. At present, the class of GMOs designed to manage pests are genetically engineered with a gene from a common soil bacteria, *Bacillus thuringinesis* (*Bt*).[8] This microbe secretes a protein that is toxic to caterpillars, as they have an enzyme in their gut that activates the toxin. Hence, *Bt* is an excellent approximation of Koch's ideal of a magic bullet—its activity is specific to the problem-causing agent.[9] By far, the most commercially significant crops engineered with the *Bt* gene are corn and cotton. However, insecticide data indicates that *Bt* corn "has had little if any impacts of corn insecticide use."[10] But *Bt* cotton has led to a significant reduction in the use of synthetic insecticides in several Western states.[11]

So, focusing on *Bt* cotton, in the recent past, cotton farmers have annually sprayed their fields with millions of pounds of highly toxic insecticides to control tobacco budworm and cotton bollworm. In addition, ever since the Sixties they have been on the pesticide treadmill. The average life for a class of synthetic insecticides has been about a decade before insects evolve resistance.[12] Replacing synthetic insecticides with GM cotton would seem to indicate progress toward addressing many of the environmental side effects associated with pesticide use in industrial agriculture. For example, *Bt* cotton only kills the organisms feeding on the plant and susceptible to the toxin, while spraying with synthetic insecticides kills a broad range of insects, including beneficial ones.[13] In addition *Bt* degrades quickly and is not toxic to mammals, birds, or fish as these animals do not have the necessary enzyme in their gut to activate the toxic protein. Therefore, this GMO addresses several of the side-effect problems associated with the use of synthetic insecticides. However, these considerable benefits could be short-lived if pests develop resistance to the *Bt* toxin. If this happens, then *Bt* cotton represents just another round on the pesticide treadmill.

This is theoretically possible, as laboratory studies have demonstrated that resistance to *Bt* can evolve if "selection pressure is strong enough.[14] Therefore, many scientists are convinced that it is only a matter of time, perhaps a decade, until *Bt* cotton will no longer be effective in fighting pests. This supposedly revolutionary technology may only be a temporary fix and not progress toward the long-term goal of environmental sustainability. Also, there is no guarantee that a new generation of GM crops can be engineered with an environmentally friendly compound like the *Bt* toxin. As medical researchers will attest, the number of magic bullets found in nature is finite. Finally, if GMOs merely perpetuate the treadmill phenomenon, then all the excitement about biotechnology will have dangerously delayed the transition to a more sustainable paradigm, such as IPM.

One reason to be pessimistic about the future of *Bt* crops is they were conceived and implemented under the narrow magic bullet model. Admittedly, *Bt* crops are a much better magic bullet for certain crops than synthetic insecticides. In addition, ad hoc provisions were made to prevent insects' from evolving resistance. However, it is doubtful that these ad hoc provisions adequately modify the magic bullet approach to prevent the treadmill phenomenon.

According to Daniel Charles's history of the biotech industry, the scientists who created *Bt* crops were attempting to make magic bullets. Charles writes that, "the genetic engineers [working on inserting the *Bt* gene into plants] spoke of 'permanent solutions' to the insect problem" (Charles, 2001: p. 82). This is similar to the way scientists once spoke of magic bullets in medicine—as permanent solutions to the problem of bacterial infection. However, biologists who study the evolution of pesticide resistance knew better. Charles comments:

Evolutionary biologists don't believe permanent solutions exist in biology. There is only adaptation, moves and countermoves, in a game of chess that never ends. For them, dreams of technologcal solutions, so common among chemical companies, are the standard object of ridicule. "Its just another silver [sic. magic] bullet," they say dismissively. Silver bullets do not work for long. (Charles, 2001: p. 82)

From Charles's remarks, it is clear there was a conflict in research paradigms between biotechnologists working for agrochemical companies and the evolutionary biologists researching resistance. The importance of this conflict is key in understanding how cultural, ecological, and evolutionary were finally included in the ad hoc management plan for *Bt* crops.

In creating these GMOs the biotech industry did not initially consider the evolution of resistant strains of insects. These concerns were only considered as an afterthought, and then reluctantly. Specifically, Charles attributes industry's acknowledgement of the potential for the evolution of resistance to the efforts of concerned academic scientists. These scientists saw in a glance that if *Bt* crops were widely planted, resistance would quickly evolve, thus rendering this highly beneficial, naturally occurring pesticide useless. In other words, industry would have squandered, for short-term profit, the long-term benefits of this unique group of proteins.

In regard to cotton, the efforts of concerned scientists resulted in a management plan requiring farmers to set aside at least 4 percent of their land as a refuge, where *Bt* cotton is not planted (Charles, 2001: p. 183). The idea, of course, is that this would prevent the evolution of resistant strains of insects. Charles summarizes how these refuges came about. He writes:

These refuges were the result of a campaign waged by scientists who believed that, without restrictions, new strains of insects would soon emerge that were resistant to Bt. Biotech companies, which wanted to sell as much genetically engineered seed as possible, pushed for smaller refuges. Many scientists believed that much larger refuges were necessary to preserve Bt as a useful tool; because once Bt failed, this gift of God would be gone forever. (Charles, 2001: p. 181)

There are at least two important points that can be learned form the way the management plan for *Bt* cotton came about. The first, as indicated above, is that the setting aside of refuges is merely an adjustment to the magic bullet approach. It bears only a superficial resemblance to the IPM paradigm. The use of an insecticide remains the primary means for controlling pests rather than preventing outbreaks by manipulating cultural and ecolgocal factors. As has been often noted, *Bt* crops "mimics the chemical-based management system."[15] The second point exposes a clash between the market model of the biotech industry, where most of the development of GMOs is taking place, and the evolutionary model used by concerned scientists.[16]

Looking more closely at this point: on the one hand, if *Bt* cotton, for example, is to be profitable, the competitive, market model indicates that the refuges cannot be too large. On the other hand, if the evolution of resistant strains is to be avoided, the evolutionary model indicates that the refuges cannot be too small. Hence, industry fought for the smallest possible refuges to maximize profits, and the concerned scientists fought for the largest possible reserves to minimize resistance. As seen above, a compromise solution was implemented.

However, many scientists felt that the size of the refuge was much too small, that at least 10 percent was needed, and some scientists argued for as much as 50 percent. The compromise of 4 percent, which was forced by microeconomics, is not necessarily sound evolutionary biology, which, of course, best tells us how to lower the probability of resistance developing. Therefore, because evolutionary factors were implemented via this ad hoc compromise, the likelihood that, sooner or later, insects will evolve resistance to *Bt* crops is much greater. Hence, the likelihood that *Bt* crops will initiate another turn of the treadmill. Significantly, for the microeconomics of the biotech industry this is not an unfortunate result. As long as the treadmill can be sustained, the magic bullet approach is justified by the competitive market model. The reason being that this approach demands maximum use of their products and when those products fail, they will supply another.

By way of summary, it should be noted that the magic bullet criticism is not a blanket critique of agricultural biotechnology. It only applies to a narrow range of GM crops that are designed along the lines of the doctrine of specific etiology in medicine. The essence of the criticism is to point out the dangers of using too narrow of a research paradigm; specifically, one that fails to adequately model cultural, ecological, and evolutionary factors. As discussed above, the most obvious place the criticism applies is at GM crops engineered to contain pesticides. However, there may be other GMOs where the criticism is appropriate. The magic bullet criticism points out flaws in a research paradigm, and not specific technologies per se. So there is no reason why GM crops engineered to contain pesticides are necessarily flawed in the same way that antibiotic technology in medicine is necessarily flawed. It is possible that when placed in the right context *Bt* crops, for example, can be a useful tool in working toward the goal of environmental sustainability. Finally, one important factor that is preventing a move away from the discredited magic bullet model is the positive relationship between this approach and the competitive market model of the biotech industry.

NOTES

1. Alteiri, 2001.
2. Dubos, 1993, p. 156, emphasis added.
3. Norell, 1984, p. 134.
4. University of California, IPM Web site, accessed 1/15/04.

5. Kuhn, 1970, p. 111.
6. Schuman, 2003, p. 85.
7. Ibid.
8. Japanese scientists identified *Bt* during an epidemic in the silkworm industry at the turn of the twentieth century. Krimsky and Wrubel, 1996, p. 57.
9. *Bt* has been safely used ever since the late 1950s as an insecticidal powder. At present six major groups of *Bt* proteins have been isolated. Their range of toxicity is small, targeting caterpillars, fly larvae, beetle larvae, and nematodes. In 1991 the worldwide sales in dollars of *Bt* insecticides represented only a tiny fraction (10^{-5}) of that of synthetic insecticides. Nonetheless, while *Bt* is not commercially that significant, it is an important tool for organic farmers. *Bt* toxins are naturally occurring, less likely to harm nontarget organisms, and they degrade quickly in water and sunlight. In sum, the *Bt* toxin does not generate many environmental side effect compared to synthetic insecticides. However, their limited range of activity and the fact that they degrade quickly has made them less attractive an option to the vast majority of insecticide using farmers. Krimsky and Wrubel, 1996, p. 56.
10. Benbrook, 2001.
11. Ibid.
12. Ibid.
13. Krimsky and Wrubel, 1996, p. 57.
14. Ibid, p. 64.
15. Benbrook, 2001.
16. Krimsky and Wrubel combine these two points to make the following observation. They write: "Agriculture would best be served by a policy of well-thought-out use of environmentally compatible control agents to conserve their effectiveness. This is in direct conflict with the competitive structure of the agrichemical and, in this case, biotechnological industry. Their purpose is to sell as much product as quickly as possible to recover the investment in research and development. Our analysis reveals, however, that one cannot separate the problem of pest control from the problem of pest resistance." Krimsky and Wrubel, 1996, p. 67.

BIBLIOGRAPHY

Benbrock, C. (2001). "Do gm Crops Mean Less Pesticide Use?" *Pesticide Outlook*. RSC Publishing, (5) pp. 204–297.

Charles, D. 2001. *Lords of the Harvest: Biotech, Big Money, and the Future of Food*. Cambridge, Mass.: Perseus Publishing.

Dubos, R. 1993. *Mirage of Health: Utopias, Progress, and Biological Change*. Rutgers, N.J.: Rutgers University Press.

Krimsky, S., and R. Wrubel, 1996. *Agricultural Biotechnology and the Environment*. Urbana, Ill.: University of Illinois Press.

Kuhn, T. 1970. *The Structure of Scientific Revolutions,* second edition. Chicago: University of Chicago Press.

Nordenfelt, L., and I. Lindhal., eds. 1984. *Health, Disease, and Causal Explanation in Medicine.* Boston: D. Reidel Publishing Co.

Norell, S. 1984. "Models of Causation in Epidemiology," in Nordenfelt and Lindahls.

Schuman, A. J. 2003. "A Concise History of Antimicrobial Therapy." *Contemporary Pediatrics.* October 2003, pp. 65–85.

Thompson, P. 1995. *The Spirit of the Soil: Agriculture and Environmental Ethics.* London: Routledge.

University of California, integrated pest management Web site: http://wwwipm.usdavis.edu/.

ETHICS AND THE LIFE SCIENCES

THE DIETARY LIMITATIONS IMPOSED BY MEXICO'S SOCIAL STRUCTURE

CHRISTINA PIÑA
SWARTHMORE COLLEGE

ABSTRACT: Blaming the individual for poor dietary habits is much easier than changing the social structure. Although society frequently assumes that the individual is able to select a particular diet amongst an array of choices, this research shows that the societal structure has quite a determinative role. This research focuses on malnutrition in Mexico and the sociopolitical and economic histories that have contributed to and maintained Mexicans' unhealthy status. The findings of this research support Weber's and Bourdieu's theories describing how individuals' choices are limited by the societal structure. With this framework in mind, resolutions include fortification, supplementation, and education. From exploratory questionnaires, Mexicans living in Monterrey showed great interest in education. This suggests a starting point for further inquiry and gives insight into possible methods of attacking malnutrition. However, effective policy changes require committed political support to ensure both short-term and long-term success.

INTRODUCTION

How easy it is to point a finger and blame the individual for an unhealthy diet. Our society turns away from complex sets of factors that contribute to unhealthy diets and prefers to simplify the problem and solution by placing the responsibility on the individual. These oversimplified explanations give the false impression that the social structure in which the individual lives need not be scrutinized. My investigation proposes that the exact opposite is true. In addition to political and economic histories giving insight into the make-up of the current social structure, theories from Weber and Bourdieu explain how an individual's choices are limited to the opportunities that the social structure provides. I apply my argument to the social context of Mexico's malnutrition problems. I examine the various aspects of this complex social structure to give insight into possible causes of the malnutrition in Mexico and to demonstrate that solutions to these problems must be multi-faceted as well. In order to match the level of structural complexity, investment should be made in both short-term and long-term projects.

Weber and Bourdieu give varying amounts of power to agency in choosing a lifestyle. Both theorists' perspectives do however, share the general idea that the individual's capacity to choose freely within the dominating social structure is limited. Weber bases the class differentiation upon consumption patterns. An example of a food item that is reflected in class differentiation in Mexico is demonstrated by the higher amount of bread consumed by the middle and upper classes while the poor populations almost exclusively eat tortillas (Ross, 1999). While wealthy individuals clearly have the power to choose which kind of grain to consume, the poor individual not only has a constrained choice of the type of food, but sometimes does not even have access to the traditional, staple tortilla. These different consumption patterns provide insight into certain lifestyles and also relate information about the life chances of that individual: "chance is socially determined, and social structure is an arrangement of chances. . . . That is, people are constrained in determining their lifestyle but have the freedom to choose within the constraints that apply to their situation in life" (Cockerham, p. 161). The makeup of the social structure is created based on the conglomeration of these chances and the boundaries that the structure creates.

Based on the research that I present in this paper, Bourdieu's perspective on the individual's life chances also apply to Mexico's social structure in regards to food consumption. He uses the idea of a "habitus" in order to explain that the individual is overpowered by the social structure: "knowledge of social structures and conditions produces enduring orientations toward action that are more or less routine, and when these orientations are acted upon they tend to reproduce the structures from which they are derived" (Cockerham, p. 163). The habitus is the individual's identification with the social structure that s/he encounters on a daily basis. The daily consumption of the tortilla is part of a Mexican's identity because the social structure provides this food item on a daily basis. As I will explain later, the nutritional quality of the tortilla has decreased from its original content due in part to political and economic influences. Because of the individual's intimate understanding of the tortilla as a part of the habitus, the individual has no choice but to continue to consume this staple food despite its inferior quality.

Diet is largely produced by lifestyle. The social structure has an extremely influential role in determining the quality and quantity of food that an individual eats. Because the tortilla is such an eminent part of the diet, I focus on this staple food's political changes and its relation to the Mexican population's health. Other elements of the diet obviously have an effect on nutritional status as well, but isolating and tracing the tortilla's political movements reveals the points at which the social structure limits important options for consumers.

Because of the vast poverty of Mexico, many individuals heavily rely on the tortillas as a source of not only caloric intake but nutrient intake as well. Thirteen million children live in extreme poverty and rely on tortillas for 80 percent of their caloric intake (Ross 1999). Balderas-Gonzales (1999) claims that corn tortillas account for over 50 percent of daily calories and protein for poor families. Poverty has an immense impact on the nutritional status as 40 percent of the Mexican population endures some form of malnutrition and about 30 percent suffer from anemia (Ross 1999). Nutrient deficiencies are especially prevalent in young children, infants, and mothers.

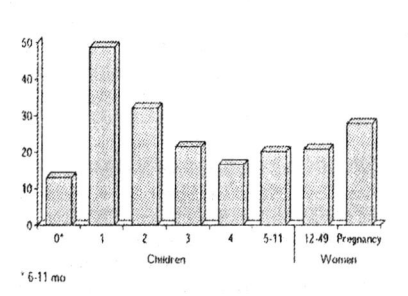

FIGURE 3. PREVALENCE OF ANEMIA IN CHILDREN AND WOMEN. NATIONAL NUTRITION SURVEY, MEXICO, 1999

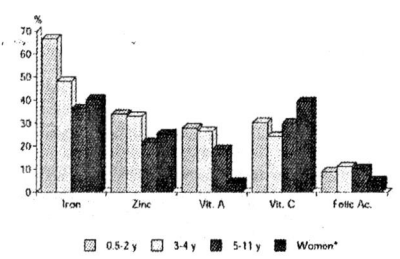

FIGURE 4. PREVALENCE OF MICRONUTRIENT DEFICIENCIES IN CHILDREN AND WOMEN. NATIONAL NUTRITION SURVEY, MEXICO, 1999

As these graphs show (Rivera and Amor 2003), the vulnerable populations are mostly children under the age of five, and women. Iron deficiency is especially widespread in children and as figure 3 shows, 50 percent of one-year old children have problems with anemia. Identifying vulnerable populations and their relative deficiencies is a starting point for policy makers. The next questions to ask are why is this the problem and how can it be solved? My research shows that the individual has less to do with the malnutrition than does the social structure that contrains the dietary options.

The low consumption of meat by poorer populations could account for some of the iron deficiency but other aspects of the diet contribute to this problem as well. The large quantities of corn, beans, and milk in the diet contain phytic acid which inhibits the absorption of the much needed iron (Rivera and Amor 2003). However, the traditional corn tortilla has many nutritious elements. When corn's nutrients are not stripped from the grain, the corn provides tortillas with a good source of calcium, protein, fiber, and potassium. Also, the corn used for the tortilla is treated with limewater which increases not only the calcium, but the phosphorous and iron content (Saldana and Brown 1984). Modern corn tortilla production incorporates more processed grains and may not provide the potential nutritious benefits. In addition to this decreased nutrient quality, corn tortillas also lack some important vitamins and thus are inadequate in fulfilling dietary needs. However, even this staple food has grown to be a difficult thing to consume for many of the poor Mexican populations.

METHODOLOGY

The sources I used in gathering my research came from a variety of disciplines in order to include basic, background information on Mexico's political and economic history. I also consulted journal articles relating to Mexico's governmental programs and sought solutions for nutrition problems in developing countries. I conducted a brief interview with a Spanish professor at Swarthmore College who is from Mexico and I also gathered perspectives from twenty-five exploratory questionnaires that were handed out in Monterrey, Mexico. The data from these surveys supports the conclusions drawn from the journals and other sources and also suggests a starting point for practical solutions.

HISTORICAL, POLITICAL, AND ECONOMIC CONTEXT

The decisions that the government makes about food policies and subsidies have a drastic effect on the availability of nutritious food. By examining how political systems interact with individuals, I hope to reveal some of the structural constraints. The Mexican Revolution of 1910–1920 is an example of the government's relationship with the Mexican citizens. Mexicans demanded land rights and equality which eventually led to the Constitution written in 1917. This document introduced the government's responsibility to intervene in cases of monopoly and to control staple food prices. Mexico was economically unstable and high inflation caused great tension between the producers and consumers. The government feared social unrest again twenty years later and in 1937, created the State Food Agency (Comité Regulador del Mercado del Trigo). The Agency intervened directly in the market of grains and made policies favoring mostly the urban consumption problems. Although the government guaranteed the purchase of the rural crops, easier railway access to larger farms left some small farmers behind. Easier access to urban areas also left many rural families without aid (Ochoa 2000).

Although the Mexican Revolution resulted in some agrarian reform, the southern regions did not take part in this and now have a much lower level of health status than the rest of Mexico. López et al. (1999) did a study that looked at the health status of children under five in a southern part of Mexico called Chiapas. As expected, children from rich peasant families have much lower rates of malnutrition than do children from poor families with small farms. While none of the agrarian families consume much meat, poorer families rely on corn and beans for protein, while richer peasant families consume about 100 more grams of milk than do poor families. These differences reflect inequalities in land and therefore income, which also relates to the options that each family has in selecting a healthy diet.

Cortés et al. (1996) come to a similar conclusion in that food consumption is related to income and social class. They look at socioeconomic status and report that while 7.7 percent of preschoolers are malnourished in the poorest stratum, hardly any malnutrition exists in the upper class. Poverty is the key factor in determining dietary options especially because of Mexico's history of inflation and decreased incomes. This lowered income occurred despite article 123 in the Constitution that assures a minimum wage that can provide for all basic needs of a family. This promise has clearly not been met because of a lack of governmental enforcement and stability.

World War II brought another period of instability. Food shortages were created because of the difficulty of importing food which led the State Food Agency to set up government stores to sell more price-reduced food. Most of the resources went into the marketing system and left the rural, agriculture producers with little aid. The government's policies during this time highly constrained the rural populations' ability to produce and consume food, much less nutritious foods. However, helping urban populations gives the outward appearance of action. The government's reaction to the social unrest appeared to be more about keeping frustrated citizens complacent than to provide effectual relief. (Ochoa 2000).

During the 1970s, staple food subsidies provided food to many poor families, but fluctuated as oil prices changed. In 1980, the government created the Mexican Food System which used money from oil revenues to invest in agricultural reform. Shortly after, Mexico underwent more economic instability, had weak foreign trade, devaluation of the peso, and uncontrollable exchange rates, making the State Food Agency's policies impossible to maintain. Consequentially, the market was carried further away from rural farmers when guaranteed prices were eliminated. These economic changes increased tortilla prices on average 73 percent each year from 1993 to 1996. The Mexicans were in outrage from such a drastic increase of price, especially because the government had led them to believe that prices would only gradually rise (Ochoa 2000; Balderas-Gonzales 1999; Smith 1996). The high price of the tortilla is a clear structural constraint that has an effect on what might otherwise be considered a simple option for poor families.

MEXICO'S ATTEMPTS AT IMPLEMENTING FOOD POLICY AND PROGRAMS

Although the Mexican Revolution demanded rights to land, jobs, salaries, education, and health, the right to nourishment is not included in the Constitution. Most of the government's intervention has been through food subsidy programs. CONASUPO created tortilla subsidies that provided 1kg/day to urban families with incomes below two minimum wages. This turned out to include 43 percent of urban families. In addition to general subsidies such as tortillas, CONASUPO provided more selective programs such as the distribution of milk. However, this distribution had many problems because it did not take into account the number of family members or the number of, or if there were, children, or pregnant mothers. The CONASUPO-LICONSA program was created to provide milk at reduced prices to children less than twelve years old and to mothers. The CONASUPO-DICCONSA program also provided corn, beans, sugar, and rice to marginalized populations. However, the proper consumption of the subsidized food is uncertain because this program lacked any educational components (Roma 1999; Barquera et al. 2001).

Levy et al. (2003) did a study that examined the effects of the tortilla subsidy on improving the nutrition of children and women in marginal areas. Three different groups were compared: families receiving tortilla subsidies, families who received tortilla subsidies five years ago, and families who had never received tortilla subsidies. The subsidies proved to be very beneficial for the families as tortillas would otherwise make up about 14 percent to 20 percent of the families' food expenses. The subsidy also shows a 7 percent improvement in the malnutrition index when comparing the families who receive the subsidy with the other two groups. Apparently, the tortilla subsidy is beneficial to the extremely poor families, but the question remains as to how to maintain funding of such an extensive program. Just as tortilla prices could not be supported by the Mexican economy and were then subject to the forces of the market, the benefits of the other food subsidy programs could not be supported. This is a clear place of tension between the needs of the individual and the limits of the social and economic structure.

In 1980, the Mexican Food System (Sistema Alimentario Mexicano or SAM) was created in order to "stimulate the production of basic foods (facilitating access to credits and improving guaranteed prices, among other strategies) to reach self-sufficiency and improve food distribution, above all among the marginal sectors" (Barquera 2001, p. 2). Although this was intended to help production, the peasant producers who needed the most assistance were not helped because of administrative organizational differences. The SAM's more technologically advanced plans did not match the peasants' outlook on production. The government's programs worked at both ends of the market system by assisting producers and consumers. However, many of the programs did not last long enough to make any notable changes.

Other programs have attempted to incorporate educational aspects into the distribution of resources in an attempt to assure effective consumption of nutrients. Progresa is a program that was created in 1997 to help alleviate poverty in rural areas. It provided financial rewards when families regularly attended health clinics and kept the children in school. This program also distributed micronutrient fortified foods to infants, children, and mothers and had reached 40 percent of rural families by 2000 (Skoufias, Davis, and de la Vega 2001; Roma 1999). By incorporating local culture into its strategies, Progresa took a structural approach at more of a micro-social level. The components of this program have much potential to improve the nutritional status of individuals because it acknowledges the power that society has on determining the health of the individual and tries to work with that structure. Another important component of Progresa is that it includes evaluation to measure its effectiveness. This program is also unique in that it only focuses on the rural populations and even wastes less money than other programs such as the tortilla subsidy programs. It is important to note that although Progresa shows more procedural sophistication than previous programs, its goal is actually to shorten the poverty gaps. Presumably this would also assist in providing healthier diets.

Other programs were created in an attempt to work at the community level as well. Community centers were established and included community kitchens in order to help women decrease their domestic workload. These centers however, did not work effectively when the lack of organization and lack of a main focus created too many inefficient activities (Roma 1999). Additional programs attempted to work with social realities by providing food at schools. In providing breakfasts at school, the government was able to provide between 20 percent and 30 percent of daily calories and protein in 1998 (Bacquera 2001).

Most of these programs either provide food directly through subsidies or through teaching self-sufficient health practices. The magnitude of financial investment however, has exceeded what should be sufficient to decrease the level of malnutrition. Although Mexico spent 758 million dollars in 1993 on food and nutrition programs, the World Bank (1994) estimates that about 405 to 675 million dollars should be able to care for 2.7 million malnourished children (Romo, Espinoza, and Salgado 1996). This ineffectiveness is due to the unstable economic and political conditions. As mentioned previously, the lack of proficient evaluations has also been a major factor in wasting resources as well as poor execution and impractical criteria for qualifying families (Romo 1996). A country with so few resources cannot afford to waste such a vast amount of money. Not only does this hurt the country financially, but it will be increasingly difficult to find the much needed political support if past attempts have failed with no documentation to show the reasons.

MEXICO'S PARALYZING INFRASTRUCTURE

Globalization has capitalized on Mexico's weak infrastructure and added health problems to the urban zones. An example of economic forces that effect daily food choices is the North American Free Trade Agreement (NAFTA) of 1994. NAFTA has made it cheaper to import corn than to grow it within Mexico and many times this imported corn is of much inferior quality and possibly transgenic. The traditional tortillas are made of corn, water, and ground limestone which provides the tortillas with a great deal of calcium. The tortillas made with imported corn and purchased flour use ground cobs and therefore lack the calcium content (Martin and Cerullo 2004). This is an obvious barrier that prevents the individual from being able to choose to be healthy.

Globalization's affect on the tortilla has not gone unnoticed. In fact, Mexicans are very disappointed with the decreased quality of the tortilla because the tortilla is of such great cultural significance. In Guadalajara, Mexico, Gabel and Boller (2003) studied the cultural importance of the tortilla through ethnographic research:

> From a symbolic perspective, tortillas, especially corn tortillas, are pride-inspiring symbols of the nation and its people. Moreover, tortillas and the corn they are made of are sacred to indigenous Mexicans and serve as a connection between modern-day Mexicans and their ancestors. (Gabel and Boller 2003, p. 135)

They found that both production and consumption of the tortilla are extremely important processes and that the Mexicans viewed the processed, plastic-wrapped tortillas to almost be a disgrace to the meaning of the tortilla. However, because of governmental forces, many of the small, local tortilla factories (tortillerías) are being closed because they cannot compete with the mass-produced tortillas sold in the supermarkets. Also, the mill owners are forced to buy Maseca flour which is a much cheaper and lower quality option than raw corn mixture.

I received similar results from the data collected from the questionnaires. Although my sample size was very small, I did notice some obvious trends as to the cultural perception of the tortilla. Seventy percent said that tortillas produced in a tortillería were much better than the tortillas sold in the supermarket. Of this group, 75 percent made an explicit reference to either the freshness of the local tortillas or to the better quality. One respondent said, "I buy them in the tortillería because they are freshly prepared and because they use corn; they process it, they grind it, and then make the tortillas. In the supermarket they use corn flour from Maseca and they aren't natural; they put in other additives." These respondents show an acute awareness about the difference in the tortillas. This shows how integral tortillas are to the Mexican "habitus." However, preference does not dictate where they actually buy their tortillas. Many of this subset said that they sometimes buy tortillas from the supermarket because they are easier to access and that many of the local tortillerías are closing down.

In 1990, President Salinas made an agreement with his family friend who was the chairman of the Maseca company. They agreed that the government would only provide flour mixtures which incidentally, only Maseca produced (Gabel and Boller 2003). The inferior quality of tortillas is disappointing to the Mexicans because the tortilla is bent and used as if it were an eating utensil. These changes in the tortillas are attributed to the forces of globalization in drastically altering the choices available for the consumer. Gabel and Boller (2003) explain that globalization is not a consumer-driven development but instead, "a process by which consumer choice is increasingly limited to a set of alternatives best allowing (agribusiness) TNC executives and major stockholders to maximize their personal and collective wealth" (p. 140). The tortilla is unfortunately subjected to the inhumane powers of globalization and has therefore lost some of its appeal and nutrient quality. This article also mentions future plans of fortifying these processed tortillas. This seems like an impractical, circular waste of resources. The traditional tortilla already had the desired quality and nutrient content. The tortilla would not need to be fortified in the first place if small mill owners were able to produce the traditional form that the consumers demand.

Instead of using the whole kernel, new technologies produce the tortillas from a corn-flour mix. The imported corn from the U.S. and Canada is of such inferior quality that some is actually made for animal feed. In 1995, the shippers marked the corn that was supposed to be sent for livestock with green dye. Shortly after, many of the tortilla factories started producing green tortillas (Ross 1999). Mexico's unstable social structure allows for incidents such as these to occur. While trying to fulfill the demand for the food supply by bringing corn into the country, Mexico could actually be adding to the malnutrition problems by decreasing the choices for the consumer to a greater extent.

A similar trend was seen with the production of oil. During the 1990s, Mexico also produced a vegetable oil that was of supreme quality. This oil had enormous health benefits, but was replaced by the very poor cottonseed oil. This cottonseed oil was imported because it cost half as much as the vegetable oil and forced the Mexican oil factories to completely close down (Camacho de Schmidt 2004). This is another example of the economic structure denying the choice of a healthier option for the consumer.

While many of the malnutrition problems are serious in rural zones where resources are scarce, the problem is growing in urban zones as well. Small rural farmers are unable to compete with the importation of grains and are forced to move into urban areas in order to find work. Many times this work is underpaid factory labor that significantly alters the lifestyle and structurally limits the workers' choices. Globalization is promoting more unhealthy lifestyles because the factory workers are only given a short lunch break instead of the traditional long lunch period. During this lunch break, cafeteria food is available and usually provides flour tortillas instead of the more nutritious corn tortillas. Although the flour tortilla may contain more vitamins, it is lower in calcium and has a much higher fat and sodium content (Saldana and Brown 1984; Martin and Cerullo 2004). Also, corn tortillas are made with water while flour tortillas are made with lard. With more Mexicans being forced into the sedentary, factory lifestyle, increased consumption of flour tortillas that are made with lard could contribute to health risks.

Not only has the lifestyle changed for rural individuals who come into the city, but for those still living in rural areas as well. About 70–80 percent of the income of small farmers comes from non-farm work (Martin and Cerullo 2004). These farmers are forced to find alternative means of income in order to survive. The effects of globalization present obvious forces upon the availability of choices for the individual. This evidence demonstrates how urban consumers and rural farmers alike are not able to access nutritious foods because of the greater political and economic forces that shape the availability of food options.

SHORT-TERM AND LONG-TERM RESOLUTIONS

The World Bank (1994) suggests that three actions need to take place simultaneously in order to decrease malnutrition: fortification of foods, allocation of supplements, and dietary change. These solutions are particularly relevant to my thesis because all of these procedures would be affecting lifestyle and would theoretically have an effect on the nutritional status of individuals. These suggestions demonstrate that the problems of unhealthy lifestyles and malnutrition are not as simple to solve as demanding that the individual consume healthier food. Although education is one of the resolutions, this alone will not suffice to improve health status. Focusing on Mexico as a case study gives ample evidence that, by limiting opportunity and access to certain foods, the structure of a society has a determinative influence on an individual's health.

While the World Bank's approach works at different areas in order to create an effective change, I want to simply mention that the resolutions do not touch on the creation and maintenance of malnutrition. Some of the approaches are indeed structural approaches but almost seem to give in to the lifestyles that are created by the social structure. These resolutions are needed in the first place because of the problems created by the broader social structure. Although the World Bank's goal is not to resolve these social problems, I only wish to mention this as a reminder that the problems stem from a much larger social system of constraints and that fortification, supplementation, and education would not be altering this structure.

Fortification of food stuffs is an inexpensive way to reach a large number of consumers. However, with globalization, diet is changing and presents the challenge in the future of being able to predict foods that many will consume. Consumer demand is important for effective fortification but if they are not educated, the demand is not realized. If the consumers do not demand the fortification, the food producers do not have any incentive to invest and sell fortified food. Because of this circular impediment, there must be structural enforcement or legislation that requires fortification of the food industry. Fortified foods must also not stray far from the original appearance and appeal of the food. Another hurdle is convincing the public to pay the extra money for the nutrient benefit (World Bank 1994). This would be quite a challenge in Mexico because so many families are struggling to buy the basic staple foods in the first place.

Another hindrance to fortification is being able to reach the truly vulnerable populations. Providing fortified food to infants might not have a significant effect if the food is not chosen carefully. Infants and children can only consume a limited quantity of staple foods and if the fortification is chosen in order to serve an adult population, then the nutritional status of infants and children may not be healthily achieved:

> The level of micronutrients needed to contribute significantly toward infant and young child nutrition requirements given the small quantities they consume would vastly exceed that needed for the rest of the population and could not be justified either by cost or safety, even if technologically feasible. (Lutter and Rivera 2003, p. 2947)

Also, for infants that are still breastfeeding, fortifying complementary foods (nonhuman, food sources) is necessary for the nutrients that are not received through breast milk. As an infant weans, the lipid intake normally decreases because the lipids come from the breast milk. These lipids are necessary in order to absorb fat-soluble vitamins. Breast milk itself is even quite variable because its quality depends on the health status of the mother which in many cases is very poor. The fortification of complementary foods was shown to decrease anemia by 6.5 percent of children under age five (Lutter and Rivera 2003). This demonstrates the potential of fortified foods to improve children's health, but would rely on the careful execution of fortification and an understanding of the vulnerable population's needs.

The World Bank emphasizes the importance of targeting vulnerable populations. As in fortification, supplementation requires the assessment of different groups' nutrient requirements and deficiencies. If a large population suffers from Vitamin A deficiency, then targeting a broader population, such as all pregnant women and children, will yield benefits. Also, utilizing pre-existing health clinics is an efficient way to dispense supplements. Regardless of the manner in which supplements are distributed, education on dosage amount and usage is critical to ensure benefits and also to prevent toxicity (World Bank 1994).

Allocating supplements would also require meticulous execution and thorough understanding of the culture. Just as choosing the correct foods to fortify is essential to the success in reaching the population, so too is the manner in which supplements are chosen to be distributed. The decisions about which supplements and how to make them available is important, but just as crucial are the mechanisms that assure consumption. The World Bank (1994) suggests that, "The actual uptake of supplements by the targeted populations requires trained, motivated health care workers who can communicate effectively with consumers to overcome fears, misinformation, and ignorance" (p. 20). Whether the supplements need to be injected or swallowed, a behavioral change will have to take place. This will require consistent communication that can only be accomplished with a thorough understanding of the culture.

Just as demand is important for fortification, it is even more essential for delivery of supplements. Other goals should be, "social marketing to raise demand and lend the program urgency, more aggressive targeting of populations, increased outreach, and improved quality of services are needed to raise and sustain the coverage of supplementation programs" (World Bank 1994, p. 21). Many of these approaches do not match well with Mexico's infrastructure because they require consistent motivation and stable political support. Mexico's history has shown that many of the programs have unstable bases and therefore end up wasting resources. Does this mean that other countries with more stable social systems would have to come in and implement the programs? This would then run the risk of misunderstandings of cultural beliefs and perhaps feelings of hostility toward outside intervention.

Structural change is useless if the policies and programs do not take into account the culture and perceptions of the population. An example of the failure to consider culture and belief systems is a small-scale feeding project that attempted to improve the nutritional status of a rural area through supplementation. This project extracted a leaf nutrient content (LNC) from alfalfa that, after being processed, could be added to foods to provide nutrients, protein, and vitamins. However, one of the program's setbacks was that the environmental climate was too harsh for the alfalfa to successfully grow during the summer months.

The characteristics of the citizens of the area in which the project took place presented other logistical barriers in carrying out the project. The residents were hesitant to accept developmental programs and the families were self-reliant and not community oriented. This made it difficult to establish and maintain local support for the project. The director also noticed that, "The traditional Mexican diet of corn tortillas, beans, a variety of fruits and vegetables and wholesome soups was being supplanted by a diet of white bread and mayonnaise sandwiches, cheap meats, watery noodle soups, flour tortillas and fewer fruits and vegetables" (Nelson 1989, p. 21). This diet probably contributed to the nutrient deficiencies of the population. Also, their attitude toward this new way of eating led them to view the supplementation as unusual and only for the poor. The failure to carefully assess the cultural beliefs of the village was the main barrier to the project's success:

> Perhaps they [the villagers] did not accept the fact that their children were undernourished and, even if they did, they might not have believed LNC to be an appropriate or realistic solution. . . . Before any new technology can be implemented in the field, it must first be ascertained if it is the appropriate one for the given culture and environment. (p. 22)

Ascertaining cultural beliefs is the first hurdle to overcome before attempting other procedures. Pelto and Backstrand (2003) discuss how power-related and belief-related factors have significant influences on nutrition in Mexico. Perhaps the more obvious factors are the economic differentials. However, because of the amount of useful data that can be extracted from economic factors, I think that belief-related factors have a tendency to be overlooked. Belief-related factors include education and psychological characteristics of human experience. The graphs below combine economic and educational factors in order to show that both have an effect on food consumption.

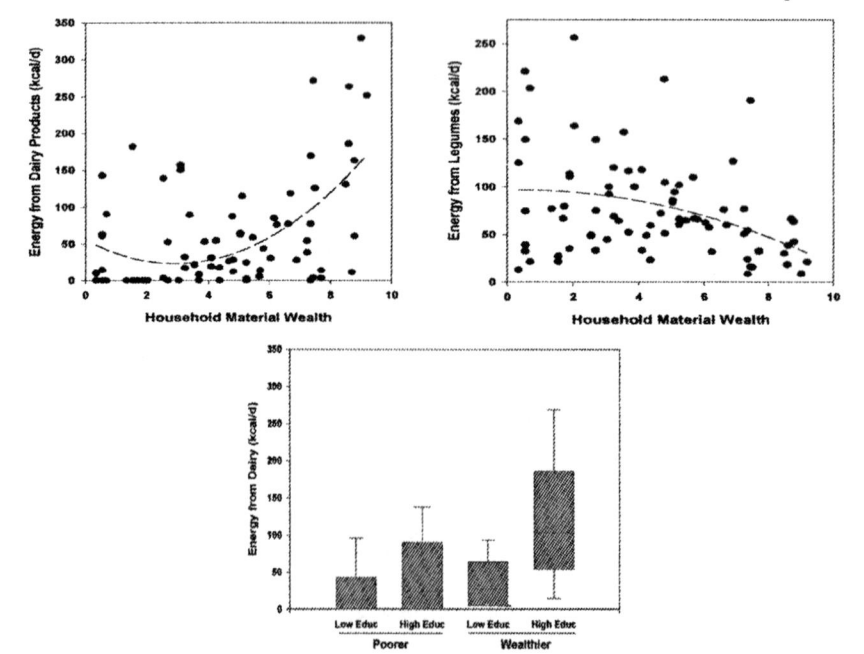

While it might be assumed that wealthier families consume more dairy products and legumes than do poorer families, the top two graphs show that this is not always the case. What accounts for this discrepancy? The bottom graph takes education into account and shows that higher levels of education will also promote increased dairy consumption. Also, in the graph on the top right, consumption of legumes is higher in families with less wealth. This could be a reflection of differences in lifestyle or in belief systems about what constitutes nutritious food. Perhaps wealthier families wish to invest economic resources in different areas of their lives, while poorer families do not even have this option and give priority to providing a healthy diet. Combining power-related and belief-related factors in research seeks to avoid reductionist conclusions about a problem and solution (Pelto and Backstrand 2003). Avoiding these conclusions will provide more practical information for policy change and nutrition programs.

Examining the culture within the household also yields applicable research. Messer (1997) provides ethnographic research based on intra-household allocation of food in order to partially account for the possible reasons for malnutrition amongst vulnerable populations. Factors that contribute to health status are, "economic and social resources, hygienic environment, cultural notions of what constitute adequate food, nutritional needs, and vulnerability to illness for different age and gender categories, and resource allocation rules" (p. 1676). Girls may be expected to be less active than boys, eat less, and be smaller. Therefore, they will consume smaller quantities of nutrients. Also, elderly individuals may receive preference in allocation of food.

Relating back to Bourdieu's theory, these characteristics of food allocation are a part of the habitus for these families because it occurs on a daily basis and therefore has become culturally acceptable. Although females may consume less than males, the cultural perspective on feeding makes the girls' diet adequate. Furthermore, the habitus could be an explanation for women's self-deprived eating patterns because they follow certain dietary rules: "females are purposely getting less but there is no discrimination or intent to deprive; rather, females perceive nutritional needs according to local understandings of nutrition, health, and development that may be nutritionally damaging from the scientific perspective" (Messer 1997, p. 1679). This would be an important thing to consider in regards to Mexico's previous subsidy programs. Perhaps the families were receiving free tortillas or other discounted staple foods, but when the food was brought into the household, were the children and mothers being deprived of the food? This could be a partial explanation as to why poor nutrition still exists. This is a prime example of the potential of anthropological research to give insight into practical health policies. If we understand that the individual's choices are limited and wish to penetrate the structure that produces the limited options, then it is necessary to understand the context of the individual's daily experiences.

Fortification requires a thorough understanding of a population's nutrition needs and lifestyle so that practical foods can be chosen to fortify. Supplementation requires an even more thorough understanding of the culture because the distribution process encompasses a great deal of personal interactions. Both of these strategies are solutions that attack the current health status. I see them as being temporary solutions because carrying them out after the vitamin deficiencies have been decreased suggests a cowardly surrender to the very limited dietary options. A long-term component needs to compliment the "quick" fixes of fortification and supplementation. Perhaps these two actions can continue on for extended periods, but energy and resources that are required to initiate the project cannot realistically last. Long-term health benefits can be accomplished through education. Provided that the resources are available, this would promote self-sufficient behaviors and help the individual to navigate the limited options in the future.

I found education to be a highly demanded way to improve health. Approximately 75 percent of those surveyed said that they would like to receive education as to what foods are healthy and what the important nutrients are. One respondent said that it would be a good idea to:

> Implement an educating culture about nutrition and making adequate, informative, and permanent campaigns so that people become conscious and know that many of their sicknesses are because they do not eat healthy food. What they eat is industrialized and with a lot of preservatives—it's like dead food.

The fact that so many are requesting education about the food they are consuming means that many individuals trust themselves more than government policies to improve their health. This exploratory research suggests that because education is desired, then it may be more acceptable than other measures such as fortification or supplementation. Globalization is introducing Mexicans with more options to junk food and processed, packaged food while limiting the traditional, healthy sources of nutrients. Education would at least create awareness that many packaged foods are not healthy.

Education alone, however, would not solve the dilemma because the traditional, dependable tortilla has been significantly altered. Fortification would help with the intake of essential nutrients but would not be easy to justify. Of the respondents who see a difference in the tortillas in the supermarket, 92 percent did not favor making changes to the tortilla through fortification ($p=.007$). This finding is not surprising because the Mexicans have had to deal with so many changes in governmental policies. Why would they accept yet another change that might worsen the quality of the tortilla even more? Perhaps the education that is so highly requested would be able to address some of the concerns that the people have regarding fortification. Detailing its use and effect on taste, quality, and price might make the fortification more acceptable. Is this a paternalistic way to improve the health of a population? Is choosing this method acceptable when the other options entail a restructuring of the political and socio-economic systems? Obviously this could not easily happen, so I argue that fortification and education are harmless mechanisms in decreasing malnutrition.

Once consumers understand and even witness the benefits from a dietary change, this would ideally be enough incentive to maintain healthy behaviors as long as access to the resources is available. Consumers must be aware of the severity of poor dietary options and/or choices in the first place and understand the implications for risks in the future. Ensuring durable change also requires policy change and enforcement of these changes. A meticulous analysis of the situation is essential before programs even start and should, "assess the coverage, quality, and cost of current efforts to remedy the problem, and it should also evaluate resources that could be marshaled in the future, including key food industries and markets" (World Bank 1994, p. 38). Implementing programs to improve diet requires investigation in many different areas from pre-existing cultural beliefs to setting realistic goals to intermittent evaluative procedures. Once again, Mexico's weak infrastructure presents some obstacles to the commitment and enforcement of what these procedures might entail.

CONCLUSION

Throughout this research, I have highlighted the ways in which Mexico's social structure drastically limits the individual's options in selecting a healthy diet. Does the acknowledgement of the power of the structure mean that the individual is eternally doomed? I do not wish to paralyze the individual anymore than the structure is already doing, but instead wish to create an awareness of its determinative power. Perhaps in the future, policies and programs will be implemented in order to decrease the level of social inequality. The suggestions for improving health status are not surrenders to this social structure, but suggestions to supersede malnutrition and intervene temporarily until real changes are made to the social structure.

The structure of society has an undeniable power over the amount of nutrient intake available for individuals. The points at which the individual has daily contact with the structure are the places to implement programs and to change dietary behaviors by creating more choice. Individuals do not have daily experiences with the officials responsible for creating trade agreements, but do have daily encounters with the tortillerías. This is an example of a location at which it is feasible to create more choices. Individuals also obviously come into daily contact with food. I highlight education as being an important strategy to make an impact on food consumption because not only is it greatly needed but is also desired.

This does not mean that the responsibility should then fall back on the individual. I can see the potential for the government to provide education and then abandon their efforts by claiming that the individuals are informed consumers. Education should instead be accompanied by adequate resources and mechanisms that make food choice practical. There seems to be a natural human reluctance to invest time and money into any project that does not show quick results. This is especially true in politics because if the public does not see any noticeable changes, then public support decreases. Policy makers' apprehensiveness to create projects that will only have results over a long period of time will be costly for the future. Fortification, supplementation, and education each have benefits and drawbacks and require varying amounts of effort, commitment, and resources. However, in order to be effective, the three strategies suggested by the World Bank need to be carried out simultaneously. Not only will it be costly because the social system will have to invest resources into trying to help sick individuals, but also will have repercussions on the outlook for national development. With a large proportion of the population suffering from illnesses and nutrition deficiencies, few will be left to reach their full potential in the education system and in the work force. I leave this open-ended question: where will the resources come from to create this change? Whoever takes charge of this large project should attack the future problems now before the future becomes the present.

BIBLIOGRAPHY

Balderas-Gonzales, Juventino. 1999. *Evaluating the Removal of the Corn Tortilla Subsidy in Mexico.* Thesis submitted for the Graduate School of the University of Colorado.

Barquera, Simón, Juan Rivera-Dommarco, and Alejandro Gasca-García. 2001. "Food and Nutrition Policies and Programs in Mexico." *Salud Pública de Mexico*, 43: September–October.

Camacho de Schmidt, Aurora. Personal Communication, May 5, 2004.

Cockerham, William C. "The Sociology of Health Behavior and Health Lifestyles." In *Handbook of Medical Sociology*, ed. C. Bird, P. Conrad, and A. Fremont. New Jersey: Prentice Hall, pp. 159–172.

Cortés, Luz María, David Cantero, and Reina Herrera. 1996. "Disponibilidad, consume de alimentos y nutrición en México y Cuba en el unbral del siglo XXI." *Acta Sociológica*, 17: pp. 135–156.

Gabel, Terrance G, and Gregory W. Boller. 2003. "A Preliminary Look at the Globalization of the Tortilla in Mexico." *Advances in Consumer Research*, 30: pp. 135–141.

Levy, Teresa Shamah, Abelardo Curiel, Lucía Nasu, Adolfo Villasana, Marco Avila Arcos, Carlota Mendoza. 2003. "El subsidio a la tortilla en México: ¿un programa nutricional o económico?" *Archivos Latinoamericanos de Nutrició,* 53: pp. 5–13.

López, Héctor, Héctor Sánchez-Pérez, Magdiel Ruiz-Flores, and Michael Fuller. 1999. "Social Inequalities and Health in Rural Chiapas, Mexico: Agricultural Economy, Nutrition, and Child Health in La Fraylesca Region." *Cuadernos de Slaud Pública*, 15: pp. 261–270.

Lutter, Chessa K., and Juan A. Rivera. 2003. "Nutrient Composition for Fortified Complementary Foods." *Journal of Nutrition*, 133: pp. 2941S–2949S.

Martin, Debra L. and Margaret Cerullo. 2004. "Feeding the Family in Troubled Times." National Science Foundation and Hampshire College. http://carbon.hampshire.edu/~nsfmexico.

Messer, Ellen. 1997. "Intra-household Allocation of Food and Health Care: Current Findings and Understandings—Introduction." *Social Science and Medicine*, 44: pp. 1675–1684.

Ochoa, Enrique C. 2000. *Feeding Mexico: The Political Uses of Food since 1910.* Delaware: Scholarly Resources Inc. Imprint.

Pelto, Gretel H., and Jeffrey R. Backstrand. 2003. "Symposium: Beliefs, Power and the State of Nutrition: Integrating Social Science Perspectives in Nutrition Interventions." *The Journal of Nutrition*, 133: pp. 297S–300S.

Rivera, Juan A., and Jaime S. Amor. 2003. "Conclusions from the Mexican National Nutrition Survey 1999: Translating Results into Nutrition Policy." *Salud Publica de Mexico*, 45: pp. S565–575.

Romo, Sara Elena, Enrique Ríos Espinosa, and Homero Martínez Salgado. 1996. "Losprogramas de ayuda alimentaria como respuesta gubernamental ante la problemática nutricional en México." *Acta Sociológic*, 17: pp. 61–89.

Ross, John. 1999. "Tortilla Wars." *The Progressive*, 63: p. 4.

Saldana, Guadalupe, and Harold E. Brown. 1984. "Nutritional Composition of Corn and-Flour." *Journal of Food Science*, 49: pp. 1202–1203.

Skoufias, Emmanuel, Benjamin Davis, and Sergio de la Vega. 2001. "Targeting the Poor in Mexico: An Evaluation of the Selection of Households into PROGRESA." *World Development*, 29: pp. 1769–1784.

Smith, James F. 1996. "Mexico Clamps Down on Tortilla Prices." *Los Angeles Times*, January 7.

The World Bank. 1994. *Enriching Lives: Overcoming Vitamin and Mineral Malnutrition in Developing Countries*. Washington D.C.: The International Bank for Reconstruction and Development.

Tuckman, Jo. 2003. "US Lifestyles Blamed for Obesity Epidemic Sweeping Mexico: Latin American Countries Succumb to Fast Food and Sedentary Behavior." *The Guardian*, August 13, p. 13.

ETHICS AND THE LIFE SCIENCES

GERM-LINE GENETIC ENHANCEMENT AND RAWLSIAN PRIMARY GOODS

FRITZ ALLHOFF
WESTERN MICHIGAN UNIVERSITY

ABSTRACT: Genetic interventions raise a host of moral issues and, of its various species, germ-line genetic enhancement is the most morally contentious. This paper surveys various arguments against germ-line enhancement and attempts to demonstrate their inadequacies. A positive argument is advanced in favor of certain forms of germ-line enhancements, which holds that they are morally permissible if and only if they augment Rawlsian primary goods, either directly or by facilitating their acquisition.

The moral permissibility of genetic intervention in humans is met with considerable skepticism. Of all the interventions, germ-line genetic enhancement is considered the most morally contentious. In this paper, I argue for the moral permissibility of some forms of germ- line genetic enhancement. I first consider several arguments against genetic intervention more generally, all of which I think either can be rejected outright or are only contingently valid given scientific limitations. I further suggest that the latter class of arguments will become impotent as the relevant technologies develop over time. I then present a positive argument in favor of certain germ-line genetic enhancements, which holds that such interventions are morally permissible if and only if they serve to augment Rawlsian primary goods, either directly or by facilitating their acquisition.

In discussing human genetic intervention, it is helpful to acknowledge two common distinctions. First, one can distinguish between somatic and germ-line cells. Somatic cells, such as skin or muscle cells, contain twenty-three chromosomal pairs and do not transmit genetic information to succeeding generations. Germ-line cells, which are the egg and the sperm cells, contain twenty-three unpaired chromosomes and provide genetic information to offspring, as well as to the future generations descended from those offspring. Second, one can distinguish between genetic therapy and genetic enhancement—alternatively referred to as negative and positive genetic engineering, respectively, although these expressions have become less fashionable. Genetic therapy aims at the treatment or prevention of a disease, whereas genetic enhancement aims at the enhancement of some capability or trait. Accepting these distinctions, there are four categories of genetic intervention: somatic cell gene therapy, somatic cell genetic enhancement, germ-line gene therapy, and germ-line genetic enhancement.[1]

Presumably, ethical enquiry could proceed independently for each of these categories. Somatic cell intervention, which affects only the subject of the intervention, is most likely less morally contentious than germ-line intervention, which will affect all future generations (in the absence of any further interventions). Germ-line intervention is consequently more extreme in scope and, furthermore, will come to affect individuals who have not consented to the procedure. One also might think that gene therapy is less morally contentious than genetic enhancement. Various arguments have been advanced in favor of this position, which include, but are not limited to: genetic enhancement is closer to "playing God," runs contrary to the goals of medicine, risks the priority of love for genome over love for child, and so forth. Although these arguments are of varying merit, it is at least worth observing that most of us feel that curing Huntington's chorea through germ-line gene therapy is less morally problematic—and more morally incumbent—than creating taller or faster children through genetic enhancement. I feel that the strength of this intuition alone provides good reason to suspect that genetic therapy is less morally contentious than genetic enhancement. Given these two results, one might postulate the following order, from least morally objectionable to most morally objectionable: somatic gene therapy, germ-line gene therapy or somatic cell genetic enhancement, and germ-line genetic enhancement. In the literature, the endorsement of these practices generally has followed the expected pattern; somatic gene therapy is the least contentious, germ-line genetic enhancement is the most contentious, and, although there are substantial debates on the intermediate practices, it seems most bioethicists, under appropriate conditions, ultimately would support both germ-line gene therapy and somatic cell genetic enhancement.

It is worth noting that the therapy/enhancement—or, alternatively, the positive/negative—distinction can be challenged. For example, an individual might be depressed either by virtue of some genetic defect, such as limited serotonin production, or because she happens to have a difficult life—e.g., death of several close relatives, loss of job, and the like. In the latter case, she could be genetically "normal" and any genetic intervention would constitute enhancement rather than therapy. However, in the former case, intervention would constitute therapy. Insofar as the therapy/ enhancement distinction is supposed to be morally relevant, one might question it on the grounds that it suggests that same genetic intervention is deemed more morally appropriate in one case than in the other. Certainly the depression could be qualitatively identical for both women and each is entitled to relief; the therapy/enhancement distinction focuses solely on the etiology of the affliction and wholly ignores the degree of suffering. Furthermore, this distinction inevitably leaves one with hard cases in which it is difficult to ascertain whether a given intervention qualifies as a treatment or as an enhancement.[2]

Nevertheless, I think it is safe to assume that most bioethicists accept this distinction, although perhaps hesitantly or in a qualified manner. In many, or even most, cases, one legitimately can view the therapy/enhancement distinction as morally relevant, and, in most cases, it will not be tremendously difficult to draw the line between what constitutes therapy and what constitutes enhancement. When difficult cases do arise, principles exist to help make the determinations. Norman Daniels, for example, has argued for the use of "quasistatistical concepts of 'normality' to argue that any intervention designed to restore or preserve a species- typical level of functioning for an individual should count as treatment, leaving only those that would give individuals capabilities beyond the range of normal human variation to fall outside the pale as enhancement" (see Juengst and Walters 2003, p. 579; also Daniels 1992). Alternatively, Eric Juengst (1997) has proposed that therapies aim at pathologies that compromise health, whereas enhancements aim at improvements that are not health-related. In most cases, either proposal will help determine what would count as a treatment and what would count as an enhancement. So, although the therapy/enhancement distinction is not above discussion, I propose to accept it for the sake of this paper, particularly because of its general intuitive resonance and its conceptual appeal.

Finally, discussions of genetic intervention frequently begin—and sometimes even end—with an evaluation of scientific limitations. There is no doubt that genetic intervention is currently in its nascent stages and, although some reasons exist to be optimistic about its future, there still are *tremendous* obstacles that must be overcome before it will be able to fulfill its potential promise. These obstacles include, but probably are not limited to, our limited knowledge regarding the human genome and the functioning of individual genes, our limited knowledge of the optimal procedural techniques, and our inability to address—or even conceive of—the inordinate economic costs of both research and practice. These issues are so daunting that it has been suggested that no "serious organized discussions" regarding genetic intervention can even take place for several years to come (Neel 1993).[3]

Although these issues obviously are critical to the ultimate success of genetic intervention, I fail to see how they impugn upon enquiry into its moral dimension. Furthermore, it is imperative that this *moral* dimension be investigated *before* the scientific limitations are overcome. As we have recently seen in the cloning debate, it is possible that the science can develop ahead of—or independently of—a healthy moral discourse.[4] When this happens, society is left to "catch up" from both a moral and a policy standpoint; advance discussion could preclude this disadvantage and ensure the appropriate readiness for scientific advance. Finally, in the case of genetic intervention, it is certainly not the case that any dialogue will be conceptually bankrupt or be wholly fruitless—we have a good idea of the relevant issues, and we can get to work on those now.

ARGUMENTS AGAINST (GERM-LINE) GENETIC ENHANCEMENT

Of the four categories of genetic intervention introduced above, I think that it is fairly clear that germ-line enhancement is the most morally contentious. Certainly, there are those who would object to the other categories. However, these objections often reflect theological concern, scientific limitations (see, e.g., Danks 1994),[5] or involved risk (see, e.g., Anderson 1990).[6] Insofar as we live in an increasingly secular society, this first type of concern may be partially allayed or marginalized. The second type of concern will abate in the wake of scientific progress, which is bound to ensue; in the event that scientific limitations do persist, this objection would remain valid, although only contingently so. Finally, no rational proponent of genetic intervention would propose that we move forward when the potential risks outweigh the potential costs, a development that one might reasonably assume to be inevitable, even if distant. (In his classic book, Jonathan Glover (1984, pp. 42–43) proposes a "principle of caution" that would hold that "we should alter genes only where we have strong reasons for thinking the risk of disaster is very small, and where the benefit is great enough to justify the risk." There might, of course, be epistemic issues but, conceptually, this is, I think, right.)

But, although there seems to be general support for somatic cell interventions and germ-line therapy, germ-line enhancement remains highly contentious, and perhaps even unpopular (see, e.g., Presidents Commission 1982; President's Council 2003). One problem with enhancement in general, as opposed to therapy, is that the scientific prospects are extremely daunting. Treatment of Huntington's chorea, for example, requires intervention at a single and known genetic locus. And, in fact, many treatments would involve intervention at single genetic loci. Enhancements, however, are far more difficult since hardly any human trait or capability is influenced by one, or even a small number, of genes.[7] Not only does the number of relevant genes frequently approach or exceed 100, the genes also contribute unequally to the development of any given trait or capability (Clark and Grunstein 2000, pp. 82–94). For an enhancement to work, not only must all, or at least many, of the genetic loci be established, but their relative contributions must be ascertained. This is a *tremendous* scientific obstacle, and one that is not likely to be overcome in the near future. Nevertheless, I am concerned with the moral, rather than the scientific, problems of germ-line enhancement. Again, I assume that the scientific limitations eventually will be overcome. Once these developments are realized, what moral objections will remain?

One argument offered to impugn the moral legitimacy of genetic enhancement holds that such practice inevitably would lead to unjust outcomes. Although different opinions exist as to what constitutes this injustice, a plausible suggestion is that genetic enhancement would be available only to a wealthy few and not to the vast majority of society. These wealthy few could augment their own abilities, or those of their progeny, through genetic enhancement and consequently widen the gap between the rich and the poor and/or leave the "have-nots" unable to compete

GERM-LINE ENHANCEMENT AND PRIMARY GOODS

in an integrated society. This injustice should be avoided; genetic enhancement, which would lead to this injustice, is therefore morally problematic.

In response to this argument, I propose that we differentiate between genetic enhancement *itself* and its *distribution*. Insofar as the preceding scenario is unjust, society could adopt a different pattern of distribution. On the pure libertarian model, the rich presumably could claim some entitlement to their resources and pursue genetic enhancement that would lead to the above effects (Nozick 1974). But other models of distribution exist. A Rawlsian, for example, could argue that the upper class may make themselves better off—as measured against a battery of primary goods that could include genetics—only insofar as they improve the situation of the least well-off class (Rawls 1999). Perhaps the genetic enhancements of the few would create a larger social product such that everyone would benefit. Or perhaps genetic enhancements for the wealthy would only be permissible if the wealthy subsidized the genetic enhancements of the nonwealthy. There are, of course, other distributive schemes as well. One could hold lotteries for genetic enhancements; nobody would be allowed to be genetically enhanced unless s/he won the lottery.[8] Perhaps part of the tax revenue could fund the procedures in the event that the lottery winner was unable to pay for the intervention.

Regardless, the obvious point is that genetic enhancement procedures alone will not lead to unjust results; there would have to be an unjust distributive scheme to enable the injustice to come about. If we can determine what constitutes a just distributive scheme, then genetic enhancement, as a good or service, can be distributed according to the principles of that scheme. So, in response to the would-be critic, I would point out that it was never the justice of genetic enhancement to which she was objecting, but rather to a specific distributive scheme. Of course, it is the case that the distributive scheme currently in place in the United States could, and perhaps would, lead to distributions of genetic enhancement that some people consider unjust. But this could change; we are many years ahead of the viability of genetic enhancement and have time to prepare for its disbursement by making appropriate policy adjustments.

There are, however, several plausible arguments that could be made directly against genetic enhancement in general or germ-line genetic enhancement more specifically. Some of these have received formulation from Erik Parens (1995; 1998) who, I think, makes the arguments about as compelling as they can be made. First, Parens considers whether genetic enhancement will compromise important facts about human existence and consequently detract from some of the aesthetic value of human experience. If, for example, genetic engineering could be used to speed up aging and circumvent the turmoil of adolescence or allay the pains of growing old, we might feel compelled to use it for these purposes. But he thinks that these processes, or our "fragility" more generally, have value and that there is at least a *prima facie* problem with interfering with them.

Clearly Parens holds a worldview that some would deny; at least it is not intuitively obvious that there is necessarily anything bad about accelerated aging, nor that there is necessarily anything good about aging, death, or human fragility. Rather, many people might think that the use of genetic enhancement to address these issues would be valuable. But other people would agree with Parens. Nevertheless, this does not impugn genetic enhancement *in general*, but rather only its use for those purposes that society, through consensus, determines to be unpalatable. In the next section, I offer a positive argument in favor of specific forms of genetic enhancement, but here it is sufficient to observe that it need not be either permitted or banned categorically; presumably, there are moral principles that can help to determine which sorts of genetic intervention should be allowed. Even Parens grants this; his point is only to establish that in *some* cases, there are reasons to think that genetic enhancement is a bad idea. This is no reason, however, to think that there are *no* cases in which genetic enhancement would be valuable.

Second, Parens wonders whether genetic enhancement would detract from accomplishments, making them less noteworthy or laudable. To use his example, we are less impressed with an athlete who performs with steroids than another athlete who turns in a comparable performance without the help of steroids. Similarly, one can imagine two athletes, one with genetic enhancement and one without. Presumably the accomplishments of the latter would be more impressive than the comparable accomplishments of the former.[9] The former accomplished the same results with fewer "resources" and is therefore more deserving of respect and awe. The question, however, is whether the potential to undermine accomplishment is a legitimate reason to avoid genetic enhancement.

I do not think so. First, successive generations always have more resources available to them than previous generations, and this fact alone does not mean that the current generation is less deserving of respect or awe than previous ones merely because its members have accomplished what they have with more resources. Consider, for example, the Olympians of classical Greece. Their athletic accomplishments, although tremendous at the time, could be duplicated now by even the most average intercollegiate athlete. In the intervening millennia, we have amassed a huge knowledge regarding nutrition, training techniques, sports medicine, and the like. This corpus of knowledge allows athletes to train in much more sophisticated manners and to perform in ways that would have been impossible long ago.

Certainly we would not want to say that current athletes *accomplish* less simply because they have advantages over their ancient Greek counterparts. More logically, one might say that the *standards of evaluation* have changed; the ancient Greeks were judged against certain criteria (relative to what was possible and expected at the time), and current athletes are judged against different criteria (relative to what is possible and expected now). Excellence and accomplishment is measured relative to some standard, and that standard is dynamic. The existence of comparative advantage cannot preclude accomplishment, it can only affect the standard against which accomplishment is measured.

Although I think that this argument alone shows that advantages do not impugn accomplishment, the same point can be made more specifically with respect to genetic—versus epistemic or social—advantages. Obviously many current professional athletes have genetic constitutions that give them tremendous advantages over the rest of us. Nevertheless, this does not diminish the respect and awe that we afford them. Certainly I do not watch Michael Jordan execute a 360° slam dunk and denigrate the feat on the grounds that his genetics made it more likely that his 6'6" frame could pull it off while my 6'0" frame had little hope. Obviously these athletes train exceptionally hard but, even if most of us were to train with comparable intensity, we would be unable to perform at their levels. The reason is that many of their talents and capabilities are genetically endowed or enabled and, as such, are beyond the reach of the average person. So I would maintain that we respect athletes *regardless* of their genetic superiority. If this is right, then genetic enhancement should not be morally contentious *merely* by virtue of the fact that it would augment performance.

A plausible response to this line of reasoning is to posit a morally relevant distinction between natural and "unnatural" (in the sense of interventionistic) genetic advantages. The reason that one scoffs at the sprinter on steroids is because he *cheated* by augmenting his actual abilities by taking drugs. Michael Jordan on the other hand, merely utilizes his natural endowments, and therefore is not deserving of any disapprobation. If this is true, then it might be a reason to disallow "unnatural" genetic enhancement. But I think that this conclusion misses an important point: most of our talents and abilities are developed through intervention. One certainly would not say that the accomplishments of a leading moral philosopher are less impressive because he was trained at the best institutions and under the best professors. Nor would one say of an athlete that her accomplishments are less valuable because she spends so much time in the gym and on the track (see Glover 1984, p. 45).

The critic would have to hold that education and exercise are morally legitimate interventions—or serve to augment our abilities and talents in a morally legitimate way—whereas genetic interventions are not. And what justification could be offered for this nonhomogenous theory regarding the moral legitimacy of intervention? That the genome is "special"? That education and exercise facilitate developments allowed by one's natural genetic endowments whereas genetic enhancement changes those endowments and is therefore immoral? Certainly these arguments would not be ridiculous, but they seem, at a minimum, to be *ad hoc* and/ or undermotivated. Furthermore, they seem to be based on the premise that "natural is good," which is probably false. For example, many natural phenomena, such as aging, disease, rape, and murder are *not* good. And, to return to a continuing theme, even *if* one thought that some genetic enhancement would undermine achievement, a conclusion I am inclined to deny, it does not logically follow that *all* genetic enhancements would be ruled out.

One final argument to consider against genetic enhancement is whether genetic enhancement is problematic because its pursuit demonstrates a failure to accept our place in nature. There is, I think, a sentiment that genetic enhancement is nefarious on the grounds that it shows a discontent with what humans have been given—whether through divine providence or through natural selection—and that this sort of dissatisfaction is unpalatable. On this line of reasoning, it is better to accept our limitations and to be content with what we have than to try to change what we are. Relatedly, we can ask what sort of people we would be if we tried to alter ourselves. Perhaps trying to do so would be demonstrative of inhumility or could be characterized as vicious in some other way.[10]

I do not think that either of these suggestions can undermine the moral legitimacy of genetic enhancement. With regard to the first, I simply do not view the pursuit of genetic enhancement as an expression of dissatisfaction with ourselves. Rather, I look at it as a chance to *improve* ourselves, and improvement is certainly not morally dubious. We undergo all sorts of processes that aim to improve the human experience: we pursue education; we exercise; we do research. Why should genetic intervention be viewed any differently? As I said above, I do not see any compelling moral distinction that can be drawn. Would it be because, most fundamentally, our genomes are what we "are," and other pursuits do not affect this fundamental fact about ourselves? Certainly I do not think that my genome is what constitutes me. I view myself as a rational, autonomous agent capable of pursuing ends that I find valuable. I also disagree that genetic enhancement would be vicious. Genetic enhancement would aim, *most fundamentally*, at the improvement of the human experience. What end could possibly be more noble than that? Genetic enhancement surely could be applied to nefarious ends, and we always must be vigilant against such abuses. But there is no reason to think that it would be intrinsically evil, the only way that genetic intervention could manifest evil would be through misapplication.

THE POSITIVE ARGUMENT

Thus far, I have attempted to reject or undermine arguments that could be made against genetic enhancement. But an important task remains: to establish a strong argument in favor of it. Probably the most obvious way to argue for genetic enhancement would be to proceed on utilitarian grounds. Look, one might say, this practice has the potential to increase human capabilities that, when exercised, will increase total aggregate happiness. If people can be made smarter, for example, they could do more good with less effort and in less time. Certainly any means to this end is, at least *prima facie*, morally laudable. Of course, the critic would want to interject that there is tremendous potential for disutility: there is scientific risk and uncertainty; there is the problem of distribution; and so forth. But, as I already have argued, these factors need not impugn the hedonic calculus. Nobody has argued that genetic intervention should proceed when the risks outweigh the benefits. Rather, we must wait, however long, until the balance sheet comes up positive, and there is no reason to think that this will not happen eventually. And, if certain distributive schemes will lead to strife and angst, then society must adopt a differing distributive scheme—surely at least one will mitigate these disutilities. Given these responses, I am quite confident that genetic enhancement could be justified quite easily on utilitarian grounds.

Nevertheless, I think that one can do better than a simple utilitarian argument. For one thing, many people are not utilitarians, so they will be unlikely to be swayed by such a line of reasoning. In addition, utilitarianism can be rather insensitive to the notions of rights and of justice, both of which should be taken seriously. To this end, one might want to consider future generations more directly, particularly as pertains to one moral feature frequently evoked against germ-line genetic enhancement: the fact that their consent is never given for the interventions (see, e.g., Lappé 1991).[11] Presumably this concern derives from Kant's moral philosophy, most specifically from the second formulation of the categorical imperative and its edict to treat individuals as ends and never merely as means or, more simply, to treat them in ways to which they would rationally consent. Technically, of course, unborn generations lack rational nature and therefore do not participate in humanity. Therefore, I think, Kant would not object to genetic intervention on the grounds that it fails to respect their (nonexistent) autonomy. Regardless, one certainly may adopt a neo-Kantian line of reasoning wherein all humanity (present and future) should be treated ways to which its members would rationally consent, were they able to do so.

Germ-line genetic enhancement will, of course, affect all future generations—although perhaps to a diminishing extent as the genetic contribution from a single generation dissipates in future generations—as long as its effects are not reversed or superceded by other interventions. So, to honor the Kantian principle, genetic intervention would be morally permissible only if *every* future generation would rationally consent to the genetic alterations made in the germ-line. Can this criterion be satisfied? Absolutely. More interestingly, however, the answer suggests which sorts of germ-line enhancements are morally permissible and which are not.

To see why some enhancements are morally permissible, I invoke John Rawls's notion of primary goods. Primary goods are those things that every rational person should value, regardless of his conception of the good: rights, liberties, opportunities, income and wealth, health, intelligence, imagination, and the like (Rawls 1999, pp. 54–55). These are the things that, *ex hypothesi*, everyone should want; it would be *irrational* to turn them down when offered. Nobody could be better off with less health or with fewer talents, for example, regardless of her life goals.[12] Building off of Rawls's concept, I propose that germ-line genetic enhancements are morally permissible *if and only if* they augment primary goods or create abilities that would lead to their augmentation.

To defend this claim, I must establish both directions of the biconditional. First, consider whether germ-line genetic enhancements are morally permissible *if* they augment primary goods. The notion of moral permissibility I am concerned with here is the Kantian one, so the central question is whether those affected by the genetic interventions would rationally consent to the enhancements, were they able to do so. Since primary goods are those that, by definition, any rational agent would want regardless of his conception of the good, *all rational agents would consent to augmentation of their primary goods*. Because rational consent is sufficient for moral permissibility, on the Kantian model, one direction of the biconditional is established.

Perhaps more interestingly, one even could argue that these enhancements would be *required* by Kant. Kant speaks of a duty to develop one's own talents; so, if genetic enhancement would consist in the development of talents, Kant might consider its pursuit to be a duty. Of course, in this case, it would not be *self*-development of talents, which is the case Kant considered, but rather someone else developing the talents of the (potential moral) agent. Nevertheless, I think that failure to develop the talents of one's progeny plausibly could yield a contradiction in will, and, if so, genetic enhancements would be required by Kant. But the prospect of a moral obligation to enhance is made even stronger by Rawls (1999), who writes: "[I]n the original position . . . the parties want to insure for their descendants the best genetic endowment. . . . The pursuit of reasonable policies in this regard is something that earlier generations owe to later ones." Although I am tempted by these lines of argumentation, I limit my investigation here to the *permissibility* of genetic enhancement.

To establish the second half of the biconditional, I must show that genetic enhancements are morally permissible *only if* they augment primary goods. In doing this, I also address a common and legitimate concern against germ-line genetic enhancement, namely that it will lead to the creation of "designer babies," such that parents might try to create, for example, the next Michael Jordan. Certainly such a use would be unpalatable. However, many of the enhancements requisite for the creation of Michael Jordanesque abilities do *not* satisfy my criterion of augmenting primary goods. For example, many great basketball players are quite tall, and prospective parents might contemplate editing their germ lines to increase the potential height of their progeny. But height is not a primary good: rational people easily could prefer not to be tall. Although height helps basketball players, it can be inconvenient with respect to entering doorways, physiological strains, and unwanted attention. Skin color, eye color, and even sex would be off limits as well for genetic intervention since none of these features constitute primary goods nor would necessarily lead to their acquisition. Rational people could disagree as to which instantiations of these characteristics would be most valuable. But why would these interventions be *immoral*? Since interventions that augmented these non-primary goods would not be desired by all rational agents, not all rational agents would consent to them. Because rational consent is a necessary condition for moral permissibility, these interventions would be immoral. Having now established the second direction of the biconditional, I have shown that genetic enhancements are morally permissible *if and only if* they augment primary goods, or create abilities that would lead to their augmentation.

For emphasis, I shall briefly reflect on the extent of interventions that *would* be morally permissible on my view. Certainly any intervention that would make progeny more healthy is morally permissible.[13] But I want to extend the argument beyond the permissibility of mere therapies (such as curing Huntington's chorea) and to authorize enhancements. Genetically engineering greater resistance to disease obviously would be morally permissible, even if the subject of the intervention already has a level of disease resistance commensurate with that of a "normal" human. There are a number of other physical enhancements that I think also would be morally permissible insofar as they would augment talents or capabilities: for example, improvements to eyesight, speed, strength, and the like. All of these characteristics have genetic bases, although environment obviously plays a substantial role as well, and I think that enhancements aimed at improving them are morally permissible. Furthermore, there are a number of mental characteristics that contribute both to our talents and to our overall intelligence. If there ever were a way to enhance mental acuity, mathematical and spatial reasoning, language faculties, creativity, musical abilities, and the like, I would propose that we should do so.[14]

Obviously these enhancements are, for now, scientifically impossible. And, to vanquish the specter of genetic determinism, I willingly concede that genetics are only part of the picture. The talents, capabilities, and physical traits of an adult cannot be read off of his or her genome; we have free will to decide which of our potential talents we choose to develop,[15] and our environments affect our physical traits as well as our opportunities to develop talents and capabilities.[16] Nevertheless, genetics can confer at least the potential to develop in certain ways and some of these potentialities are more valuable than others, both prudentially and morally.

In conclusion, I think a strong argument can be made to support the moral permissibility of certain types of genetic enhancement in general and germ-line genetic enhancement in particular. Specifically, such interventions are morally permissible if and only if they serve to augment Rawlsian primary goods, either directly or by facilitating their acquisition.

NOTES

This paper originally was presented at Monash University's Department of Philosophy and Centre for Bioethics, and I thank its faculty and graduate students for helpful comments and challenging criticisms. I also thank two anonymous reviewers for their comments and suggestions.

1. Beauchamp and Walters (2003, pp. 454–456) provide a good introduction to genetic intervention.

2. For a an excellent and detailed discussion of the therapy/enhancement distinction, see Buchanan et al. (2000, pp. 104–155).

3. Neel is particularly concerned with the dialogue over germ-line therapy, although I think that he would not object to the extension of his view to genetic enhancement as well. See also Walters and Palmer (1997, chap. 3) for another response to Neel.

4. More specifically, I am thinking here of the surprise announcement of Dolly's creation and the quick and uncritical condemnation issued against cloning both by the public and by politicians. If there had been a more open forum prior to this announcement, I suspect that the reaction would have been quite different. Alternatively, the open forum might have generated a critically informed indictment of cloning in which case there might have been legitimate pressures to abandon research.

5. For an assessment of the existing scientific possibilities, and consideration of some moral arguments, see Gordon (1999). Gordon's article is, however, a few years old and scientific possibilities are growing rapidly.

6. Anderson also argues against the moral dangers of genetic intervention.

7. The fact that (most) enhancements would require intervention at more genetic loci than (most) therapies is even intuitively obvious. Consider a car: there are a vast number of singular malfunctions that could cripple a car's normal functioning, yet there are very few, if any, singular interventions that could enhance it. To substantially improve a car most likely would require redesign of at least one entire system. Similarly, with human genetics, a vast number of malfunctions could substantially lower the welfare of an individual (and therefore open the door for genetic therapy), yet there are few singular interventions that could substantially *enhance* an individual's welfare—to use the vernacular of the biologist, most traits are highly multigenic and quantitative.

8. This idea has been proposed by Maxwell Mehlman and Jeffrey Botkin (1998) and further developed in Mehlman (2000).

9. Although I am concerned particularly with whether genetic enhancement can be said to undermine accomplishment, Juengst and Walters (2003) offer a related discussion as to whether genetic enhancement constitutes a form of cheating. They are inclined to think that it does not, but admit that it might force institutions—be they athletic, educational, and the like—to reconceive their standards of evaluation, to redesign the "game," or else to prohibit certain enhancements.

10. In an article that I very much like, Thomas Hill (1983) argues for environmental consciousness on the grounds that we would not be good people if we did not care about our environment. If, for example, we decimate forests to build strip malls, we display an arrogance and an attitude that is morally blameworthy. Although I agree with his argument as pertains to the environment, I do not think that it can be extended to impugn genetic enhancement.

11. Juengst and Walters (2003) offer a good discussion of other rights-related issues for germ-line intervention.

12. Someone might object to the categorical value of some primary goods, such as intelligence or talents, on the grounds that their possession brings about higher expectations and greater pressures. Although such an objection risks regression to Mill's swine, one, nevertheless, could grant that, insofar as expectations and pressures are undesirable, the goods that would lead to them are not primary since some rational agent would disapprove of them. Even though this concession might undermine certain proposed goods, such as intelligence, it is unlikely to impugn others, such as health. But, more fundamentally, I am tremendously skeptical that *any* of Rawls's proposed primary goods could make someone worse off. Although one might not like the pressures associated with increased intelligence, for example, the intelligence *itself* is certainly desirable.

13. Walters and Palmer (1997) also argue that interventions aimed at health are morally permissible. However, I think additional enhancements are legitimate as well.

14. Although some people might be reluctant to accept the relationship between genetics and mental abilities, such a relationship most certainly exists. (As importantly, there is no moral cost to acknowledging such a relationship, since doing so has no normative implications.) For discussion of the genetic basis of cognitive ability, see Clark and Grunstein (2000, pp.

221–238); Bouchard (1998); Bouchard and McGue (1981); Plomin et al. (1994); and Daniels et al. (1998).

15. For discussion of the relationship between genetics and free will, see Clark and Grunstein (2000, pp. 265–270); Pinker (2002, pp. 174–185); Ridley (1999, pp. 301–313).

16. For discussion of the interactions between genes and the environment, see Clark and Grunstein (2000, pp. 218–220, 253–265) and Ridley (1999, pp. 65–75).

BIBLIOGRAPHY

Anderson, W. French. 1990. Genetics and Human Malleability. *Hastings Center Report* 20 ([No.?]): 21–24.

Beauchamp, Tom L., and LeRoy Walters, eds. 2003. *Contemporary Issues in Bioethics*. 6th ed. Belmont, Calif.: Wadsworth.

Bouchard, Thomas. 1998. Genetic and Environmental Influence on Adult Intelligence and Special Mental Ability. *Human Biology* 70: 281–296.

Bouchard, Thomas, and Matthew McGue. 1981. Familial Studies of Intelligence. *Science* 212: 1055–1059.

Buchanan, Allen, Dan Brock, Norman Daniels, and Dan Wikler. 2000. *From Change to Choice: Genetics and Justice*. New York: Cambridge University Press.

Clark, William R., and Michael Grunstein. 2000. *Are We Hardwired?: The Role of Genes in Human Behavior*. Oxford: Oxford University Press.

Daniels, J., P. McGuffin, M. Owen, and R. Plomin. 1998. Molecular Genetic Studies of Cognitive Ability. *Human Biology* 70: 281–296.

Daniels, Norman. 1992. Growth Hormone Therapy for Short Stature: Can We Support the Treatment/Enhancement Distinction? *Growth: Genetics & Hormones* 8 (Supplement 1): 46–48.

Danks, David. 1994. Germ-Line Gene Therapy: No Place in Treatment of Genetic Disease. *Human Gene Therapy* 5: 151–152.

Glover, Jonathan. 1984. *What Sort of People Should There Be?* London: Penguin Books.

Gordon, John W. 1999. Genetic Enhancement in Humans. *Science* 283: 2023–2024.

Hill, Thomas J. 1983. Ideals of Human Excellence and Preserving Natural Environments. *Environmental Ethics* 5: 211–224.

Juengst, Eric. 1997. Can Enhancement Be Distinguished from Prevention in Genetic Medicine? *Journal of Medicine and Philosophy* 22: 125–142.

Juengst, Eric, and LeRoy Walters. 2003. Ethical Issues in Human Gene Transfer Research. In *Ethical Issues in Modern Medicine*, 6th ed., ed. Bonnie Steinbock, John D. Arras, and Alex John London, pp. 571–584. New York: McGraw-Hill.

Lappé, Marc. 1991. Ethical Issues in Manipulating the Human Germ Line. *Journal of Medicine and Philosophy* 16: 621–639.

Mehlman, Maxwell J. 2000. "The Law of Above Averages: Leveling the New Genetic Playing Field." *Iowa Law Review* 85: 517–593.

Mehlman, Maxwell J., and Jeffrey Botkin. 1998. *Access to the Genome: The Challenge to Equality*. Washington, D.C.: Georgetown University Press.

Neel, James. 1993. "Germ-Line Therapy: Another View." *Human Gene Therapy* 4: 127–128.
Nozick, Robert. 1974. *Anarchy, State, and Utopia*. New York: Basic Books.
Parens, Erik. 1995. "The Goodness of Fragility: On the Prospect of Genetic Technologies Aimed at the Enhancement of Human Capacities." *Kennedy Institute of Ethics Journal* 5: 141–153.
——— , ed. 1998. *Enhancing Human Traits: Ethical and Social Implications*. Washington, D.C.: Georgetown University Press.
Pinker, Steven. 2002. *The Blank Slate: The Modern Denial of Human Nature*. New York: Viking.
Plomin, R., G. E. McClearn, D. Smith, et al. 1994. "DNA Markers Associated with High versus Low IQ." *Behavior Genetics* 24: 107–118.
President's Commission for the Study of Ethical Problems in Medicine and Biomedical and Behavioral Research. 1982. *Splicing Life*. Washington, D.C.: U.S. Government Printing Office.
President's Council on Bioethics. 200X. *Beyond Therapy*. Washington, D.C.. Available at www.bioethics.gov. Accessed January 2005.
Rawls, John. 1999. *A Theory of Justice*. Rev. ed. Cambridge, Mass.: Harvard University Press.
Ridley, Matt. 1999. *Genome: The Autobiography of a Species in 23 Chapters*. New York: Perennial.
Walters, LeRoy, and Julie Palmer. 1997. *The Ethics of Human Gene Therapy*. New York: Oxford University Press.

ETHICS AND THE LIFE SCIENCES

TELOMERES AND THE ETHICS OF HUMAN CLONING

FRITZ ALLHOFF
WESTERN MICHIGAN UNIVERSITY

ABSTRACT: In search of a potential problem with cloning, I investigate the phenomenon of telomere shortening which is caused by cell replication; clones created from somatic cells will have shortened telomeres and therefore reach a state of senescence more rapidly. While genetic intervention might fix this problem at some point in the future, I ask whether, absent technological advances, this biological phenomenon undermines the moral permissibility of cloning.

TELOMERES AND AGING

James Watson, who, along with Francis Crick, discovered the famed double-helix structure of DNA, also observed that polymerases (which copy DNA) are unable to begin the transcription process at the very end of a DNA strand. Rather, transcription must begin *within* the genetic code and, consequently, the end is not replicated. If the DNA that was not replicated possessed valuable genetic data, then this would obviously be quite bad; each successive replication would delete part of our genome. But, as it turns out, natural selection found a clever way around this problem. At the end of a strand of DNA, we (and all other living organisms) are endowed with telomeres which are like the protective are like the protective aglets on the ends of shoelaces; the telomere serves no function other than to protect our genome against the imperfection of polymerases. Given that our polymerases fail to replicate some segment on the

© 2007 Philosophy Documentation Center   pp. 231–237

end of our genome, it is obviously better that those segments are non-coding DNA than valuable genes.

Each time a cell divides and the genetic material in the cell nucleus is copied, the telomere shortens as the polymerase copies only (an interior) part of the chromosomes. Telomeres, while long, are nevertheless constantly growing shorter. Interestingly, telomeres only shorten in somatic cells (e.g., hair, skin, etc.), and not in germ line (e.g., sperm and egg) cells. Why? Humans have a gene (on the fourteenth chromosome) called TEP1 which codes for the production of a protein that forms part of a "biological machine" called telomerase (Ridley 2000). Though the process is not yet fully understood, telomerase repairs shortened telomeres by re-lengthening them. In most human tissues, the genes that create telomerase are deactivated which consequently preordains the shortening of telomeres. In germ line cells, however, the genes for telomerase are not deactivated. (Notably, malignant cancer cells reactivate the telomerase genes, a process which allows the cancer to reproduce without telomere shortening.)

Why does any of this matter? There is little controversy that telomere shortening is the central reason that cells grow old and die—cell division stops once the telomeres become sufficiently short and the cells consequently begin to senesce. But there are also good reasons to think that the shortening of telomeres is one of the reasons that the entire organism ages (and dies); research has also shown that this sort of cellular aging can lead to degenerative diseases and conditions. For example, the chromosomes in arterial cells typically have shorter telomeres than the chromosomes in venous cells. This is no doubt because arterial cells are under higher pressures and become damaged more often; consequently, they have to repair themselves, which involves cell copying and telomere shortening. Arterial cells therefore reach a state of senescence faster than venous cells, which is why we die from arterial hardening rather than venous hardening (Chang and Harley 1995).

Despite these findings, there are certainly causes of senescence other than the shortening of telomeres. Rather, it is far more likely that telomere shortening is one of the contributing factors to senescence, of which there are likely to be many (Austad 1997).[1] Therefore, I grant that telomere shortening is not the only factor that contributes to aging. Nevertheless, it is uncontentious to claim that telomere shortening leads to aging (on both the cellular and organism levels), and that the relationship between telomeres and aging is quite important.

It should also be observed that the prospect of genetic engineering could solve the problem of telomere shortening: if it becomes possible to reactivate telomerase (or insert genes that create it), organisms will be able to able to repair frayed telomeres and cells will be, at least theoretically, immortal. The Geron Corporation (www.geron.com), for example, has done extensive research on telomeres and telomerase and has been able to insert genes for telomerase into cells that otherwise lacked those genes; the cells were then able to divide indefinitely. Whether an active telomerase gene is inserted or whether the current deactivated ones can be reactivated, science offers the hope of being able to respond to senescence induced by telomere shortening. What is equally exciting is that this research is likely to also yield ways to deactivate the telomerase in cancers which would consequently limit cancerous growth. Despite a cautious optimism on these fronts, it is not currently technologically possible to engineer wide-scale reactivation of telomerase in the human body (or to insert the gene which would code for telomerase production), nor is it likely that this breakthrough will come in the immediate future. So, for the moment, we are stuck with senescence once our cells cease to replicate.

CLONING

Now, we can consider cloning. Though the point may now be obvious, allow me to make it explicit. Imagine that a thirty-year old woman wished to create a genetic clone. She would have to acquire a denucleated egg and insert the nucleus from one of her *somatic* cells (remember that germ-line nuclei have unpaired chromosomes) into this egg. The DNA contained within the nucleus of her somatic cell would have shortened telomeres because it would have been generated after several generations of cellular replication. The clone would therefore *begin* its existence with shortened telomeres; its constitutive cells would have fewer replications in their futures than those of a zygote created by germ line cells whose telomeres would have been re-lengthened by telomerase. The clone would therefore senesce more rapidly (or, perhaps more accurately, *earlier*) than a child conceived through sexual reproduction and this senescence would result in heightened susceptibility to degenerative conditions and diseases, as well as shortened life expectancy (Associated Press 2003).[2]

What sorts of normative conclusions can be informed by this biological consideration? It seems obvious that there is at least something wrong/bad/undesirable with cloning given these consequences, but what is it? Laura Purdy has argued that reproduction is immoral if the child will not lead a "minimally satisfying life"; she argues that this criterion can be defended on either a consequentialist or contractarian approach (Purdy 2000). Accepting Purdy's suggestion, we could ask whether cloning would be immoral given the biological considerations that we have been discussing. Clearly there is no reason to think that a clone with shortened telomeres would fail to have a minimally satisfying life. His life would be comparatively less desirable than a "normal" life in virtue of an earlier onset of senescence and, presumably, a shortened life span, but it is wildly implausible to think that this life would not be one worth living (especially from the point of view of the clone).

Another potential response would be to argue that cloning *harms* the clone by subjugating him to various undesirable propensities (such as earlier onset of degenerative conditions). Some philosophers have argued against the logical coherence of this notion (as applied to "wrongful birth" more generally), and it is instructive to look at the argument. One plausible account of harm is to apply a counterfactual (or comparative) criterion: X harms Y by doing A if Y would be better off had X not done A (Parfit 1984; Kagen 2002). For example, I harm my friend by kicking him because he would have been better off had I not kicked him. While this account of harm is not perfect (there are problems with over-determination), it is certainly one that is widely considered and constitutes the starting point for many "advanced" versions (Nagel 1979; McMahan 2002). So, we could ask, does cloning harm the clone? If so, the clone would have to have been better off *had cloning not taken place*. However, this condition can obviously not be met; had cloning not taken place, the clone would not exist at all, much less have a higher level of overall welfare. Therefore, cloning cannot harm the clone (Parfit 1986; Robertson 1994).[3] While some non-comparative accounts of harm have been proposed (Shiffrin 1999; Woodward 1986), I nevertheless take the received view to be consistent with the general spirit (if not the details) of the above proposal. Therefore, I think it is fairly uncontentious to *deny* that cloning harms the clone.

Nonetheless, we could adopt an impersonal comparative account, which would hold that cloning is wrong because the life of a clone would be worse (in some way) than that of a non-clone.[4] Parfit, for example, proposes *The Same Number Quality Claim* (Q): "If in either of two outcomes the same number of people would ever live, it would be bad if those who live are worse off, or have a lower quality of life, than those who would have lived" (Parfit 1986).[5] To take his example, consider a fourteen year-old girl who chooses to have a child now rather than wait until she would be able to provide a better life for the child. Insofar as it is metaphysically impossible for *that* child to have been born substantially later, we must locate the wrongness of the girl's act not in its effect upon *that* child, but rather in the fact that she *could* have created some child with higher welfare had she waited.

I think that Parfit's line here is extremely compelling. Turning back to cloning, we might apply (Q).[6] If parents were to have children, we might think that they should produce the children, to the best of their ability, that would be maximally well off. Children would obviously be better off being born with normal, as opposed to stunted telomeres, so parents should do whatever they can to avoid this problem. Obviously sexual reproduction would not transfer shortened telomeres to offspring so, all else being equal, sexual reproduction is (for now) better than cloning.

But for many of those who would consider cloning, sexual reproduction is not an option. We might, for example, imagine a single person, a sterile couple, or a homosexual couple who is trying to reproduce. In these cases, is cloning morally permissible? If we take Parfit's principle seriously, cloning would only be morally permissible if it were to maximize the welfare of the potential offspring. *If*, for example, the option were to clone or to not reproduce, then cloning would still be morally permissible so long as the clonant would live a minimally satisfying life (which, I think, we have every reason to believe that s/he would).

But, more likely, there would be other options. For example, we can wait to see whether genetic engineering will be able to reactivate telomerase in somatic cells or to insert a gene that would code for its production. If the technologies do develop, then we could have cloning without moral hazards. Given the potential risks of cloning now and their potential abatement at some time in the future, it seems that we should wait and see if science can fulfill its potential. Alternatively, prospective cloners might seek sperm or egg donation for sexual reproduction. If one parent is sterile, the other could reproduce sexually with a third party (through

IVF, of course). Similar options would be available for single parents or homosexual couples, though males would obviously need to obtain gestational surrogates (which is not without moral problems).

While I am inclined to support Parfit's principle, I nevertheless have one concern. Namely, I worry that a full endorsement of (Q) might serve as an indictment against all sub-optimal reproductions; this indictment would follow from a commitment to any maximizing consequentialism. For example, my daughter might be worse off given her acquisition of half of my wife's genes than a daughter that could have been created had I mated with someone who was more genetically fit. Does this mean that it was wrong to reproduce with my wife as opposed to someone else? I would certainly want to resist this conclusion.

Perhaps, however, the consequentialist need not have this commitment: it is certainly plausible to think that utility is not maximized if reproductions aimed solely at maximizing the welfare of the child. If every reproduction were to be certified (either morally or legally) as maximizing the welfare of the child, there would be obvious effects upon the relationships of the parents, the relationships between the parents and the children, etc. If, for example, my wife were to inform me that "our" child's welfare would be maximized if she were to be inseminated by a donor (of high genetic worth) rather than reproducing with me, there could obviously be negative effects. So, while we might grant that, *prima facie*, the welfare of children should be maximized, there might be significant countervailing considerations that would allow for reproductions which would not maximize the welfare of the child.[7] Finally, it is worth observing that the consequentialist approach would only commit parents to producing the best children that they were *able* to. While many reproductions might be sub-optimal insofar as children's lives would not be maximally rewarding, we could nevertheless observe that the parents were constrained, to some extent, with the lives that they could offer their offspring.

CONCLUSION

In conclusion, I have suggested that there is a biological issue inherent in human cloning: the clonant will senesce earlier than someone who was created through sexual reproduction. While there is cause for cautious optimism that genetic engineering will be able to address this problem, the solution is, at best, still a few years away. Investigating the normative implications of this biological phenomenon, I proposed that we adopt an impersonal comparative approach, which would hold that we should reproduce so as to maximize the welfare of our offspring (to the best of our abilities). This is similar to principles argued for by Derek Parfit and Julian Savulescu and, hopefully, has intuitive resonance and conceptual appeal. It is unlikely that cloning (at the present time) will satisfy this criterion given the existence of alternative means of reproduction and/or given the potential technological developments in the future. Therefore, I suggest that we have located at least a *prima facie* problem with human cloning, though I grant that this problem is contingent upon scientific limitations that might dissolve.

NOTES

I would like to thank Matthew Hanser and two anonymous reviewers for helpful comments on an earlier version of this paper.

1. Genetic disposition is likely to be another substantial contributor to senescence. For example, George Martin, a geneticist and pathologist, has estimated that as many as 7,000 of our 100,000 genes may influence some aspect of our aging.

2. It should be said that this is not mere speculation: Dolly, the first mammalian clone, was euthanized after developing both lung cancer at an early age (as well as arthritis); scientists observed that she "had started to show signs of wear more typical of an older animal." (Most sheep live to be eleven to twelve years old, whereas Dolly lived to be six.).

3. Parfit argues that this consequence leads to the "non-identity problem": we cannot say that the decision to reproduce was worse for that child because, absent reproduction, that child would not exist. (Let's assume the child has a minimally satisfying life.) This line has also been taken by John Robertson in several articles, as well as in his book *Children of Choice: Freedom and the New Reproductive Technologies*. For a contrary view, see Melinda Robert's "Human Cloning: A Case of No Harm Done?"

4. This is, of course, a consequentialist approach that might not be supported by non-consequentialists. Nevertheless, the principle I offer will hopefully be benign enough so as to alienate very few people. Parfit does consider (and reject) deontic solutions.

5. See also Savulescu 2001; Savulescu proposes a similar principle of "procreative beneficence": "couples (or single reproducers) should select the child, of the possible children they could have, who is expected to have the best life, or at least as good a life as the others, based on the relevant, available information."

6. Similarly, we might apply Savulescu's principle of procreative beneficence; either principle will yield the same result.

7. Or, alternatively, we might reasonably think that a child's life would go better if he were to actually be related to both of his parents rather than score higher on some detached barometer of genetic fitness. Certainly an unrelated parent would not feel as close to a child as a related parent would; these feelings could be manifested as utility considerations. This response would also allay my concern.

BIBLIOGRAPHY

Associated Press. 2004. "Cloning Pioneer Dolly Put to Death." 14 February.

Austad, S. J. 1997. *Why We Age?* New York: John Wiley & Sons.

Chang, E., and C. B. Harley. 1995. "Telomere Length and Replicative Aging in Human Vascular Tissues. *Proceedings of the National Academy of Sciences of the U. S. A.* 92:11190-94.

Kagen, S. 2002. *The Limits of Morality.* Oxford: Oxford University Press.

Kass, L. 1997. "The Wisdom of Repugnance." New Republic 216:17-26.

McMahan, J. 2002. *The Ethics of Killing: Problems at the Margins of Life.* Oxford: Oxford University Press.

TELOMERES AND THE ETHICS OF HUMAN CLONING 237

Nagel, T. 1979. "Death." In *Mortal Questions*, pp. 1-10. Cambridge: Cambridge University Press.

Parfit, D. 1986. *Reasons and Persons*. New York: Oxford University Press.

Purdy, L. M. 2000. "Genetics and Reproductive Risk: Can Having Children Be Immoral?" In *Biomedical Ethics*, ed. T. A. Mappes and D. Degrazia. Boston: McGraw Hill.

Ridley, M. 2000. *Genome: The Autobiography of a Species in 23 Chapters*. New York: Perennial.

Roberts, M. 1996. "Human Cloning: A Case of No Harm Done?" *Journal of Medicine and Philosophy* 21:537-54.

Robertson, J. A. 1994. *Children of Choice: Freedom and the New Reproductive Technologies*. Princeton: Princeton University Press.

Savulescu, J. 2001. "Procreative Beneficence: Why We Should Select the Best Children." *Bioethics* 15:414-26.

Shiffrin, S. 1999. "Wrongful Life, Procreative Responsibility, and the Significance of Harm." *Legal Theory* 5:117-48.

Woodward, J. 1986. "The Non-identity Problem." *Ethics* 96:804-31.

ETHICS AND THE LIFE SCIENCES

SAVING SEVEN EMBRYOS OR SAVING ONE CHILD? MICHAEL SANDEL ON THE MORAL STATUS OF HUMAN EMBRYOS

GREGOR DAMSCHEN AND DIETER SCHÖNECKER
UNIVERSTÄT HALLE-WITTENBERG AND UNIVERSITÄT SIEGEN

ABSTRACT: Suppose a fire broke out in a fertility clinic. One had time to save either a young girl, or a tray of ten human embryos. Would it be wrong to save the girl? According to Michael Sandel, the moral intuition is to save the girl; what is more, one ought to do so, and this demonstrates that human embryos do not possess full personhood, and hence deserve only limited respect and may be killed for medical research. We will argue, however, that no relevant ethical implications can be drawn from the thought experiment. It demonstrates neither that one always ought to let the embryos die, nor does it allow for any general conclusion concerning the moral status of human embryos.

1. A THOUGHT EXPERIMENT

Philosophical thought experiments hardly ever make it into a greater public arena. Most of them are quite obscure, and only very few are of broader interest. However, in spring 2001 there was a heated and public discussion in Germany about such a thought experiment after Reinhard Merkel, a professor of legal studies at the University of Hamburg, had introduced it to a wider audience in the high-circulation weekly *Die Zeit* (Merkel, 2001). Only a few readers, though, realized that the thought experiment actually goes back to an article published in the *Hastings Center Report* by George Annas who in turn took up a suggestion by Leonard Glantz: "If a fire broke out in the laboratory where these seven embryos are stored, and a two-month-old child was in one corner of the laboratory, the seven embryos in another, and you could only save either the embryos or the child, I doubt you would have any hesitancy in saving the infant" (Annas, 1989, p. 22). Recently, Michael Sandel (2003) has reformulated this thought experiment: "Suppose a fire broke out in a fertility clinic. You had time to save either a five year old girl, or a tray of ten embryos. Would it be wrong to save the girl?" (pp. 9–10).[1] As Merkel points out in his version of this thought experiment, one should assume that the child is already unconscious to block any distorting considerations about the baby feeling pain etc. Also, Merkel is right in saying that the number of embryos is irrelevant; so it could be seven, seven hundred or seven thousand, it does not matter. It does not matter because according to Annas, Merkel, Sandel, et al., the normative conclusion would always be the same: "Saving the infant, however, acknowledges that the child is not equated with the embryo" (Annas, 1989, p. 22). Similarly, Sandel uses the thought experiment to argue against a view he calls the "'equal moral status' view," i.e., "to attribute full personhood to the embryo" (p. 9).

Since Annas makes little of the thought experiment, and because Merkel's position is not known beyond Germany, we will concentrate on Sandel's version of the thought experiment which actually puts it in the context of considerations about the moral status of human embryos.[2] Sandel's position is all the more interesting since he is a member of the President's Council on Bioethics. In this council, the thought experiment was used as an argument against the position that embryos must not be used for stem cell research.

We will argue, first, that Sandel has not made up his mind about the epistemological status of his thought experiment, and that, second, *nothing* morally relevant follows from this thought experiment; despite its initial plausibility, no valid moral conclusion can be drawn from it. Note that the purpose of this article is not to demonstrate directly what the moral status of human embryos actually is, and it will also not discuss the numerous ethical, epistemological, and ontological problems that arise in the debate about the moral status of human embryos.[3] This paper is limited in scope; it is primarily critical.

2. THE EPISTEMIC ROLE OF THE THOUGHT EXPERIMENT

Sandel begins with the claim that there "are three possible ways of conceiving the moral status of the embryo—as a thing, as a person, or as something in between" (9). It is not entirely clear what he means by 'something in between.' In any event, he seems to hold that the human embryo is not merely a thing, and even if it were, one could not conclude that it did not deserve any respect. For personhood, Sandel argues—quite correctly, we think—is not a necessary condition for something to deserve (at least some) respect. However, he attacks the view which he calls the 'equal moral status' view. According to this position, embryos have to be attributed "full personhood" (p. 9). What that ('full personhood') really means also is not clear. But it is Sandel's strategy to play out the "full implications" (p. 9) of this view and then to show by applying a kind of *reductio ad absurdum*—we call it later an *ethical modus tollens*—that this view is untenable. Thus, he attempts to demonstrate that if embryos had full personhood, we needed to save those ten

SANDEL ON THE MORAL STATUS OF HUMAN EMBRYOS

embryos rather than the child, but that actually we must not save the embryos which is why it cannot be true that embryos have full personhood.[4]

Let us now have a closer look at the thought experiment. We are to suppose that a fire broke out in a fertility clinic and that we only had time to save either a five year old girl, or a tray of ten embryos. The question then is: Would it be wrong to save the girl? Clearly, Sandel suggests that not only would it not be wrong to save the girl, but rather it would be wrong to save the embryos and let the child die. This might be true or false, but the first question is: What epistemic role does this thought experiment have? Is Sandel saying that we gain ethical insight about the question whether embryos have 'full personhood' by thinking about the thought experiment inasmuch as it elicits some kind of *moral intuition* such that we just know (*see*) that saving the embryos would be wrong and hence know that embryos are not persons? Or is the thought experiment just an *illustration* of an ethical insight that is grounded in some non-intuitive argument? Since Sandel does not provide a non-intuitive argument one has to assume that in some sense the thought experiment *itself* is the argument. But that in turns means that Sandel bases his conclusion that embryos do not have full personhood and, *a fortiori*, his claim that they may be killed for medical research, on a moral intuition. There is, we agree (though we cannot show this here), nothing wrong with relying on moral intuitions simply because we have to. But one should methodologically be clear about this, and Sandel is not.

Sandel bases his conclusion on a moral intuition because he really avails himself of what might be called an *ethical modus tollens*. By this we mean arguments of the following form:

1. If some ethical theory x is true, the action y is obligatory.
2. Action y is not obligatory.

Therefore,

3. Ethical theory x is not true.

Of course, this argument-form is valid. But it is also obvious that the soundness of actual arguments that are based upon this argument-form depends on the truth-claim in the second premise.[5] The thing is that the truth of this premise is taken for granted, i.e., taken as *intuitively true*. It is just assumed as (self-)evident that some actions are obligatory, prohibited, or permissible. But this assumption itself, the truth of the second premise itself (i.e., the truth of a basic moral or ethical proposition), is taken for granted.

Here is Sandel's ethical modus tollens:

1. If the embryo has full personhood, it is obligatory to save ten embryos rather than a child if a fire broke out in a fertility clinic and we had only time to save either the embryos or the child.
2. But it is not obligatory to save the embryos.

Therefore,

3. the embryo does not have full personhood.

Sandel's two other arguments run just like that:
1. If the embryo has full personhood, we have to punish harvesting stem cells as severe as we punish any murder of born humans.
2. But we ought not punish harvesting stem cells so severely.

Therefore,

3. the embryo does not have full personhood.

And then again:

1. If the embryo has full personhood, we must regard the natural death of embryos due to miscarriage as morally equivalent to infant mortality.
2. But we ought not regard the natural death of embryos due to miscarriage as morally equivalent to infant mortality.

Therefore,

3. the embryo does not have full personhood.

We do not submit that it is improper to refer to basic moral intuitions. As a matter of fact, we believe that we cannot do away with moral intuitions in moral philosophy, and hence not with the ethical modus tollens either. The ethical modus tollens is indeed indispensable and inevitable. Ethics has to start and to end somewhere, and there is no *Münchhausen-trick* that somehow allowed us to prove what is right or wrong without already presupposing that we have ethical insight. But again, one needs to be clear what this means and implies, and at least in his paper Sandel does not elaborate his understanding of intuitions.

3. THREE ALTERNATIVE VERSIONS OF THE THOUGHT EXPERIMENT

How important it is to reflect on the power and limits of moral intuitions will show if we reflect upon other versions of the thought experiment. In Sandel's thought experiment, it is tacitly assumed that we will *always*, under *all* circumstances and quite independend of the context, prefer to save the child rather than the embryos. But is this really true? Let us consider other versions of the thought experiment.

Suppose there is a young man who is, as far as he knows, the last survivor of a devastating catastrophe; he is the last living member of the species homo sapiens sapiens. Let us also suppose that in this version of the thought experiment the child is a boy, and let us also suppose that the young man can either save this boy or 100 embryos. These embryos, let us further assume, could develop into adult members of the species homo sapiens sapiens, namely not only boys, but also girls (for there are already artificial uteri and the young man could then raise these children). If the man saves the child, the human species will be made extinct; if he saves the embryos, however, there is a good chance that mankind will survive. How is the young man to decide? We actually believe that his duty to save mankind overrides his duty to save the child (provided there were such a duty at all).[6] However, that is not the point. The point is that in such a case one has to admit that it is, at least, *not self-evident* whether the child should be saved rather than the embryos. This has two implications: First, it is not intuitively clear that if one either can save a child or embryos one ought to save the child. Second, it is clear that even if we all agree that one ought to save the embryos and hence mankind, this decision, in and of itself, does not imply that the moral status of the child is such that it may be killed for medical research.

Or how about this version: The rescuer who rushes into the fertility clinic is actually the biological father of the two embryos in danger to burn; these embryos are supposed to be implanted the next day into his wife's uterus. Both of them have long been looking forward to this, they already have chosen names for their embryos and have established an emotional contact with their descendants, and the embryos are known to be genetically healthy. (In the meanwhile, artifical fertilization has made such great progress that the chance of these embryos developing into healthy fetuses is just as good as that of naturally conceived embryos.) Is it really 'intuitively clear' that the father must let his embryos die in order to save the child? Would we all blame him for saving the embryos? What 'intuition' do 'we' have? Again, it is far from *self-evident* that the child should be saved rather than the embryos. And whatever our decision might be, nothing follows from this regarding the question whether embryos may or may not be killed for medical research.

Obviously, many cases and situations can be construed as that what appeared to be intuitively clear (the child is to be saved, the embryos must die) is not so self-evident, after all. Here is another version worth reflecting upon: Let there be a tray of ten embryos on the one side of the room, but instead of a child, let us assume there is (for some reason) a mortally ill and unconscious woman on the other side who we know will only live for a day or so. So again we ask: Is it intuitively clear that we must let the embryos die and save the woman?

4. CONCLUSIONS

So what follows morally from the thought experiment? Well, nothing, really. According to Sandel, our reaction to the thought experiment is to rescue the baby rather than the embryos, and this allegedly proves that human embryos do not have full personhood, which in turn justifies, following Sandel, the conclusion that it is morally permissible to kill human embryos for medical research. First, however, it is by no means *self-evident* that we always really react to this thought experiment in the way Sandel suggests (saving the child and not the embryos)—or *ought* to react, for that matter. With regard to each of the three versions of the thought experiment we presented, there is good intuitive evidence to believe that most or all of us would rescue the embryos rather than the born human—or *ought* to, for that matter. In these versions of the thought experiment the embryos rather than a born human being have to be rescued, and still no one would reasonably argue that the born human being (therefore) does not have full personhood. It is only in *these* situations that embryos, babies, or mortally ill humans deserve less protection *because* they are embryos, babies, or mortally ill humans, respectively. Nothing can be inferred for other situations. (It is, by the way, one thing to *actively* kill human embryos to conduct research with them, and quite another to be in a tragic situation in which only embryos or whoever can be saved.)

Second, even if Sandel's arguments *were* sound, they would not warrant the conclusion he draws, to wit, that "stem cell research to cure debilitating disease, using six days old blastocysts, cloned or uncloned, is a noble exercise of our human ingenuity to promote healing and to play our part in repairing the given world" (p. 10). On Sandel's account, the embryo has a moral status that is apparently *not* sufficient to pay it the amount of respect that is sufficient not to kill it for the purposes of medical research. But how does Sandel know this? Even if Sandel were successful in showing that embryos do not have full personhood—hence not an 'equal moral status'—he still needed to demonstrate that killing them for stem cell research is morally permissible. But from the alleged fact that embryos do not have full personhood, one cannot jump to the conclusion that they may be killed for medical research. That conclusion would only be valid if it had been shown that full personhood is a necessary condition for a moral status that renders protection from being killed for medical research. But Sandel himself denies that personhood is such a necessary condition.[7] He has offered no argument, no criterion by strength of which we would be able to determine when something or someone does have such a low moral status that it might be killed for medical research. From the fact that someone or something is not a full person it simply cannot be concluded that he, she, or it may be killed for medical research, just as one cannot imply from the fact that someone or something must not be killed for medical research that he, she, or it is a full person. At least this does not go without saying. But Sandel suggests it does because this is the underlying assumption of his argument.

The crucial question to answer here is not what the general moral status of human embryos is (whatever such a 'general' status might be), whether they are full persons, or whether they have the same moral status as we do. Even if embryos do not have the same moral status as we do, they might be due such a great amount of respect that we must not kill them for medical research. So the question really at stake simply is: Is it permissible to kill human embryos in order to conduct stem cell research?[8]

So even if, in fact, the thought experiment could show that in such a situation we would always react (or ought to react) in the way suggested by Sandel, nothing could be inferred from it with respect to the general moral status of human embryos. Just because a human being is mortally ill, it has not lost its personhood, and the fact that a human being is mortally ill does not give anyone the right to exploit it for medical reseach. Certainly, we have not proven here that embryos possess full personhood. But from the alledged fact that they rather than the child have to die in a burning lab, it does not follow that they do not, and even if they did not, it does not follow that they may die for medical research. Let's bid farewell to this kind of thought experiment.

NOTES

1. Plain numbers in parenthesis refer to the article by Sandel, 2003.
2. For an analysis of Merkel's thesis cf. Damschen and Schönecker 2003b.
3. We have tried to show elsewhere (Damschen and Schönecker 2003a) that killing embryos for stem cell research is morally not permissible. Using an argument from potentiality, we were neither dependent on a concept of personhood nor on a general definition of what a 'moral status' is.
4. Later he offers two similar arguments that come to the same conclusion.
5. One gets analogous argument-forms in combination with deontic concepts such as "prohibited" and "permissible"; hence there need not be a negation in the second premise.
6. Cf. Jonas's imperative that "there ought to be humanity" (Jonas, 1979, p. 91, our translation).
7. "Personhood is not the only warrant for respect" (p. 9). If personhood were a necessary condition for deserving respect, Sandel could not consistently believe that "we consider it a failure of respect when a thoughtless hiker carves his initials in an ancient sequoia—not because we regard the sequoia as a person, but because we consider it a natural wonder worthy of appreciation and awe" (p. 9).
8. Again, it is not the purpose of this paper to answer this question. It is only our intention to prove Sandel's argument to be untenable.

BIBLIOGRAPHY

Annas, George. 1989. "A French Homunculus in a Tennessee Court," in: *Hastings Center Report*, pp. 20–22.

Damschen, Gregor, and Dieter Schönecker. 2003a. "In dubio pro embryone. Neue Argumente zum moralischen Status menschlicher Embryonen," in *Der moralische Status menschlicher Embryonen. Pro und contra Spezies-, Kontinuums-, Identitäts und Potentialitätsargument*, ed. G. Damschen and D. Schönecker. New York/Berlin: de Gruyter, pp. 187–267.

———. 2003b. "Zukünftig ϕ. Über ein subjektivistisches Gedankenexperiment in der Embryonendebatte," in: *Jahrbuch für Wissenschaft und Ethik*, vol. 8, pp. 67–93.

Jonas, Hans. 1979. *Das Prinzip Verantwortung. Versuch einer Ethik für die technologische Zivilisation*, Frankfurt am Main: Suhrkamp.

Merkel, Reinhard. 2001. "Rechte für Embryonen?" in *Biopolitik. Die Positionen*, ed. Chr. Geyer. Frankfurt am Main: Suhrkamp, pp. 51–64.

Sandel, Michael. 2003. "The Ethical Implications of Human Cloning," in: *Jahrbuch für Wissenschaft und Ethik*, vol. 8, pp. 5–10.

ETHICS AND THE LIFE SCIENCES

A CLONE BY ANY OTHER NAME: THE DELAWARE CLONING BILL AS A MODEL OF MISDIRECTION

KATHERIN A. ROGERS
UNIVERSITY OF DELAWARE

ABSTRACT: The possibility of cloning human beings raises the difficult question: Which human lives have value and deserve legal protection? Current cloning legislation tries to hide the problem by illegitimately renaming the entities and processes in question. The Delaware cloning bill, (SB55 2003/2004) for example, permits and protects the creation of human embryos by cloning, as long as they will be destroyed for research and therapeutic purposes, but it adopts terminology which renders its import unclear. I show that, in the case of cloning legislation, the burden of proof is on those who would adopt new terminology, and it has not been met.

The possibility of cloning human beings has raised, in a new context, the very difficult philosophical question which has been central to public policy debates over issues such as euthanasia and abortion: Which instances of human life have value, count as members of the moral community, and deserve the protection of the law? I argue that at least in some cases current cloning legislation has tried to hide the problem by illegitimately renaming the entities and processes in question, and I take the proposed legislation of my home state of Delaware as an example. It aims to permit and protect the creation of human embryos by cloning as long as such embryos will subsequently be destroyed for research and other medical purposes, but it adopts terminology which makes this consequence extremely unclear to someone not versed in the issue. I show that, while language changes constantly and renaming is often legitimate or at least harmless, in the case of cloning legislation, the burden of proof is on those who would adopt new terminology, and this burden has not been met.

It is traditional in the United States to assume that there is a moral order which the legal code ought to conform to, or at least not violate. It is standard to test legislation against ethical considerations—standard, but certainly not easy. And one of the hardest issues arising from the intersection of legislation and ethics in the last thirty years is the very vexed question of how to carve out the moral community. The law should, at least to some extent, reflect our moral obligations towards others, but which others are to be included within the fold of beings with moral status? Different ethical theories are likely to render different answers. For some species of utilitarian, for example, what counts is the ability to experience pleasure and pain, and so, lower animals may fall within the moral community. Some Kantians hold that only one who can act as a moral agent is a member of the moral community, and so the child below the age of reason is left outside the fold. Natural Law ethicists see membership in the natural kind, humanity, as vital. This is the crucial criterion for many in the Judeo-Christian tradition as well, since it accords with the understanding that human beings are somehow special, having been made by God in His image. But then the questions arise: What does it take to be a member of the kind? And when does one actually become one? These questions in turn rest on more fundamental, and perhaps more difficult, metaphysical questions about the very nature of kinds.

It is no wonder that legislatures and courts seem eager to avoid any head-on attempt to tackle the issue of the criteria for moral status. *Roe v. Wade* (1973) is probably the most famous case of recent decades in which the scope of the moral community was a central issue. A major question was whether or not the fetus might be the sort of being which could deserve the protection of the law. The Supreme Court recognized the importance of the question, but explicitly declined to answer it. "We need not" wrote Justice Blackmun giving the majority opinion, "resolve the difficult question of when life begins." Presumably he means when "moral status" begins, since no one doubts that the human organism is *alive* from conception. (I will use the term "human organism" to refer to the living human at all stages in an effort to avoid terms with morally-charged connotations like "human being" or "human individual."). What is at issue is whether or not the living organism could deserve legal protection. He goes on, "When those trained in the respective disciplines of medicine, philosophy, and theology are unable to arrive at any consensus, the judiciary, at this point in the development of man's knowledge, is not in a position to speculate as to the answer."[1] Thus, they leave open the possibility that the embryo and fetus might be members of the moral community, deserving some or all of the protections accorded to later stages of human development.[2]

Roe v. Wade chose not to rule on the moral status of human embryos and fetuses, and whatever one thinks of the decision, it recognizes the importance of the question, and explicitly adopts a stand on it, albeit the negative one of finding it too difficult and hence refusing to try to answer it. The question has been open and under discussion for the last thirty years, and comes to the forefront again in a new way now that it is possible to clone human embryos. To date the issue has

DELAWARE CLONING BILL AS A MODEL OF MISDIRECTION   249

been one for legislatures to deal with, and the legislation now being proposed must be charged with failure to face it. Take for example Delaware Senate Bill 55, "The Cloning Prohibition and Research Protection Act."[3] This legislation adopts nonstandard terminology and renames the entities and processes in question in a way which seems to bury the facts and so avoid the moral question.[4] (Here I intend to criticize only the legislation, not the legislators. I do not know the authors or the history of the bill and believe it entirely possible that it was crafted without any intention to misrepresent.)

SB55, sometimes referred to as the "Anti-cloning" bill, seeks two aims. First, as its official name suggests, it seeks to prohibit human cloning, but, as we will see, not all human cloning. As amended it states that "No person shall clone or attempt to clone a human using any cloning technique." And then it defines "Clone a human" to mean "creation of a fully-developed human that is genetically identical or substantially genetically identical to a fully-developed, living or previously living human using any cloning technique." That is, cloning a human organism is prohibited if the organism is allowed to reach "full development." The phrase "fully-developed human" is not further defined, which seems to make a practical understanding of what exactly is prohibited almost impossible. From the scope of the discussion one assumes that the bill is intended to prohibit, for example, the creation of a human organism which will be allowed to mature to twelve years from birth. But a twelve-year old is not a "fully-developed" human, on one, very commonplace, understanding of the phrase. It has many years to go before it matures completely physically and mentally and reaches the fullest extent of its powers. Presumably someone who had a use for the cloned twelve-year-old's body parts might argue that he has not, under SB55, "cloned a human." On a different understanding of "fully-developed" the human newborn, assuming it is normal and healthy, is a fully-developed human newborn, and so a fully-developed human. But a human embryo, on the same assumptions, might be considered a fully-developed human embryo, and hence a fully-developed human, with the consequence that all human cloning is prohibited, which, as we will see, is not the intent of the bill. Granted, the bill would have an almost impossibly difficult time defining "fully-developed human" in a way that would suit its purposes, given that it wants to protect some sorts of human cloning while it prohibits others.[5] But surely it is a fair criticism of a piece of legislation if it is worded in a way that makes its practical import opaque.

So it is not clear exactly what human cloning SB55 intends to prohibit, beyond the creation of cloned adults. But it is clear what it intends to permit. "Nothing in this Chapter shall restrict other areas of biomedical and agricultural research including, but not limited to, important and promising work that involves the use of somatic cell nuclear transfer or other cloning technologies."[6] What this means is that human embryos may be created through cloning. But the bill never says this explicitly. It speaks of "the product of a nuclear transfer of a human somatic cell into an enucleated human oocyte" and "the result of the cloning technique known as somatic cell nuclear transfer" and, for example, prohibits the implantation of such a "result" in a human or artificial uterus. But it never speaks of the "product" or the "result" as a human embryo.

The "result" certainly is a human embryo. If it were not, then you could not harvest its human embryonic stem cells, a process the bill intends to protect. If it were not, then it could not, if implanted in a human or artificial uterus, grow into the "fully-developed human" whose creation the bill intends to prohibit. Dolly, the cloned sheep, was certainly a sheep. She was a sheep with an anomalous history, having had only one genetic parent. But a successfully cloned plant or animal has all the essential attributes of the non-cloned plant or animal. In terms of what it is, it is exactly the same kind of thing as its non-cloned cousin. A cloned human adult, would be a human adult with all that that entails, and a cloned human embryo is a human embryo.

Use of the language of scientific technique, rather than the term "embryo," in SB55 promotes misunderstanding. For example, in an opinion piece to the *Wilmington News Journal* (February 9, 2004), Robert Venables, the state senator who authored SB55, seems to find a radical difference in kind between the embryo created the old-fashioned way, and the cloned embryo. He writes that, "In human conception, a male sperm cell is united with a female egg in the uterus to form a human fetus. Many people, myself included, believe that life begins and a human soul comes into being at the moment the sperm fertilizes the egg to form what will become a human fetus."[7] The author of SB55 seems to suggest that the cloned embryo is radically different from the embryo created by the fertilized egg because it would not have a soul. Probably the most puzzling consequence of this view is the entailment that, should some law-breaker succeed in "cloning a human," the result would apparently be a fully-developed human without a soul, perhaps a soulless adult human being. But if the soul does not contribute anything to the development of the human being, such that one can be a mature adult without one, then it is hard to see why it should be a necessary element for inclusion in the moral community. And since the soulless, cloned, human adult would, in all likelihood, be indistinguishable from the ensouled human adult, it is hard to see what evidence there could be for thinking that any human beings *have* souls. Certainly there are many views of the nature and activity of the soul, but we had better be clear on our metaphysical assumptions before we allow them to generate our life and death moral and legal decisions. He makes the moral point that, "If in order to save another child's life through this new technology it would be necessary to kill a human fetus or a sperm fertilized human egg, I would say 'Never!' I would never approve of taking of one life to save another life. . . . But the type of research that Senate Bill 55 seeks to promote does not involve the killing of a human fetus or a sperm fertilized human egg." He goes on to argue that the "product of somatic cell nuclear transfer" is radically different. "Somatic cell nuclear transfer is fundamentally different from the process of human conception. [It] does not involve a sperm cell. It takes an egg . . . removes its DNA and combines it with a cell taken from an adult person. . . . [T]he DNA from the adult cell combines with the egg to grow cells that can then be used for medical purposes for human benefit."

The phrase, "the adult cell combines with the egg to grow cells" is very telling. It implies that the "product of somatic cell nuclear transfer" is just a bit of cell-growing tissue, but not a discrete, living, human organism the sort of thing which can grow into one of us. An editorial supporting SB55 in the same issue of the *News Journal* seemed to draw this mistaken conclusion when it opined that those who opposed the bill hold the strange view that "all embryonic tissue constitutes human life." But of course, that's not it at all. The argument is that human *embryos* are members of the moral community deserving of protection, and hence, bringing them into being for the purpose of harvesting their stem cells and destroying them in the process, is morally wrong. Many people, apparently including the author of SB55, are of this opinion. In order to promote an informed discussion among the legislators of Delaware and their constituents, SB55 should state clearly that what it intends to protect is the cloning of human embryos for research and therapeutic uses, and not hide behind the unfamiliar language of "somatic cell nuclear transfer."

But perhaps this is a bit too fast. The bill does not lie when it uses the term "product of somatic cell nuclear transfer" rather than "embryo." What justification is there for insisting on the term "embryo"? Why should the terminology of the anti-SB55 side be adopted over the terminology of the pro-SB55 side? After all, language changes all the time. New technologies introduce new terms. Sometimes the same events and entities can rightly be described in very different ways bearing very different moral entailments. Must the old, familiar, terms be retained just because they are old and familiar? True, the voting public in general has some idea of what "embryo" means and probably very little of what "somatic cell nuclear transfer" means, but the framing of bills relevant to technological advances cannot be held hostage to the public's lack of knowledge. If we are to criticize SB55 for making an illegitimate move in renaming the embryo we need to offer some argument.

First, a case must be made that the burden of proof is on those who would abandon the older term and adopt the new and unfamiliar term.[8] In the cloning debate, as in others involving the question of moral status, at least two reasons can be adduced why the older term is preferable in the absence of good reason to abandon it. The first has to do with the fact that the older term was the term in use to name that sort of thing before the moral question arose. The charge against the newer term is that it is an ad hoc usage introduced precisely to prejudice the outcome in the new moral debate. The same claim cannot be leveled against the older term since it was the standard term before the moral issue ever came up.

It might be argued that the older term prejudices the case as well, since, as we have noted, many people are likely to consider a "human embryo" something of value, whereas they will not respond the same way to a "product of somatic cell nuclear transfer." But this point supports the contention that the burden of proof is on those who would replace "embryo" with "product, etc." Assuming that neither side can command a consensus on the issue, legislators should probably treat the moral status of the embryo as an open question. Presumably, *ceteris paribus*, it is far worse to mistakenly *fail* to value and protect something which is a member of the moral community, than it is to mistakenly value and protect something which is not.[9] Since we have to choose one term or the other, in the absence of consensus, we should err on the side of caution and choose the term which slants towards the ascription of value rather than the one which slants away from it. So the burden of proof is on SB55 to justify use of "product etc." over "embryo."

It cannot be argued that everyone must use the same name for the same thing, nor that we must always retain the names we used in the past. According to SB55 one has "attempted to clone a human being" if one produces a cloned human embryo with the intention of allowing it to reach full development, and this is to be prohibited. But if one engages in exactly the same action of producing a cloned human embryo with the intention of harvesting its stem cells, then one has not attempted to clone a human being, but simply created a "product of somatic cell nuclear transfer" and one's actions are permitted. The critic of SB55 may argue that it is illegitimate to describe the same action and entity so differently based simply on the intent of the agent. But again, this is a bit too fast. There are cases where it seems standard and acceptable to use different names and descriptions for two actions or entities which are essentially the same, the difference lying in the intent of the agent. And this even in cases where the different characterizations entail different moral consequences. Suppose, for example, that Peter and Paul buy essentially similar pigs. Paul's intent is to make his pig a pet. He calls his action, "Buying a new pet" and calls his pig "Hamlet." Peter intends to make his pig dinner. He calls his action, "Buying a pig for dinner" and refers to his pig as " pork chops *in potentia*." These differing characterizations based on intent seem entirely legitimate, and this even though it is at least arguable that people have certain moral obligations towards their pets, and so by characterizing his action and his pig as he does, Paul incurs moral obligations which Peter does not.

Can SB55 defend its very different characterizations of cloning for reproduction and cloning for research by arguing, along the lines of the pig analogy, that in this case, too, different intentions justify different names and descriptions? No. The reason that Peter may characterize his act as "dinner-buying," is that there is a solid consensus (though it is certainly not unanimous), that use of pigs as dinner is morally acceptable. Calling his pig "pork" is not an attempt on Peter's part to win in a widely debated issue by redefining the terms. The same is not true of the cloning debate. If it were entirely clear to everyone that it is legitimate to use human embryos for research, then the "product" language might be justifiable. But in that case presumably there would be little motive to adopt it.

It might be argued that the renaming in SB55 is important because cloning is a scientific issue involving new technology. The old terms do not capture the scientific and technological advances and so are inaccurate. And certainly there are many instances in which our language and concepts have rightly changed due to such advances. New processes and entities have come into being, and new methods of amassing information have given us new insights into the same old beings, and often all of this progress has sent ripples through our moral thinking, including our thinking about membership in the moral community. For example, new information has shown that the old idea of "quickening," the fetus "coming to life" around the fourth month from conception, is entirely wrong. The human organism is alive from conception. In the past it could have been argued that there is a morally significant difference in the fetus before and after quickening since before that stage its destruction does not mean the killing of a living human organism. Now we recognize that "quickening" is a misnomer and has no moral significance.[10] At the other end of life, new methods of establishing brain activity have allowed us to introduce the category of brain death such that we can now distinguish between what is still a living human organism, and what is better considered a cadaver with a heartbeat.

Instead of focusing on new information, could the argument be that contemporary technology has produced a new kind of thing, the cloned embryo, which requires a new name? For the sake of analyzing the connection between moral status and the "new kind of thing," allow a little science-fiction. Suppose in the future certain beings which began as computers "wake up." They become self-conscious, intelligent, thinking things, the subjects of experience, able to suffer and enjoy, evaluate and choose. One can foresee that the legislative debate over those who would protect the thinking computers and those who would not might include disagreement over what name to give them. The pro-protection side would insist upon their likeness to us, and perhaps coin a new term in hopes of influencing public opinion: maybe something like "compu-people." The anti-protection side would likely insist that these intelligent computers are "just computers" and should be referred to as such. But here the new terminology seems justified. Although there is a presumption in favor of the term used before the debate arose, in this case the principle that, when the moral status of an entity is debatable, we should ere on the side of caution and use the term more likely to suggest moral status would support "compu-people." More importantly, here we have a genuinely new entity. A self-conscious "computer" is a radically different kind of thing from "just a computer" and so the nature of the thing demands a new name.

Does the scientific and technological progress involved in cloning justify the use of "product of somatic cell nuclear transfer" rather than the term "embryo"? Has some new information about the entities in question come to light? No. We are discovering new things about human embryos all the time, but there is no new information about the embryo produced through normal conception or about the embryo produced through cloning which should lead us to regard the latter as less a human embryo than the former. Has some new thing been brought into being? In a sense, yes. Deliberately cloned human embryos are a recent phenomenon. But is this new thing sufficiently different from the old-fashioned embryo as to be a new *kind* of thing? Is the difference relevant to the moral standing of the thing such that we are justified in renaming it in a way which will encourage people not to think of it as something deserving of protection? I do not see an argument for the position that, with respect to determining moral status, having one genetic parent rather than two is very significant morally. In prohibiting cloning for reproduction, SB55 recognizes that the entity in question is a living human organism which, if allowed to mature, would become one of us. Surely if any embryo has moral status it is because of what it actually is, not because of how it came to be what it is. Perhaps there are other possible justifications for renaming processes and entities, but none which is relevant to the cloning of human embryos springs to mind. At this stage it is safe to conclude that the renaming in SB55 is illegitimate.

The status of the human embryo, and the ultimate moral and legal conclusions to be drawn concerning it, are extremely complex and difficult issues which contemporary technology has forced us to face as a society. In this country the presumption is that the public will participate with our elected legislators in discussion about laws which affect us all. But informed debate requires that, insofar as reasonably possible, laws should be written to make their intent clear, not to hide it. SB55 intends to outlaw cloning human embryos for reproduction and permit it for research and therapeutic purposes. It should say so.[11]

NOTES

1. *Roe v. Wade*, 410 U.S. 113 at 159 (1973).

2. Their ultimate conclusion, that women have a constitutionally guaranteed right to abortion pretty much on demand throughout pregnancy, is puzzling given this position, but not completely indefensible. One might, as Judith Jarvis Thomson has famously argued, hold that the fetus is a person in the full moral sense, and yet conclude that in most circumstances the pregnant woman has no duty to care for it, and that this entails her right to abort it ("A Defense of Abortion," *Philosophy & Public Affairs*, (1) 1971). There is much to criticize in her argument, but the fact that it can be made could defend *Roe v. Wade* against the charge that it is flatly and obviously contradictory when it holds that the fetus may be the sort of thing that should be protected, while arguing that constitutionally it is not to be protected, at least against the woman carrying it.

3. I use the version of SB55 which was under consideration by the Delaware legislature when the first draft of the present paper was written in the winter of 2004. SB55 has since died, to be replaced by SB80 which also permits cloning human embryos for research purposes, but emphasizes use of "unwanted" and "donated" embryos from invitro fertilization. As of the final revision of the present paper (Fall 2005) the Delaware legislature is scheduled to vote on SB80 in January 2006.

4. The New Jersey Bill permitting cloning human embryos for stem cell research in New Jersey (Senate, No. 1909, Introduced September 30,2002) flanks the entire cloning issue by embedding it in a paragraph about stem cell research generally: "2.a. It is the public policy of this State that research involving the derivation and use of human embryonic stem cells, human embryonic germ cells and human adult stem cells from any source, *including somatic cell nuclear transplantation* [my italics], shall: (1) be permitted in this State; . . ." The bill repeatedly mentions that stem cell research raises "ethical concerns . . . [which] must be

DELAWARE CLONING BILL AS A MODEL OF MISDIRECTION 255

carefully considered; . . . " (1.g.). But, not only does it not consider them, it does not even hint at what the concerns might be.

5. Its best bet might be to draw the line after the embryonic stage at which stem cells can be harvested. This, however, would prohibit growing older embryos and fetuses, and perhaps those who want to see some human cloning protected would be unwilling to lose these potential sources of material for research and medicine.

6. To a philosopher, not restricting does not necessarily entail protecting, but the title of the bill and the discussion surrounding it in the Delaware legislature, make it clear that the legislators understand the bill as a positive step to protect cloning.

7. Mention of a soul highlights the need for some work in background metaphysics. First it should be noted that there are a great many philosophers who hold that human beings do have moral status, but do not have souls under most philosophical definitions of "soul." The question of the moral status of a human organism can be debated between people of diverse opinions on the issue of the existence of a soul. But assume that human beings do have souls. The suggestion here seems to be that fertilization by a sperm is a requisite for the new embryo to have a soul. But why should this be so? The Roman Catholic Church, as a result of centuries of debate, eventually concluded that the soul is the informing principle of the body, and so "it is because of its spiritual soul that the body made of matter becomes a living, human body; . . . " Further, "every spiritual soul is created immediately by God—it is not "produced" by the parents." (*Catechism of the Catholic Church*, (Urbi et Orbi Communications, 1994) Sections 365–366). According to this doctrine, the sperm does not play a role in the creation of a soul. And this view of the soul entails that the very fact that a human embryo is alive and can grow and develop is conclusive proof that it does indeed have a soul, since it is the soul which gives life and guides the human development of the organism. That the "product of a nuclear transfer" is a living human organism, proves it to have a soul on this theory.

8. The simple expedient of clarifying the situation by using both terms, describing the organism in question as "a human embryo produced by somatic cell nuclear transfer," would in practice be the same as retaining the term "embryo" and is a move which SB55 carefully avoids.

9. On most imaginable scenarios, shooting the child in the bushes on the belief that it might have been a deer, is far worse than failing to shoot the deer, on the belief that it might have been a child.

10. Those who hold that it is up to the pregnant woman to decide if the fetus is to be granted moral status might find some moral significance in the first stage at which the woman can feel the fetus. The assumption of this paper is that judgements regarding what entities are to be included in the moral community are not merely matters of private sentiment.

11. I would like to thank Rae Stabosz for her considerable help in researching and writing this paper.

ETHICS AND THE LIFE SCIENCES

WRONGFUL LIFE, WRONGFUL DISABILITY, AND THE ARGUMENT AGAINST CLONING

DAVID K. CHAN
UNIVERSITY OF WISCONSIN-STEVENS POINT

ABSTRACT: Philosophical problems with the concept of wronging someone in bringing the person into existence, especially the non-identity problem, have been much discussed in connection with forms of assisted reproduction that carry risks of harms either greater than or not otherwise present in natural reproduction. In this essay, I discuss the meaning of claims of wrongful life, distinguishing them from claims of wrongful disability. Attempts to conceptualize wrongful disability in terms of either the harmed existence of the offspring, or the possibility of less harmful alternatives, are found unsatisfactory. A contractualist approach that provides an account of wronging that is independent of harming is considered. Finally, I present a new approach that necessitates an account of *reasons* for procreation that could justify harm to the offspring. These reasons are not the kind that require or prohibit actions of certain types, but reflect what the agent sees as intrinsically valuable in acting.

Is it always a good thing to bring new human life into existence? If not, how do we decide whether it is morally right to procreate? It is sometimes argued that it is wrong to create offspring that come into being in a less than satisfactory condition, or in adverse circumstances. This form of argument is seen in cases described as 'wrongful life' in legal action taken against parents and physicians. Many philosophers have pointed out problems with the concept of wronging someone in bringing the person into existence. The need to properly evaluate such arguments has become urgent with forms of assisted reproduction such as in-vitro fertilization and surrogate motherhood that carry risks of harms either greater than or not otherwise present in natural reproduction.[1] A similar type of argument can be formulated to challenge the use of cloning technology as a form of human reproduction.

In this essay, I will first discuss the meaning of claims of wrongful life, and distinguish them from claims of wrongful disability. I will suggest that attempts to conceptualize wrongful disability in terms of either the harmed existence of the offspring, or the possibility of less harmful alternatives, are unsatisfactory. I will then consider a non-consequentialist approach that provides an account of wronging that is independent of harming. Finally, I will present a new approach that necessitates an account of *reasons* for procreation that could justify harm to the offspring. I show that this approach helps answer questions not just about the ethics of human cloning but also natural reproduction that risks harm to the offspring.

Two caveats are needed at the outset. First, it is the wrongful life and disability arguments that are the subject of this paper, and discussion of reproductive cloning is relevant insofar as it illustrates my points in a particularly insightful way. Whether or not the wrongful life and disability arguments are effective objections to cloning is obviously not going to be decisive regarding the ethics of cloning, and I do not make any claims here to settle the larger issue. Second, some ethicists focus on wrongful life as a legal concept, but my discussion in this paper seeks to provide moral guidance on the issues that go beyond what can be legislated. My broader application of the concepts of wrongful life and wrongful disability, which may appear to change the subject, can be defended on the grounds that medical professionals should receive moral guidance beyond the law, and that as far as the violation of rights involved in claims of wrongful life is concerned, both legal and moral rights should be relevant.

I. LINKING WRONGING WITH HARMING

What does it mean for the offspring to come into existence in a less than satisfactory condition, such that it is entitled to a claim of being wronged by those responsible for its having been born? There is a range of possibilities.[2] The offspring may inherit a disease, the risk of which its parents were aware, or should have been aware. The offspring may be born to parents who are unable to care for it. The offspring may be born illegitimate and as a result suffer from rejection and discrimination. Finally, the offspring may be created with reproductive technologies that deprive it of normalcy. On the other hand, it may have imperfections due to the *failure* of those responsible to make use of genetic screening (and gene therapy when this becomes viable).

Many kinds of wrong are found in the above examples. Failure to care for the offspring is the kind of wrong associated with child abuse and neglect. If legal action is taken against the parents, the wrong that is alleged is not the kind that falls under wrongful life or disability. The latter wrongs are the kind that are intrinsically connected with the child's existence, and that arise in cases in which the only way a harmful condition can be avoided is for the affected individual never to exist at all. Philosophers have found that this kind of wrong cannot easily be accounted for using "conventional, commonsense,

THE ARGUMENT AGAINST CLONING 259

and philosophical accounts of harm and harm prevention." The problem is that "there is no unharmed condition, because there is no unharmed individual with whom to make [a] comparison."[3]

In philosophical accounts of harm, harm is usually measured in comparison with a standard of what is normal, and what is normal is dependent both on social norms and on the state of technology. If, for instance, it were not possible to screen for Down's syndrome, babies born with the disease are defective but cannot be said to have suffered harm. But with the possibility of screening, a question arises as to whether it is acceptable to make a choice not to know, or to refuse to abort after the disease has been detected. The answer depends on our sense of just how unsatisfactory the condition of the child has to be before it can claim to have been harmed by being brought to birth, which in turn depends on society's attitudes and practices.

A number of arguments have been used against any claim on behalf of the offspring to having been harmed in being brought into existence. First, it may be suggested that mere existence has value great enough to offset any amount of unhappiness or suffering contained in the life. Second, the life of a person born with disease or handicap often has compensating benefits that make the life as a whole worth living. In other words, the person is not better off not having been born, unless her suffering is so severe that non-existence is preferable. And if she has not been made worse off, she has not been harmed, even though she is "born in a condition extremely harmful to it."[4] Third, there is an argument to show that it is not possible to harm a person if the imperfection is necessary for the person to exist at all.[5]

In order to respond to these arguments, it is necessary to be clear about the difference between wrongful life and wrongful disability.[6] In wrongful disability cases, but not wrongful life cases, the harm suffered in being brought into existence, though significant, is not so serious as to make life not worthwhile for the offspring.[7] In wrongful life cases, it is thought to be better for the offspring never to have existed at all. In wrongful disability cases, the offspring has not been made worse off by the disability as its life is on the whole worth living. The distinction is important because wrongful disability cases are probably much more common than wrongful life cases, and the former cases are of special philosophical interest because they have difficulty with what is known as 'the non-identity problem,'[8] of which more will be said below.

The first argument above is directed at wrongful life and wrongful disability cases alike. But what it maintains is the strikingly perverse idea that the value of mere existence can outweigh the disvalue of a life full of suffering, even a life that is anguished and not just flawed. Not only would it mean that it is beneficial for persons to be brought into existence regardless of their condition, it seems to follow that we have an obligation to procreate in vast numbers without concern for the quality of life.[9] Consider now the second argument against the offspring's claim to be wronged in being brought into existence. The argument denies any justification for wrongful disability but leaves wrongful life claims unchallenged. A reply can be made as follows. Even if a person's life is overall worth living, the fact that she has more suffering than normal means that she exists in a harmed condition, and if so, then she has been harmed in being brought into existence.[10] This reply runs up against the third argument considered below. In addition, whether it follows from her being harmed to her being wronged is a further question that I will be discussing in the next section.

The third argument raises the interesting philosophical issue of the non-identity problem. If it can be sustained, it is especially pertinent to forms of assisted reproduction that carry a risk of abnormality. The philosophical argument is that if it is not possible for the person to have existed without the imperfection, then it cannot be claimed that the person is harmed in comparison to a normal person.[11] As a normal child could have been brought into existence only if it had not been conceived in the way that the imperfect person was, the imperfect person could not possibly be a normal person. The imperfect person can only avoid her disability by not coming into existence at all. The problem is that the harm that accounts for a wrong that is suffered is a person-based harm. And if there is no person-based harm, there is no wrong.

I mentioned earlier that the non-identity problem is more of a problem for the concept of wrongful disability. In a wrongful life case, the person is so disabled as to not have a worthwhile life. She has a life with terrible burdens and no compensating benefits. "There is nothing in it to make it a good for the person whose life it is; instead, its nature makes it only a burden and torment."[12] There is no need to compare her life with that of a normal person to show that she has been harmed. All that we need to know is that she has been given a life of such awful quality that it is worse than non-existence.[13] And we know that it is worse because no one would choose such an existence over permanent extinction.[14] On the other hand, a person whose disability does not make her life so bad that it is not worthwhile has a life that is preferable to non-existence. She is badly off only in comparison with the life of a normal person. But, as she can avoid her disability only by not coming into existence at all, the latter kind of life is not an alternative that was ever open to her. Thus, she has not been harmed in being given a life.

In the next section, I begin to explore a response to the third argument that grants that the imperfections necessary for a human to exist are not harms that she as a particular person suffers. My aim is eventually to show that it is one thing for a person to be *wronged*, another for her to be harmed.

II. SEPARATING WRONGING FROM HARMING (I)

Various ways have been suggested to distinguish between harming and wronging. On a person-based account of doing wrong, no wrong has been done unless the consequences *for the person* who suffers are so severe as to render her life not worth living. The problem is that this criterion makes it most unlikely that anyone, even someone deliberately brought into existence in a harmed condition, is ever wronged in being harmed. A more attractive line to take is to say that it is wrong to choose "deliberately to increase unnecessarily the amount of harm or suffering in the world."[15] For instance, if a woman who is pregnant with five embryos, having had them genetically screened, gets her doctor to selectively terminate the healthy ones and to allow the disabled ones to come to birth, she clearly has done a wrong in "wantonly introducing a certain evil into the world."[16] This is so, no matter what we decide concerning whether she has harmed those born disabled instead of being terminated.[17]

Since doing a wrong is not necessarily or not just a matter of harming a person, any question of being made worse off becomes moot in cases where wrongs can be established on the basis suggested above. What we have is an account of wrong that moves away from person-based harm to a kind of 'aggregative consequentialism' or 'non-person-affecting principle.'[18] If it is *wrong to choose deliberately to increase unnecessarily the amount of harm or suffering in the world*, then one has done wrong in bringing defective babies into existence even if one has benefited them more than one has harmed them, given that they have lives worth living. The mother in our example could have had the normal babies instead of the disabled ones. Hence the suffering she introduced into the world through her choice in selective termination constitutes unnecessary harm, and she is wrong in so choosing. On the other hand, parents who must have disabled children if they are to have children at all would *wrong the children* only if the children would find their lives so severely harmed as to be not worth living.[19]

Is this account, one that in its essential details is the one given by John Harris,[20] of the wrong in wrongful life and wrongful disability fully acceptable? Intuitively, we want to say that it is not always a wrongful act to bring persons into existence in a less than perfect condition or circumstances. With wrongness understood solely in terms of person-based harm, one may be forced to claim that one does no harm at all in bringing into existence persons in a harmed condition (a counter-intuitive claim), or that it is always acceptable to conceive a disabled child so long as the child has a life that is worth living (rendering the selective termination case permissible). With wrong and harm separated, we can distinguish necessary from unnecessary harm. Necessity of the harm is understood in terms of whether there are alternatives in which the harm is absent. In the selective termination case, since the woman could have terminated the disabled embryos and kept the healthy ones, there is an alternative without the harm.

The problem I see for Harris's account concerns whether it is a sufficient condition, for there being no wrong done in bringing into existence children in a harmed condition, that *there are no alternatives without the harm*.[21] Is it acceptable to deny both claims of wrongful life and claims of wrongful disability, given that such harm does not render the life not worth living but cannot be avoided if the parents are to have children at all? Since one can be very severely harmed before one reaches a stage when one's life is not worth living, the account's permissiveness in this respect indicates that another moral consideration has been left out. I shall try to say what this is. But first there is another account of wronging to discuss.

III. SEPARATING WRONGING FROM HARMING (II)

So far, we have attempted to separate wronging from person-based harming but not from the amount of harm in the world. We moved from what Parfit calls 'same person' choices to 'same number' choices. In same number choices, "the same number of persons exists in each of the alternative courses of action from which an agent chooses, but the identities of some of the persons ... is affected by the choice."[22] Non-person-affecting principles enable us to avoid the non-identity problem when we attempt to show that a person is wronged, who has a worthwhile life and who can be without her disability only if she had not existed. Such a person was not harmed in being brought into existence. But someone, who had chosen deliberately to increase unnecessarily the amount of harm or suffering in the world, such as the woman who chose to bring to birth a defective embryo, had wronged her in bringing her into existence.

To avoid the non-identity problem we could instead take a non-consequentialist approach, detaching the concept of wronging from that of harming altogether. Wronging, it could be argued, has to do with the wrongdoer's agency, not the harm to someone that is merely the result of a wrongful act. The possibility of someone not being harmed, or even being benefited when she is wronged, as exemplified by the wrongful disability cases, shows that "what one does has an intrinsic significance in moral reasoning that is independent of what happens as a result of what one does."[23] The idea is to focus on the wrongdoer's agency, not the person-based harm that may or may not be inflicted on the object of his action. According to Kumar, the advantage of this approach to wronging is that "considerations concerning the fixity of psycho-physical personal identity can be shown to be of no moral relevance."[24]

Kumar draws on Scanlon's contractualist approach in his argument that "a claim to have been wronged requires that certain *legitimate expectations*, to which one is entitled in virtue of a valid moral principle, have been violated."[25] What is morally relevant to someone's being wronged is the culpable failure of the wrongdoer to fulfill legitimate expectations arising from the character of the relationship that she stands in with respect to the wronged party. Harm may or may not result from the latter's being wronged, but harm is not the basis of a claim to have been wronged. In the cases of wrongful disability that we have been discussing, it is argued that someone who intends to conceive a child "has reason to take it to be the case that the intended, but yet to be conceived, child will be of the type required for her to owe it to the child to take appropriate measures to protect its welfare, regardless of what its particular token identity turns out to be."[26] She thereby wrongs the child in not taking such appropriate measures.

The move to base wrongfulness on agency rather than the effect on the victim has the advantage not just of avoiding the non-identity problem. By showing how wronging a person does not in general require that the person be harmed, it turns out that the wrongful disability cases are not really exceptional at all and that conventional ethical theories do have the resources to explain how a wrongful disability claim is justified. But there is more than one way of basing wrongfulness on the character of agency. Kumar's way focuses on the types of action that a person in a particular relationship is expected to fulfill or to avoid. I will now present an account of wronging that takes into account the agent's *reason* for choosing actions, such as to conceive a child who is more disabled than normal. I will suggest that this type of action is not always wrong and does not always wrong the child. If so, Kumar's account will not be satisfactory.

IV. LIMITS TO PROCREATIVE LIBERTY

One may well take the view that the permissiveness of Harris's account is not a problem. The right to procreative liberty is a given in American society.[27] While the legal right to terminate a pregnancy is seriously challenged, there are no legal obstacles to would-be parents having a child even if they are not financially able to provide for it, or if they risk passing on a genetic disease. It is only when reproductive technologies are introduced to assist infertile couples that the right to use such methods may be questioned. We have seen concerns raised about the risk of harming offspring produced by such methods. But if we accept the arguments in Section II that such harms count against procreation only when there are less harmful alternatives, then the right to procreate entitles couples who have no other way of conceiving a child to use technology that is risky but not likely to cause the offspring so much harm as to render its life not worth living.

Advances in cloning technology may eventually overcome concerns about the harm to cloned offspring,[28] and clearly, scientists have a duty to improve the technology to reduce the risk of harm. But based on Harris's account of wrongful life and disability, if the possible harm to the clone is not so great as to render its life not worth living, those who want to use the technology can use it without doing wrong to the clone. This seems to be too permissive. To determine what other consideration could be relevant to the permissibility of cloning, consider a fantastic example:

The Last Panda

I am Ming, the last male panda in the world. I live in a nice zoo in New York. I am getting rather old, but every year, my human attendants fly me to London, where Ling, the last female panda in the world, lives. They hope that we will mate and have little pandas. Ling and I have a little problem getting it done the natural way. Last year, our human attendants decided to try in-vitro fertilization. It was not successful. They found that Ling's ova are defective and not viable. This year, they want to try something that has not been done with a panda before, only with sheep. They want to clone me! They have asked to have DNA taken from a cell in my body transferred into the nucleus of an enucleated egg cell from another species of bear. They cannot guarantee success, and there is a high risk of producing deformed pandas. If they wait a few more years, the technology would be safer to use. But by then, I may be dead. They plan to do the same with Ling, so that there will be male and female pandas after we are both dead. I am consulting my bio-ethicist about this.

Assuming that pandas were persons with moral rights, would the little pandas have a wrongful disability claim if they turned out to be somewhat disabled, but to still have lives worth living? On the basis of Harris's account, they would not. There is no other way of bringing into existence Ming's offspring, and despite their harmed condition, their lives are still worth living.

Harris's standard for a claim of wrongfulness should apply in similar fashion to an example with humans: Morris and Lois are unable to have children. They try IVF without success. Morris wants to be cloned, but the use of cloning on human beings risks harm to the offspring, but *not enough to make the clone's life not worth living*. I think that we should not agree with Harris that Morris's defective clone has no case for being wronged. It is true that Morris and Lois (like the pandas) have no other way of having children of their own. But they are not resorting to cloning in order to preserve the species! We have not been given reason to think that they should have children one way or another, even going to such lengths as using technology that may be severely harmful to the offspring (but short of rendering its life not worth living).

We need an account of wrongful life and wrongful disability as a way of limiting the right to procreative liberty. For it is difficult to make a case for limiting that right by appeal to the rights of non-existent or yet-to-exist beings, given philosophical disputes as to whether there are such rights, and if there are, whether these have the same weight as the rights of already existent beings. But Harris's account of the wrong in wrongful life and disability seems to be *too permissive* if it shows that Morris and Lois do not wrong the clone in bringing it into existence in a harmed condition.[29] On the other hand, an account that considers any disability suffered by the clone as sufficient for wrongful disability is too restrictive. It is a dilemma for such a restrictive account that: Either we apply a double-standard in not restricting natural reproduction, in the way that we restrict assisted reproduction, to cases where the offspring are perfectly adequate; or we should impose unduly coercive requirements that have to be met by all prospective parents.

V. REASONS AND WRONGFUL DISABILITY

The in-between standard that I propose is one that balances the likelihood and severity of the harm to the offspring with the reasons that parents have for seeking to reproduce. In the example of *The Last Panda*, avoiding extinction is a good enough reason for the panda to justify some harm to its offspring that can result from using cloning at the present level of technology. On the other hand, Morris and Lois were not attributed with a good reason that could be used to justify the harm suffered by the resulting clone.[30] But this does not mean that there can never be good enough reasons. Such reasons may be easier to find as cloning technology becomes safer. We know that natural reproduction also involves a risk that the offspring will come into being in less than satisfactory condition. Yet we do not restrict it to parents who can guarantee that their offspring would not inherit genetic defects and that the children would grow up adequately provided for. This is because we believe that having children of one's own contributes to human happiness, and achieving this is a legitimate life-goal of most people. It is an important goal even if, unlike for the pandas facing extinction, it is not so important as to justify risking the level of harm to the offspring associated with cloning at the present level of technology. But some degree of risk is morally acceptable. The widely shared desire of human individuals to procreate is a good enough reason to justify some harm should that eventuate in natural reproduction and the same justification should apply to assisted reproduction, including cloning.

Admittedly, claims about whether a reason is good enough seem to be rather subjective. First, is it from the human perspective or the panda's perspective that saving the species is a reason for cloning?[31] Second, if Morris seeks to perpetuate his family lineage, is that not equally as good a reason for cloning as that of saving the species?[32] Third, the value of procreation and the acceptability of a means of reproduction are culturally variable, and are also dependent on religious beliefs. For instance, some believe that there is good reason to procreate only by natural means. It is clear then that my approach requires me to insist that the evaluation of reasons can be given an objective basis, but it would be too large a project to embark on here.

Let me instead address another objection to my account, and do so in a way that would help to clarify what I mean by reasons that justify actions that involve harm. It may be held that an action that is wrong is wrong regardless of the reason that someone has for doing it. If cloning is too risky in terms of bringing into existence persons in a harmed condition, then it would be wrong to use the technology, no matter that it is done to save the species or to satisfy a desire to procreate or to make a contribution to science. But such an absolutist view of right and wrong cannot always be sustained. For instance, is it right or wrong for a physician to treat a hypochondriac with a placebo? Let's say that receiving the placebo is a benefit for the hypochondriac who will feel better because he thinks he is being treated. Let's say that there are no alternative ways of providing him with such a benefit. I do not think that we can say that the treatment is always right because it benefits the hypochondriac, or that it is always wrong because it involves deception and the dishonesty of charging for medication that is not real. Instead, it is right if the physician's reason for the 'treatment' is to prevent the hypochondriac from taking medication that she knows will harm him, but it is wrong if her reason is to collect payment for the 'treatment.'

Could it be that the physician had wronged the hypochondriac, even though the physician's action was not itself wrong? If the hypochondriac had been wronged, what would his grievance against the physician be? That he paid for something (medication) that he did not receive? On an analogue of Harris's account of wrongful life and disability, the physician may deny wronging him since there was no alternative way to benefit him and the harm that he suffered did not render his life not worth living. But I am suggesting that if the physician's reason for treating the hypochondriac is to collect payment, there is justification for the latter to complain of being wronged. But if he were told that the physician's reason is to prevent him from harming himself with real medication, it would be odd for him to complain.[33]

My point here about wrongful action is not an application of Kantian ethics, where intention is taken into account in the formulation of a maxim describing a type of action that either is or is not to be performed. The type of action in the previous example is that of making a hypochondriac feel better by treating him with a placebo. The type of action in the examples of the Last Panda and of Morris and Lois is that of creating a genetic offspring by cloning. *These action types do not yet determine for us whether it is right or wrong to do them.* The reason I take to be a possible justification for the action is not the intention attached to the description of the action that is used to test maxims, but the agent's larger goal that reflects what he, she, or it cares about or sees as intrinsically valuable in the action. Reasons of this kind are described in Aristotelian ethics as the agent's final ends, or the constituents of the agent's happiness.[34]

Similarly, the contractualist account of Kumar would decide right and wrong by the type of action performed, given the relationship between agent and person affected by the action. If the caretaker owes it to the child not to let the child "suffer a serious harm or disability or a serious loss of happiness or good,"[35] then, given the current state of technology, to create a genetic offspring by the risky procedure of cloning would be wrong for both the pandas and the humans. This is contrary to what I have said about the example. Maybe we are too unsure of our intuitions about cloning for my example to be decisive. Consider then the example of Smith, an African-American, who is refused an airline ticket. Suppose that the plane that he would have boarded crashes, killing all aboard.[36] Now it may be thought an advantage of Kumar's account that the action of refusing to sell a ticket to Smith wrongs him even if he is not harmed (and in fact benefits) as a result. Given the relationship and the legitimate expectations entailed, the clerk wronged him in refusing to sell him a ticket. It seems to me that Smith is wronged because the clerk's reason is a racist one. Compare the clerk's action to that of Dobby the house-elf who prevented Harry Potter from boarding the train to Hogwarts.[37] The reason he acted was to protect Harry Potter, whom he knew would be in danger if he were to return to school. It seems clear that

THE ARGUMENT AGAINST CLONING 267

Dobby did not wrong Harry Potter, and that Harry had no cause for complaint when he discovered what Dobby had done. So it is not the type of action, but the reason for doing it that determines whether someone is wronged when the agent does it.

Finally, I should consider the account of wrongful life and disability given by Feinberg and by Steinbock, in their suggestion that children can be wronged by being brought into existence if they are deprived of the minimally decent existence to which all citizens are entitled: "It is a wrong to the child to be born with such serious handicaps that very many of its basic interests are doomed in advance."[38] Such wrong can be done without the child's life being so seriously impaired that it is worse than no life at all.[39] On this account, parents who do not have the alternative of having children without the handicaps could still wrong a child that is born with the handicaps, even if its life is worth living.

Although we agree on the last point, there are two ways in which my account of the wrong in wrongful disability works better than Steinbock's account. First, Steinbock has to assume that the 'basic interests' of human beings are determinate and can be agreed upon. This is unlikely to be the case. Any attempt at drawing up a list of basic interests is bound to be controversial and provide insufficient confidence in the account of wrongful life that is based upon them. Second, even if a list of basic interests can be agreed upon, the standard it provides would be too rigid. Either cloning technology threatens basic interests of the offspring or it does not. In my view, whether or not it is wrong to clone depends on the reason for cloning, that is, the agent's end that is served by cloning. Steinbock's view would insist on saying the same thing for both the pandas and the humans Morris and Lois, whereas I suggest that there is a reason that justifies only the cloning of the last panda, and for this reason, it is not a wrong to the baby panda if it turns out to be the less than perfect product of cloning.

It is true that compiling a list of final ends for persons, as needed by my account, may be no less difficult to do than it is for Steinbock's basic human interests. But Steinbock has the much larger task of determining all the basic interests of the clone that are affected by the defects that can result from cloning. On the other hand, where wrongful disability as an argument against procreating is concerned, all that is required for me is agreement that both procreation and avoiding harm to other persons, especially innocent offspring, are appropriate goals that constitute a flourishing human life.[40] There is indeed room on my account for disagreement concerning the relative value of each of these two goals. But I think it is an advantage of my account that whether the achievement of each goal is a justifying reason for or against cloning depends on the circumstances. What may be a good reason for cloning the last panda is not a good reason for cloning Morris, but it could be a good reason for cloning Morris when the risk of defects in the clone has been sufficiently lowered by technological advances.

VI. CONCLUSION

The concepts of wrongful life and wrongful disability are puzzling. Under what circumstances may we say that a parent has wronged her offspring in not trying to avoid passing a disease to her child? If we cannot draw a line in cases of natural reproduction, can these concepts be used to argue against new reproductive technologies such as cloning? Surprisingly, thinking about wrongful life and disability in the context of assisted reproduction actually helps us think about wrongful life and disability in natural reproduction. Recognizing the need to avoid a double standard, we seek to draw the same line between what is wrongful and what is not, regardless of how the offspring is brought into existence. My account justifies some harm to the offspring on the basis of the reasons we have for reproducing, and it is able to provide a line between harms that are wrongful and harms that are not which applies to cloning as well as natural reproduction.

I would argue that, in a sense, my discussion of the reasons a person might have in risking harm to the offspring in choosing to reproduce in a certain way or under certain circumstances either fills in a gap or draws out an element implicit in the legalistic accounts of wrongful life such as Feinberg's. It fills in a gap because Feinberg has a problem dealing with cases where a child's handicapped existence is preferable to no existence at all, leading him to allow exceptions to his liberal doctrine that there should be no criminal liability without a victim.[41] And he makes his case for an exception with examples where the reason for bringing into existence an impaired child is "deliberate, malicious, and sadistic."[42] On the other hand, in arguing against a tort claim for a child who is not so handicapped that his life is not worth living, he compares holding the mother liable with "holding a rescuer liable for injuries he caused an endangered person that were necessary to his saving that person's life."[43] What is crucial but left unsaid here is that the rescuer has a good reason for risking harm to the person he rescued. The thrust of my argument in this paper is that, provided we can establish what are good reasons for risking harm to the offspring in reproducing naturally or artificially, we can use the reason a person has for bringing a child into existence to either justify the harm risked or articulate a moral claim of either wrongful life or wrongful disability.[44]

NOTES

1. Bonnie Steinbock, "Surrogate Motherhood as Prenatal Adoption," *Law, Medicine, and Health Care* 16 (1988), brings up the issue of wrongful life in a discussion of surrogate motherhood.

2. If the offspring is handicapped or suffers poor health due to negligence by the mother during pregnancy, such as drug taking, the wrong suffered does not count as wrongful life or disability. The wrong here is not that of bringing a child into existence, but is the wrong of harming an already existent child.

3. Allen Buchanan, Dan W. Brock, Norman Daniels, and Daniel Wikler, *From Chance to Choice: Genetics and Justice* (Cambridge: Cambridge University Press, 2000), pp. 224–225.

4. Joel Feinberg, *Harm to Others* (New York: Oxford University Press, 1984), p. 102, takes this line of argument.

5. This argument has been used by Steinbock, "Surrogate Motherhood as Prenatal Adoption," in relation to surrogacy, and by John A. Robertson, "The Question of Human Cloning," *Hastings Center Report* 24 (1994), in relation to cloning.

6. I follow Buchanan, et al., *From Chance to Choice*, p. 225, in separating wrongful disability from wrongful life. Many other philosophers treat both kinds of cases as wrongful life cases.

7. Another way to mark the difference is to distinguish between lives that are simply flawed but not anguished, and lives that are anguished, as Melinda A. Roberts does in "Can it Ever Be Better Never to Have Existed at All?" *Journal of Applied Philosophy* 20 (2003), pp. 159–185.

8. The problem is famously associated with the work of Derek Parfit in *Reasons and Persons* (New York: Oxford University Press, 1984), pp. 351–379.

9. As Melinda Roberts, "Can it Ever Be Better Never to Have Existed at All?" pp. 169–170, points out, the idea that an anguished existence is not worse than non-existence opens the door to a conclusion that is not just repugnant (the situation imagined by Parfit, *Reasons and Persons*, pp. 381–390) but beyond repugnant (the situation imagined in the text here).

10. John Harris, *Clones, Genes, and Immortality* (Oxford: Oxford University Press, 1998), p. 110.

11. Melinda A. Roberts, "Human Cloning: A Case of No Harm Done?" *The Journal of Medicine and Philosophy* 21 (1996), pp. 547–549, has, however, argued that even if the argument is valid in some cases, it is not in general true: for a cloned human being can be harmed in being brought into existence as a clone.

12. Buchanan, et al., *From Chance to Choice: Genetics and Justice*, p. 235.

13. Roberts, "Can it Ever Be Better Never to Have Existed at All?" p. 168, argues that, because in not existing a person's level of well-being is zero, a level of well-being that is negative (a life not worth living) is worse than non-existence.

14. Joel Feinberg, "Wrongful Life and the Counterfactual Element in Harming," *Social Philosophy and Policy* 4 (1986), p. 164, points out that faced with being born again after death as a Tay-Sachs baby, most people would opt for immediate permanent extinction.

15. Harris, *Clones, Genes, and Immortality*, p. 107. Feinberg, *Harm to Others*, p. 103, discussing a case from Parfit, recognizes this as a morally wrongful act, but not as a wrong to the child in the sense that it can have a "personal grievance" against anyone.

16. The example and point is made by Harris, *Clones, Genes, and Immortality*, pp. 106–107. Harris disagrees with Feinberg who thinks that a child can be wronged only where nonexistence is preferable to her present condition.

17. Harris and Feinberg hold opposing views regarding whether harming or wronging is the broader category. For Feinberg, a child caused to exist in a severely harmed condition

has not been harmed if she has not been made worse off, but the mother, as in the selective termination example, could still have done a wrong. Harris reverses the order of the distinction, holding that harming is not necessarily wronging because someone might be harmed in order to benefit them, as in medical treatments with adverse side-effects, and perhaps as in voluntary euthanasia.

18. Roberts, "Can it Ever Be Better Never to Have Existed at All?" contrasts aggregative consequentialism with person-based consequentialism. Buchanan, et al., *From Chance to Choice: Genetics and Justice*, pp. 248–250, contrasts non-person-affecting principles with person-affecting principles.

19. Harris, *Clones, Genes, and Immortality*, p. 111. Harris specifically says that *the children would be wronged*, not just that the parents did a wrong. In other words, the children have a wrongful life claim since their lives are not worth living, but no wrongful disability claim otherwise.

20. When Harris deems it wrong to introduce avoidable suffering into the world, even if there is a net benefit, his account of wrong seems to go beyond a utilitarian weighting of benefits and harms.

21. Harris, *Clones, Genes, and Immortality*, p. 118, makes a distinction between legal wrong and moral wrong. Taking his account as concerned with the former, it will *not* be a problem for Harris's account that it does not show what is morally wrong in those cases where severely harmed children are brought into existence and there is no alternative without the harm. One could, however, take issue with his suggestion that legal claims are not appropriate in these cases.

22. Buchanan, et al., *From Chance to Choice: Genetics and Justice*, p. 248. Parfit discusses the distinction in "Future Generations: Further Problems," *Philosophy and Public Affairs* 11 (1982), pp. 113–172.

23. Rahul Kumar, "Who Can Be Wronged?" *Philosophy and Public Affairs* 31 (2003), p. 105. Kumar provides examples of wronging without harming (or without additional harming) on pp. 103–104.

24. Ibid., p. 101.

25. Ibid., p. 106. Scanlonian contractualism is drawn from T. M. Scanlon, *What We Owe to Each Other* (Cambridge, Mass.: Harvard University Press, 1998).

26. Ibid., p. 114. In other words, the caretaker-child relationship that gives rise to legitimate expectations tracks the child 'type,' not its token identity.

27. For a discussion of procreative liberty, see John A. Robertson, *Children of Choice: Freedom and the New Reproductive Technologies* (Princeton: Princeton University Press, 1994). John Stuart Mill in *On Liberty* sees the freedom to reproduce as morally limited by the prospects for the offspring to be properly educated and cared for.

28. I also think that some of the alleged harms are in fact based on misconceptions or false assumptions about cloning, in particular the thesis of genetic determinism. In my paper, "Human Dignity and the Human Genome," I criticize the objection that reproductive human cloning would violate human dignity.

29. Despite agreement with Feinberg on many matters, Bonnie Steinbock, "The Logical Case for 'Wrongful Life,'" *Hastings Center Report* 16 (1986), p. 19, also challenges his

THE ARGUMENT AGAINST CLONING 271

view that it is necessary, for the child to have been wronged, that it be better off never having been born. She presents a dilemma: "Either we have to maintain, implausibly, that a rational person would prefer nonexistence to, say, being born deaf, or we must dismiss the suits of less seriously impaired children, who may still have expensive medical and educational needs."

30. The reasons Raelians and rogue scientists have for carrying out human cloning are clearly not good enough! Given the high risk of harm, if Morris and Lois believe that their desire to have a genetic offspring is a good enough reason to produce a clone of Morris, they can be criticized for an insufficient concern for the offspring's well-being.

31. On my account of reasons that justify the risk of harm, it has to be the panda's choice to procreate by cloning, chosen for the panda's own reasons. The worry here is whether we are reading human reasons, such as preserving an endangered species, into the panda's decision-making about bringing into existence an offspring by cloning.

32. Thanks to Mark Nowacki for bringing this point to my attention.

33. Of course, there may be other things that the physician does, besides treating him with a placebo, that gives him cause to complain.

34. This paragraph serves the purpose of clarification, not the purpose of arguing for the superiority of Aristotelian ethics over Kantian ethics, which would take a much longer paper going beyond the subject of this paper.

35. This clause comes from principle M, originally from Buchanan, et al., *From Chance to Choice: Genetics and Justice*, p. 226, and reproduced in Kumar, "Who Can Be Wronged?" p. 112.

36. I adapt the example from James Woodward, "The Non-Identity Problem," *Ethics* 96 (1986), p. 810, who uses it to make a different point.

37. J. K. Rowling, *Harry Potter and the Chamber of Secrets* (New York: Scholastic Inc., 1999).

38. Feinberg, *Harm to Others*, p. 99; Steinbock, "The Logical Case for 'Wrongful Life,'" p. 19.

39. It is here that Steinbock, ibid., and I would part company with Feinberg, *Harm to Others*, p. 102, who insists that a child is wronged only if it is negligently or deliberately permitted to be born into a life not worth living. That is, the only wrong is wrongful life, not wrongful disability. As I am contrasting the account with my own, I shall henceforth refer to it as Steinbock's account. Steinbock uses wrongful life to include wrongful disability.

40. Of course, if it turns out that it is necessary to identify basic human interests before the appropriate goals of a flourishing human life can be established, my account will not have the advantage on this point that I think it has. Once again, it seems that I need an objective account by which to evaluate the goals that constitute reasons for acting. Although I do not have the space to defend such an account in this paper, I will mention that a natural law ethics would list both procreation and avoiding harm to others as basic human goods.

41. Feinberg, "Wrongful Life and the Counterfactual Element in Harming," p.172.

42. Feinberg, ibid., p. 171, imagines a mother who "wants the experience later of mothering a child that will be more completely dependent on her, and for a longer period, than a normal

child would or, even worse, because she wishes to glory sadistically in a child's frustrations and sufferings."

43. Ibid., p. 169.

44. Thanks to Fred Adams for organizing the Ethics and the Life Sciences Conference at the University of Delaware in October 2004, to Stephen C. Taylor for serving as commentator on my paper at the conference, and to the audience for their questions. An earlier version of the paper was presented at a colloquium at the National University of Singapore in June 2004, at which I also benefited from comments and suggestions from the audience.

ETHICS AND THE LIFE SCIENCES

WRONGFUL LIFE, SUICIDE, AND EUTHANASIA

JAKOB ELSTER
UNIVERSITY OF OSLO

ABSTRACT: "Wrongful life" claims are made by persons born with a disease to the effect that they should not have been born. I ask whether we can say that if someone claims that he would have been better off if he were not born, he would be better off if he died. I examine the relationship between the following propositions:

(1) It would have been better for me if I were not born.
(2) My life (as a whole) is not worth living.
(3) It would be better for me if I died.
(4) I desire to die.
(5) I should commit suicide/ ask for euthanasia.

If a person claims that he would have preferred not to be born, this normally implies that it would be better for him if he died. But this does not necessarily imply that he desires to die, or that he should commit suicide.

1. INTRODUCTION

The many advances in the technology of prenatal diagnosis in recent years—and more progress will certainly be made in the years to come—has made it possible to detect with increasing accuracy an increasing number of diseases and disorders before a child is born, thus making it possible to abort foetuses with these diseases.[1] Furthermore, the testing of embryos before implantation and carrier testing of couples before they conceive make it possible to identify possible diseases after *in vitro* conception, but before pregnancy starts, and even before conception.

© 2007 Philosophy Documentation Center   pp. 273–282

This technology raises a large number of ethical problems, one of which is the possibility that children who are born with a disease which the parents or the doctor knew about or could have known about during or before the pregnancy might sue either their parents or the doctor for having been born. In these cases, called "wrongful life" cases, the child typically claims either that the doctor should have given the parents the information about the disease, so that the parents could have decided not to have the child (either by choosing another embryo, by not conceiving, or through abortion), or, if the parents had access to this information, that the parents should have chosen not to have the child.

The development of new forms of reproductive technology might also lead to wrongful life-claims, since, at least when these technologies are not fully developed, they may involve risks for the future child, who nevertheless could not have been created without these technologies.

The wrongful life-claim "I would rather not have been born" must be distinguished from the claim "I would rather have been born healthy." There are certain situations in which the latter claim cannot meaningfully be made.[2] These are situations where the disease or disability which makes the life unbearable for the person making the claim is inseparable from that person's existence. In these cases, a person cannot wish to have been born without the disease or disability, since he could not have been the person he is without the disease. This follows from what Derek Parfit calls the "time-dependence claim": "If any particular person had not been conceived when he was in fact conceived, it is *in fact* true that he would never have existed."[3] It is this inseparability of the person's existence and his disease or disability which accounts for the particularity of wrongful life-cases.

2. EXTENDING OUR INTUITIONS FROM EUTHANASIA AND SUICIDE TO WRONGFUL LIFE CASES

Wrongful life-claims raise several ethical and philosophical problems. Some of these problems, as well as the paradoxes involved in wrongful life claims, have been treated by, among others, Derek Parfit and David Heyd.[4] In this paper, I will leave most of these problems aside, in order to examine one particular problem which has not received sufficient attention, viz. the relationship between the claim made in wrongful life- cases and the desire for euthanasia or suicide.[5] I will ask the following question: If someone claims, as they do in wrongful life cases, that they would have been better off if they were not born, would they not then be better off if they were killed? This question can be divided into two sub-questions:

1. Does the wrongful life claim imply that it would be better for the person making the claim if he died?
2. If the answer to the preceding question is positive, does this give grounds for offering euthanasia to the person, or for the person killing himself?

The practices of euthanasia and suicide are problematic in several respects; they raise ethical problems concerning the morality of the acts in question, psychological problems both for those who commit the acts (in the case of euthanasia) and for the relatives of the deceased, and social problems for society as a whole. These problems (which I will not discuss in any detail) provide, as we shall see, one reason why a wrongful life claim need not entail a desire for euthanasia or suicide. But these problems, serious as they are, should not be allowed to obscure the first sub-question: does a wrongful life claim imply that it would be better for a person that he died?

One could separate the general question of what a wrongful life claim implies from the practices of suicide and euthanasia by rephrasing the question as follows: Does a wrongful life claim imply that, if the person making the claim were to suddenly and painlessly die accidentally, this would be better for that person than if he continues to live? The reason I nevertheless choose as my starting point the very problematic ideas of suicide and euthanasia, is that wrongful life cases are often hard to get an intuitive grasp on, whereas we have somewhat clearer intuitions concerning suicide and euthanasia, if only because we have thought about them more. So if we can show that wrongful life claims sometimes entail that suicide or euthanasia are appropriate, this can help us get a better grasp on wrongful life claims. I do not intend to propose euthanasia or suicide as a practical solution to wrongful life cases: this would seem abhorrent to many. But precisely because of the abhorrent character of such a solution, if one could show that euthanasia was a natural consequence of wrongful life claims, this could perhaps constitute a sort of *reductio ad absurdum* of wrongful life claims. Most people accept euthanasia, if at all, only in cases of terminal illness and intense suffering: only in these cases is one said to be better off dead. The question then is if someone who is not in such a situation can claim that he would have been better off not to be born (as one does in wrongful life cases) and at the same time not claim that he would have been better off dead. If the latter claim (I would be better off dead) is entailed by the former claim (I would have been better off not to be born), our intuitive negative reaction to the latter claim might extend to the former claim as well. And vice versa, if we accept euthanasia and suicide to a certain degree, this acceptance might give us a reason to accept wrongful life claims in similar cases.

My general strategy is thus to see if we can draw on our intuitions about suicide and euthanasia in order to judge the legitimacy of wrongful life claims. In order to do this, we must see what further claims are implicit in a wrongful life claim.

The explicit claim one makes in a wrongful life claim is:

(1) It would have been better for me if I were not born.

The flip-side of this claim is:

(2) My life (as a whole) is not worth living.

The question I wish to ask is whether this claim implies a third claim:

(3) It would be better for me if I died.

Also important is the question whether this third claim implies:

(4) I desire to die.

or the stronger claim:

(5) I should commit suicide/ask for euthanasia.

Examining whether claim (1) and (2) imply claims (3) to (5), might be a way of examining if, and in what cases, claims (1) and (2) can be reasonably made.

The focus of this article is somewhat different from the focus of many discussions of wrongful life-cases. I take the claims (1) and (2) to be two complementary aspects of a wrongful life claim: (1) and (2) entail each other, but their focus is somewhat different. (1) is backward-looking, comparing a person's actual life with non-existence, whereas (2) is only concerned with evaluating the actual life. (2) is not necessarily forward-looking, but it easily invites questions about whether it does not follow that the person making the claim would be better off if he died. The main focus in the discussion of wrongful life-cases has been on the first claim, because of the philosophical and logical problems involved in comparing one's present quality of life with one's non-existence. For example, David Heyd argues in his book *Genethics—Moral Issues in the Creation of People* that such a comparison is impossible and so wrongful life claims do not make sense.

I will, however, focus on the claim (2) that life is not worth living. This claim is directly entailed by the first claim, so that if one wanted to invalidate a wrongful life-claim, one could either follow Heyd and say that (1) makes no sense, or one could refute (2). My guess is that most people's immediate negative reaction to wrongful life claims is a reaction against (2), not a reaction based on the impossibility of comparing an actual life with nothing. We are used to the concept of a life not being worth living from discussions of suicide and euthanasia, but in those cases it is usually just a phase of life which is not worth living, not a life as a whole. (I am supposing here that the quality of life of the person making the wrongful life claim is more or less stable over his lifetime. I will return to the question of what follows when this is not the case.)

3. WHY (3) NEED NOT IMPLY (4) OR (5)

Imagine a case where someone—let us call him Peter—makes a wrongful life claim. Let us ask Peter if he is contradicting his own wrongful life claim since he does not commit suicide. There are several answers Peter might give:

a) Peter might answer by giving arguments related to the particular nature of suicide, which he might be opposed to for religious or philosophical reasons. This kind of answer does not deny that (3) can entail (4), but it denies that (5) follows from (4).

b) Or Peter might claim that *he* would be better off if he killed himself, but he does not desire to die, since his death would make the people who cared about him worse off.[6] This would deny that (4) follows from (3).

c) Peter might also say that the people who cared about him would be worse off not because he died, but because he committed suicide (or because they participated in the euthanasia), and they would have difficulties living with the guilt caused by this. They would then be better off if he did not commit suicide (or ask for euthanasia), but they would have been even better off if he died from an accident, so they had nothing to feel guilty about. This answer is another way of denying that (5) follows from (4).

This discussion, in particular answers b) and c), points to what appears to be a major difference between saying one would prefer not to have been born, and saying one would prefer to die. Once a person has already been born, he has a history and is related to other human beings, for whom he might matter very much. So even though it might be better for Peter if he died, he would not want this, because it would be worse for the people who cared about him. Similarly, Peter might have important life projects which he wants to finish.[7] Therefore, he might not want to die, even though he recognizes that he would be better off if he died. Of course, this objection raises the question of whether Peter can even be said to be better off dead, if he has a life project or relations with other people which are so important that they make him not want to die. In order for Peter to say that he would be better off dead, but that he nevertheless does not desire to die, because he desires that certain preferences concerning people he cares about or concerning his life project be satisfied, we need to claim that what is best for Peter can be separated from the satisfaction of his preferences. And even if this is the case, the question remains whether Peter in this situation can be said to have a life not worth living. I will return to this question.

We have seen that there can be reasons related to Peter's relationships and convictions which stop him from going from claim (3) to claim (4) (although it would be better for him to die, the thought of how those who care about him would react or about his life projects stops him from wanting to die), and from claim (4) to claim (5) (though he might even want to die, his religious or philosophical convictions, or considerations of the guilt those who care about him would feel, makes Peter opposed to suicide and euthanasia). I will now consider if there are reasons hindering the implication from claim (2) to claim (3).

4. IF ONE'S LIFE IS NOT WORTH LIVING, WOULD ONE BE BETTER OFF IF ONE DIED?

I want to claim that—with two exceptions, which I will come back to—there are no reasons stopping us from going from claim (2) to claim (3): if one's life really is not worth living, one would clearly be better off dead. We can see this by looking at the inverse case, when life is very much worth living. Why is death in this case a bad thing, if death is just non-existence, which will not be experienced by any subject? Thomas Nagel writes, in his article "Death," where he tries to explain why death sometimes is undesirable: "If we are to make sense of the view that to die is bad, it must be on the ground that life is a good and death is the corresponding deprivation or loss, but not because of any positive features, but because of the desirability of what it removes."[8] Nagel's point is that death is not a state which involves a negative experience in itself, but is negative because it entails the loss of something good, viz. life. If this view is correct, it should follow inversely that if life is bad (and that is what is claimed when one says it is not worth living), death is a positive and desirable[9] thing, because it entails the loss of something bad.

It might be useful to examine further on what grounds we can make the claim (2) that life is not worth living. The claim needs to be made more precise. One can see what it means for someone to say about a person's life, after his death, that it had not been worth living: it means that the *sum* of that person's experiences turned out to be negative.[10] It is less clear, however, how this claim can be made while a person is still alive.

This idea that a man's life can only be judged to be good or bad after he is dead was a commonplace in ancient Greece, expressed notably by Sophocles:

count no man happy till he dies, free of pain at last.[11]

The quote from Sophocles could be interpreted in a more pessimistic manner, implying that only when a person is dead, is he happy, but I do not think that is a correct interpretation. Indeed, Aristotle, commenting on a passage from Herodotus, where Herodotus puts the same idea in the mouth of Solon,[12] writes: "We do not say, then, that someone is happy during the time he is dead, and Solon's point is not this [absurd one], but rather that when a human being has died, we can safely pronounce [that he was] blessed [before he died], on the assumption that he is now finally beyond evils and misfortunes."[13]

But since one's future is always open for new negative, or positive, events, how can one say *in the middle of one's life* that life is not worth living? One can certainly say about one's life *so far* that it was not worth living. But can one, before one is dead, say it about one's life as a whole, and thus come to the conclusion that one would be better off dead? I think that if one can reasonably expect nothing better of future life than what one has had in the past, such a claim can indeed be made. Such a claim would not be based uniquely on a person's present situation, but on an estimation of what the final judgement of this person's life will be after his death. If a person estimates that his life will very probably be considered, after his death, as not worth having been lived, this can give him grounds for making the claim (2)—that his life is not worth living—even in the middle of his life. And if he does makes this claim, it follows, as we saw earlier, that he should claim (3) that he would be better off if he died.

5. OBJECTIONS AGAINST INFERRING (3) FROM (2)

I will now consider three objections against going from (2) to (3). One possible objection could be that precisely because a person's life so far has been awful, this person has more than other people a reason to go on living in order to try to lead a life so happy that his future happiness will weigh up for his past suffering, so that his life as a whole will come out positive. Although this argument is psychologically understandable, it would be irrational, if it is really the case that the person has no reason to expect that his life will go better. It is the equivalent of a gambler who, having lost a lot of money on the horses, considers this to be a reason to keep on playing, since, having already spent so much, he ought to give himself a chance to finally win.

There is another, more serious objection against my claim that (2) entails (3).[14] I have assumed that a person's quality of life is more or less stable over his lifetime. But of course this need not be true. In particular, someone's life might overall be so full of suffering that (2) is true, but the person may be at a point in life where his future is nevertheless worth living. One can imagine for example that Peter is born with a horrible disease which makes him suffer terribly every waking minute. When Peter is fifty years old, a cure is invented, so that he will no longer suffer, but the pain of his first fifty years was so bad that he still wants to make claim (2). He need not however make claim (3), since the rest of his life will be worth living.

I do not want to deny that such cases might exist, and in these cases (3) does not follow from (2). But I do think that such cases are rare. This is not because I believe that the cases where the first part of someone's life is bad and the second part is good are rare. It is rather because I think that in most such cases, the life as a whole is worth living, even if the total of suffering in the bad part of the life exceeds the total of good experiences in the good part of the life. This is because the correct way of judging whether someone's life is worth living might not be by summing up his positive and negative experiences. Rather, a person's life should be seen as what G. E. Moore calls an "organic unity," that is a whole such that "The value of a whole must not be assumed to be the same as the sum of the values of its parts."[15] (Here I simply suggest this reply to the objection. Developing a full-fledged view of a human life as an organic unity would require further development.)

A third objection is related to a point we saw earlier:[16] The person making the wrongful life claim might have a life project which he wants to finish. We saw earlier that this possibility permits us to block the inference from (3) to (4). But it might also permit us to block the inference from (2) to (3). If the quality of my life is not simply dependent on my subjective experience, but also on certain preferences I have being satisfied, it may be the case that I would not be better off if I died no matter what my subjective experience is. This idea explains why we sometimes can say that things which happen after a person's death affect the quality of that person's life, if the things which happened affect his life projects or other important preferences, in particular preferences concerning persons he cares about.

I suspect that in most cases where one has a life project which makes one accept great suffering in order to finish it, this project is of such value that it does make the life as a whole worth living. However, I do recognize that there may be cases where someone has a life not worth living and yet would not be better off dead because his life project would then not be fulfilled. One such case would be where Peter has wronged someone terribly and his life project is to repair this wrong and redeem himself. Although this project is so important for Peter that it makes it the case that Peter would not be best off dead, it remains true that Peter's life is not worth living, because if Peter had not lived, there would no wrong to redeem, and so the project would have no value.[17] There are therefore two important, but I believe, rare exceptions to my claim that (3) always follows from (2).

6. WHAT LESSONS CAN WE DRAW FROM THIS?

As mentioned above, people often have clearer intuitions about when to accept euthanasia than about how to consider wrongful life claims. But what lessons can we draw from our discussion concerning the possibility of extending our intuitions concerning euthanasia and suicide to cover wrongful life claims? Indeed there is a parallel between euthanasia and wrongful life claims, but the parallel is not absolute, so we will not have a one-to-one-relationship saying that in the cases when one accepts (5) one always accepts (1) and (2) or vice versa.

It seems to me that, if people accept the practice of euthanasia at all, the claim (3) that one would be better off dead is a necessary but not sufficient condition for the acceptability of euthanasia. (We must still overcome the objections against going from (3) to (4) and (5).) (3) also plays a central role in most wrongful life claims (although we have seen two possible exceptions.) One way of refuting a wrongful life claim is therefore to refute that (3) holds (and show that this is not one of the exceptional cases where (3) does not follow from (2)). But if in a given case we accept euthanasia, we cannot reject the wrongful-life-claim on the ground that (3) does not hold. That is perhaps the main positive conclusion which we can draw.

But in wrongful life cases one also makes the claim (2), where one says something about the quality of one's whole life, not just about the quality of the remaining part of one's life. This claim need not follow from (3), so we can reject a wrongful life-claim even in cases where we would accept a euthanasia claim, because we accept (3), but not (2). This would typically be the case where someone suffers at the end of a long and good life.

Furthermore, it is not the case that we should only accept wrongful life claims for the cases in which we accept euthanasia. When we accept euthanasia, we go from claim (3) to claims (4) and (5). We thus count as insufficiently important the reasons which might hinder the inferences from (3) to (4) and (5)—viz., that suicide and euthanasia may be morally wrong, or that the person's death would have negative consequences for that person's relations or life projects. Accepting euthanasia thus involves counting the badness of life to be so great that it outweighs the *prima facie* reasons against wanting to die and against taking a life.[18] Accepting the claim made in wrongful life cases, however, does not involve a judgement about the force of these *prima facie* reasons, even if it does involve accepting that one can make the claim (3) that it would be better for a person if he died. So there can be cases where we refuse to accept (5), yet we accept (3).

7. CONCLUSION

If a person makes the claim that he would have preferred not to be born, this normally implies the claim that it would be better for him if he died. This does not necessarily imply that he desires to die, or that he should commit suicide, since there are other considerations, related to philosophical and religious convictions, to his life projects and to his relationships with other people, that may stop him from drawing such conclusions.

This discussion does not seek to solve the paradoxes involved in wrongful life claims, but it can give an indication of when such a claim can even be considered. Saying that it would be better to be dead is indeed a very strong claim, and it is not sure that everyone making a wrongful life claim would be prepared to make such a strong claim.

NOTES

1. I want to thank David Heyd, Kai Draper and Lene Bomann-Larsen for useful comments to earlier versions of this article.

2. Of course, one might also claim, as e.g., David Heyd does, that the former claim might not meaningfully be made either.

3. *Reasons and Persons*, p. 351.

4. David Heyd, *Genethics—Moral Issues in the Creation of People* especially chapter 1, and Derek Parfit, *Reasons and Persons*, part 4.

5. Euthanasia and suicide are two different practices and although they share many features, they also raise different moral issues. For the purposes of this article, however, I will treat the two as morally equivalent.

6. The claim that his death would cause them pain would not be a sufficient reason against suicide if the person thought that in the long run they would be better off after his death.

7. I owe this point to Kai Draper.

8. Reprinted in *Mortal Questions*, p. 4.

9. At least if one ignores the possible reasons hindering one from going from claim (3) to claim (4).

10. Some people would claim that such a calculation is unfeasible, and in practice, this might be the case; in principle, it might be feasible however, at least in extreme cases where it seems clear what the outcome is.

11. *Oedipus the King* (last line of the play).

12. *Histories*, book I, pp. 31–36.

13. *Nicomachean Ethics*, 1100a 15-20.

14. I owe this objection to Kai Draper.

15. *Principia Ethica*, p. 28.

16. I owe this objection too to Kai Draper.

17. A similar objection could be made about personal relationships. If someone has such close relationships with other persons that these persons would suffer greatly if he died, and if furthermore his concern for them is so great that it stops him from thinking that he

would be better off if he died, does it not follow that his life is worth living? Mutual caring relationships are precisely a central part of what makes life worth living. Although this is often the case, there can be cases where someone can deny (3) on the grounds that he would be worse off if his loved ones suffered because of his death, yet affirm (2). An example would be if Peter is born with a disease and his mother has sacrificed everything else to take care of him. Peter loves his mother and recognizes that she would be devastated if he died. Nevertheless, he recognizes that it would have been better for her had he not been born at all. In this case, the mutual loving relationship which makes it the case that Peter does not think he would be better off dead might not be enough to make Peter's life worth living.

18. Most people consider that there are such reasons, even if the person involved thinks it would be better for him to be dead. However, the reasons against taking a life in the case of suicide may be less strong than the reasons against taking a life in the case of euthanasia. Many people would nevertheless consider suicide to be in some degree morally problematic.

BIBLIOGRAPHY

Aristotle. 1985. *Nicomachean Ethics*, trans. Terence Irwin. Hackett Publishing Company

Heyd, David. 1992. *Genethics—Moral Issues in the Creation of People*. University of California Press

Moore, G. E. 1960. *Principia Ethica*. Cambridge University Press

Nagel, Thomas. 1979. *Mortal Questions*. Cambridge University Press

Parfit, Derek. 1987. *Reasons and Persons*. Clarendon Press

Sophocles. 1984. *Oedipus the King* in *The Three Theban Plays*, trans. Robert Fagles. Penguin Classics

ETHICS AND THE LIFE SCIENCES

ARE WE GOOD ENOUGH?
THE PARADOX OF GENETIC ENHANCEMENT

LISA BELLANTONI
ALBRIGHT COLLEGE

ABSTRACT: If we can enhance ourselves genetically, should we? A plague of recent works in bioethics insist that we should not. Bill McKibben, for example, joins a chorus of theorists who oppose enhancement efforts not because they might harm individuals or undermine social practices, but because they imperil a human nature that is already "good enough," and threaten to catapult us into a "post-human" world. But what's wrong, exactly, with being post-human? These positions never answer that question. They fail biologically, sociologically, and ethically, I argue, because just as ethics aims to improve us, bioethics aims properly to direct biotechnologies to enhance human life prospects. To refuse that injunction is to assert that we can, but will not, improve ourselves. It is to undercut any basis for bioethics, and for human progress, moral or material. It is, to re-work Leon Kass's singularly apt phrase, moral hubris run amok.

I

If we can genetically engineer better people, should we? That question might ask if we are virtuous, or just, or principled enough to use such technologies wisely. But it might ask instead, following Leon Kass,[1] Hans Jonas,[2] and Gerald McKenny[3] among others, if we aren't "already good enough." While these theorists decry a brave new "post-human" world, they do not defend their operative premise: that human nature is already of such over-riding moral worth that we ought not to modify it, whether for therapeutic or for eugenic purposes. In contrast, Bill McKibben's recent book *Enough: Staying Human in an Engineered Age*, explicitly maintains that we should refuse any genetic interventions that alter our nature at the germ-line.[4] McKibben maintains that our pursuits bear no point, pose no challenge, earn no admiration, if we are engineered, for example, with the enhanced athletic or artistic capacities to perform them. Such feats prove "inauthentic" because we select neither those capacities nor how we pursue them. Those activities offer no means for self-discovery; as the "mysterious" processes by which we now fashion our identities "from the pieces around us" give way to future parents' "complete power" to design their children, and even to create a genetic over-class so "superior from the human past" that they "won't be [recognizably] our kids" (p. 61). Above all, McKibben laments that we "may engineer ourselves out of existence" (p. 8). Running that risk implies that there is "nothing particularly significant about the human present," "nothing unique, special, even sacred." In contrast, two facets under- write that significance: our struggle with finitude, including our genetic limitations, and our capacity to choose how we face that finitude: Rejecting the "techno-zealots" efforts to enhance us, he affirms instead that, "What makes us unique is that we can restrain ourselves. We can decide not to do something that we are able to do. We can set limits on our desires. We can say, "Enough" (pp. 203–204).

These two aspects of human experience, however, are unaffected by any variant of our physical nature. Biological beings, however well engineered, could never secure biological invulnerability. Human organisms, for instance, face an ever-evolving rush of bacteria, viruses, and environmental pathogens, and if none of those kill us, the incessant cell division that sustains our lives ultimately ends them. Genetic enhancement might modify these vulnerabilities, but could never eliminate them. Moreover, holding that genetic modifications alone may render us "post-human" denies basic principles of evolutionary biology. Humans have, after all, undergone unceasing genetic modification since climbing down from the trees. If genetic modification per se is the criterion for demarcating "post-human" as a biological category, we are already post-human. Indeed, while we were "good enough" at each point along our evolutionary path, as evidenced by our species' survival, we were equally not "good enough" at that same time, and on the same basis: biological existence abhors genetic stagnation.

McKibben might mean instead that we are "good enough" not in our biological identity, but in our capacities. He advances such a case when decrying enhancements to athletic or musical ability as diminishing the value of human achievement in those areas. But here he mistakes genetic potential for realized skill. Lengthened muscle fibers or expanded lung capacity will not, by themselves, unfurl the habits and skills—and drives—that fuel exceptional athletic performance. However much we learn about the genetic bases of such accomplishments, we are unlikely to discover, or manufacture, a "Michael Jordan gene." McKibben might grant that but insist that far future modifications could produce children, for example, with the hand-eye coordination of a Barry Bonds, and the drives to use that capacity in particular ways. But genetic potential, however expanded, is no more likely to exercise such a determinative impact on our activities than it does today. Many people with extra-ordinary talents do not develop them, some for lack of opportunity, others through deliberate refusal. Conversely, some people with less genetic potential to succeed doggedly pursue their favored activities. Are those talent-challenged people in the grips of an as yet undiscovered baseball or basketball gene that drives them to pursuits for which they are otherwise unsuited? If we hold, as most current theorists do, that environment and heredity both under-write these interests, we have no basis for expecting that the proportionate influence of those two facets will change, i.e., that it will be any easier to engineer motivation than it is at present.

Moreover, if our range of genetic potentials expands, so too will the realized value McKibben attributes to our mastery of the activities those potentials permit. He dismisses enhanced athletic capacities as futile if other competitors are comparably enhanced. But if such enhancement improves the competitors' aggregate achievement, surely it doesn't undercut the prospect that those with superior habits, training, skill, and drive will still excel beyond that aggregate level. And if such enhancements are widely distributed, they may raise the aggregate and exceptional levels of performance while also broadening the base of those who could enjoy such activities at some still considerable level of skill and accomplishment. On these counts, if we follow McKibben in locating these activities' primary value in our struggle to perform them, we could enhance that struggle both (a) by broadening the range of activities we can perform, and (b) by raising the standards of human accomplishment that those pursuits permit.

McKibben might object here that the sheer contingency by which those capacities are now distributed lends them some of their human value: their unexpected occurrence, and their openness to unanticipated development. Yet even if the proportion of selected traits rises, and the distribution of capacities becomes more homogenous, the impact of genetic, as opposed to social or personal, influences over any individual's interests and vocations, is unlikely to increase. For that reason, it's equally unlikely that the novelty that any individual introduces into his or her social milieu will diminish. If many more individuals have many more potential talents, then we may expect the opposite effect. Individuals would have more rather than less genetically mediated choice, because they would have more genetic potentials to select among. Similarly, if such genetic potential were more broadly distributed across the population, and the norms for accomplishment expanded, we might just as readily predict greater rather than less individual diversity, because carving out distinctive niches would come under greater individual control.

The points above indicate several critical weaknesses in McKibben's position. First, it espouses a genetic determinism inconsistent with how deeply environmental influences shape our genetic potentials. Second, it ignores how genetic enhancement may expand individual choices. More broadly, it never establishes why genetic enhancement changes human nature in ways that other enhancements of our capacities do not. To take a simple example, would expanding the human hearing range change our communicative capacities more than writing, telephones, or the Internet have? Such a modification seems trivial compared to the latter cases. Yet for McKibben, an altered biology changes us in ways that modifying how we transmit information does not. He is hardly alone in this belief. But if we embrace such a patent genetic determinism, his criticism of genetic enhancement is wholly unfounded. If we are our genes, they will determine our behavior whether we acquire them through natural accident, parental choice, or the state lottery. Yet if we reject that genetic determinism, we must grant that we can refuse our genetic potentials no less than we can refuse contemporary technologies. Indeed, if we affirm with McKibben that the struggle with finitude defines human activities, then we should refuse technological enhancements on precisely the same ground. The only way to resist that stark conclusion is to insist that genetic modification changes our nature in ways that other technological modifications do not, and that those changes are morally verboten.

But just what is that nature that McKibben and like-minded theorists such as Kass aim to protect? Despite their impassioned defense of human dignity, their opposition to genetic enhancement takes a quite different tack. Kass laments such efforts as emblems of human egoism, hubris, and choice run amok. McKibben never tells us what's so great about human nature, apart from its enduring struggle against finitude, but warns against the despotisms and injustices that present people will heap upon future persons if given half a chance. These concerns thematize basic elements of human nature long criticized by ethicists: egoism, selfishness, greed, racism. Nevertheless, McKibben says, whatever we hold about intractable human behavioral patterns such as "group solidarity" and our "short-term focus," the humans who practice these patterns century after century "are still the same people" (pp. 59–61, 63). In contrast, genetically enhanced post-humans would seem so different that we could not even envision their interests, because they would not "be like us" (p. 65). That lament, however, begs two critical questions: First, why we would want future persons to remain so much like us, and second, whether we should endorse that desire, and adopt it as an over-arching ethical mandate.

McKibben answers that query in the affirmative. If we have a nature, we may describe it as variously egoistic or altruistic, hostile or beneficent, clannish or sociable. We might even accept that this mix is "good enough" to make us not only who we are, but who we should aspire to be. Yet, McKibben asks, given the option of modifying this nature, "Would we ever have enough strength to choose life as we've always known it?" (p. 223). Even if sustaining this nature is the harder choice, he says, it is the more morally valuable choice, an acceptance of human finitude, an achievement of human maturity, of "Life at peace with itself" (p. 223). But as a moral mandate, this aspiration is fatally flawed. Biologically, life has never been a peaceful affair, and gives no indication of becoming such. More tellingly, if we isolate not the biological but the human element of human nature, the vast majority of our moral inheritances counsel us not to live by or to affirm, but to refashion that nature, to redirect it in pursuit of more humanly worthy aims, to improve it. Indeed, any ethical evaluation of our behavior presupposes (a) that a gap obtains between what we are and what we can be, (b) that we have some capacity to move from what we are to what we should be, and (c) that given (a) and (b), we are obliged to modify that nature in service of that ideal.

Ethics is a crucible. It is not a celebration of human nature so much as a remedy for its deepest flaws. Its point is emphatically not to justify or to idolize human nature, but to modify or recreate it. For their part, McKibben and Kass depart starkly from the ethicist's habit of locating human worth in human capacities. They do not ascribe the worth of our human nature to our rationality, our language, our moral capacity, or any other feature historically used to confer moral worth on us alone. But what's missing without these referents is any recognition of the social role that ethics has traditionally occupied. For the ancients, it aimed to reform our habits, to accord our lives with the cosmos; for the Medievals, it aimed to accord us to God's will; for moderns such as Kant, it aimed to bring our actions under reason's dominion; for Utilitarians, it aimed to maximize our happiness. Unanimous among these approaches is their form: they describe human nature, identify its flaws, and propose ways to hew that nature more closely to an appropriately human life. Human nature is thus less a factual repository of moral worth, then an index of our challenges, an amalgam of what we are and what we might become. Indeed, that is its moral function even within an extra-ordinarily anthro-pocentric approach such as Kant's, which locates human dignity not in human nature, but in an ideal of humanity pursued in struggle with, and even in opposition to, that nature.

Human nature, that is, has never been presented as an unalloyed good, nor as an object of deliberate moral maintenance; quite the contrary. More importantly, I suggest, what Kass and McKibben describe and aim to defend as human nature, i.e., our enduring struggle with finitude, seems more indicative of what older authors called "the human condition." That distinction is important because it's not clear (a) that altering human nature would essentially alter that condition, or (b) that we would "lose" anything of over-riding ethical value by altering that nature. Let's put the matter directly: If we enhance human strength, or endurance, or memory, are we improving human beings? With memory enhancement, we expand a human capacity, change the conditions under which we act, and maybe even modify an aspect of human nature. But does modifying our memory capacities differ in kind from upgrading computer memory capacities? McKibben insists that whereas the first changes how we manage information, the second changes us. But that distinction assumes wrongly that we are our biology in ways that we are not our culture or technology, and that changing our biology changes us in ways that changing our technology does not. That assumption is false first, I suggest, because it changes nothing of substantive moral import to us. It does nothing to modify the human traits that ethicists most lament, among them egoism, aggression, group solidarity, and the like. Indeed, on those measures, the striking element of those enhancements is not how much they improve human beings, but how little they do so—at least in ways that arouse substantive moral concern.

I stress that point because it illustrates perhaps the most basic shortcoming of McKibben's position. He insists both that genetic enhancements such as improved strength or disease resistance change us, and that our current capacities are so morally valuable that any efforts to modify them, even those aimed at ameliorating biological vulnerabilities, unethically imperil human nature. Let's take him at his word. If he correctly holds that we are essentially our genetic inheritances, then we will presumably find genetic bases for all human behaviors, among them endemic patterns like racism and aggression. So, let's suppose, however implausibly, that we discover the genetic complex that yields racism, and find further that we could easily disable that complex, preventing its occurrence in future humans. McKibben's position entails not only that we refuse to act upon this knowledge—just as we must refuse to engineer people genetically resistant to cancer, or Alzheimer's, or congenital idiocy—but also that we equally endorse all of these genetic propensities of human nature as essential components of "who we are," as morally valuable because they are human. This implication raises two critical questions. First, would changing our propensity to Alzheimer's and to racism equally change our nature? And second, would modifying either propensity eliminate something of moral value in our nature? We can affirm the first point only if we say that Alzheimer's disease and racism are equally biological components of our nature, and equally essential to who we are, such that we are equally enjoined to retain both if we are to remain human.

But if we reject such genetic determinism and say instead that these propensities are valuable not merely because they are inevitable but because they are ours, we endorse the egoism McKibben excoriates. Indeed, if we are, by nature, moral, then the positions he and Kass advance are deeply incoherent: we warrant such moral concern, they say, that we cannot be trusted to enhance our capacities, even on therapeutic grounds, because we will invariably turn such efforts to egoistic ends, eliminating human nature in the process. Ethicists of all stripes, of course, have tried to identify what makes us uniquely morally valuable, to convert our egoism into a measure of our virtue. That this assertion is extraordinarily egoistic escapes defenders of the faith. But the belief is less troublesome than its implication: If we are already good enough, and should remain so though we could be so much better, we are akin to those who remand their children's care to faith healers. They would have us tell future persons that they will suffer enduring ailments and weaknesses, even though we could have prevented them, because "that's what being human is." But surely we can hope for more for those future persons, and surely we are obliged to improve their life prospects, not merely because we can, but because it diminishes us—morally—when we can do better, but refuse to. Surely that is not the moral heritage we wish to pass on to our successors: that you can do and be better than we were, better than you are, but that you should not choose to do so. That demand is beyond hubris run amok—it is a bioethic devoid of life. Such a position is unethical in the extreme, and refutes the conditions under which we may claim to live as ethical beings, chief among them the hope, if not the expectation, that the human future can be better than the human past, better even than the human present. McKibben's demand heralds instead a perverse form of post-humanity, the newly minted "good enough" type that we emphatically ought not to sanction.

NOTES

1. Leon R. Kass, *Life, Liberty and the Defense of Dignity: The Challenge for Bioethics*, San Francisco: Encounter Books, 2002.

2. Hans Jonas, *The Imperative of Responsibility: In Search of an Ethics for the Technological Age*, Chicago: University of Chicago Press, 1984.

3. Gerald P. McKenny, *To Relieve the Human Condition: Bioethics, Technology and the Body*, New York: State University of New York Press, 1997.

4. Bill McKibben, *Enough: Staying Human in an Engineered Age*, New York: Henry Holt and Co., 2003.

ETHICS AND THE LIFE SCIENCES

LEARNING TO COPE WITH AMBIGUITY: REFLECTIONS ON THE TERRI SCHIAVO CASE

BRAD F. MELLON
BETHEL SEMINARY OF THE EAST

ABSTRACT: The present study, "Learning to Cope With Ambiguity: Reflections on the Terri Schiavo Case" looks at the many complexities of dealing with *Persistent Vegetative State* (PVS). By its very nature PVS is ambiguous. It is difficult to diagnose and, even when the diagnosis appears to be certain, there is a multiplicity of ethical issues and treatment options to consider. There are four high profile PVS court cases that can help us understand the Schiavo situation. They are Karen Ann Quinlan, Nancy Kruzan, Helga Wanglie, and Daniel Fiori. These cases share many common features with each other and with Schiavo. In the final analysis, the judicial decisions inevitably point us to the ongoing need to live and cope with ambiguity.

INTRODUCTION

The recent case of Terry Schiavo in Florida has again lifted the issues surrounding PVS before the public eye. Ms. Schiavo's medical condition began in 1990 when she went into cardiac arrest resulting in brain asphyxia. Some physicians labeled her condition as PVS, while others disagreed. Her parents produced videos demonstrating what they believed to be Terri's ability to respond to her environment. This case was so hotly debated that Robert Cranston refered to it as a 'debacle.'[1]

One of the main problems in dealing with PVS is that it can defy analysis. Patients experience sleep-wake cycles, can open their eyes, appear to smile, and cry out, but have no awareness of self, others, or their surroundings. They cannot respond to stimuli. However, "functions of the hypothalamus and brain stem are sufficiently retained to allow the patient to survive with medical and nursing care."[2] Many PVS patients can live indefinitely with the aid of life support, thus physicians can have a difficult time offering a diagnosis, and family members and surrogates find themselves agonizing over treatment options.

PVS IN THE COURTS

As mentioned above, there are four high profile court decisions involving PVS that can offer a perspective on the Schiavo case.[3] In chronological order they are Karen Ann Quinlan, Nancy Kruzan, Helga Wanglie, and Daniel Fiori. The following is a summary of these cases.[4]

Karen Ann Quinlan was just seventeen years old in 1975 when she lapsed into a coma at a New Jersey hospital. Her brain had been deprived of oxygen on two separate occasions for at least fifteen minutes each, and when her condition deteriorated into a PVS the medical staff could offer no hope of recovery. Karen was placed on life-support that included a respirator and feeding tube.

After consulting with their priest, Karen's parents requested that the respirator be removed. The primary-care physician refused to comply based on personal moral concerns, so the case went to court. During that time the hospital decided to wean Karen off the respirator in the event the judge should decide in the Quinlans' favor. The New Jersey Supreme Court did decide to allow removal of the respirator and Karen lived another eight years with the aid of the feeding tube.

Nancy Kruzan, a twenty-five year old resident of Missouri, was involved in a motor vehicle accident. Her brain had been deprived of oxygen for about fifteen to twenty minutes, leaving her in a PVS. After five years on life-support her family began to observe a decline in her condition. Believing that Nancy would not have wished to continue indefinitely in this state, they asked for the removal of the feeding tube. A further consideration was the cost of treatment that had swelled to $130,000 annually.

Although a lower court granted the request, the State of Missouri appealed the decision. The Missouri Supreme Court reversed the decision on the basis that Nancy's diminished quality of life did not outweigh the state's interest in the sanctity of life. The case was subsequently taken before the U.S. Supreme Court, which affirmed Missouri's right to set standards. Once back in Missouri, however, Nancy's case took an interesting twist. A number of her friends surfaced to testify in court that Nancy had indeed spoken with them years before about her wishes should she become incapacitated. In December of 1990 a judge ordered the removal of the tube and Nancy passed away several days later.

Although quite similar to Quinlan and Kruzan, the case of eighty-six year old Helga Wanglie of Minnesota moves in a different direction. In May 1990 Helga was on a respirator following surgery when she suffered a heart attack. Like Quinlan and Kruzan she experienced anoxia for several minutes, leading her doctors to offer a PVS diagnosis. At that time a feeding tube was inserted. She had no living will (LW) or advance directive (AD). Unlike the Quinlan and Kruzan situations, however, it was Helga's medical staff that suggested the possibility of taking her off of life support, while her family was opposed to the idea.

The hospital and staff took the case to court, arguing that medical personnel should not be forced to treat if it is not in the patient's best interest. The Wanglie family countered with the request that Helga be kept alive by any and all possible means. In a July 1991 decision, the judge rejected the hospital's stance and gave full guardianship to Helga's husband, Oliver. Helga died three days later. The cost of her treatment came to about $750,000.

Like Karen Ann Quinlan, Daniel Fiori of Pennsylvania was in his teens when severe head injuries suffered in an accident left him in a PVS.[5] Daniel received nutrition and hydration through a feeding tube. After seventeen years with no apparent improvement, his mother and court-appointed guardian filed a petition with the facility to have the feeding tube removed. She admitted that Daniel had never expressed his wishes about this with her previously, but affirmed that based on his love of life he would not want to continue in this condition.

The skilled nursing facility (SNF) refused the request and Daniel's mother brought the case to court. A judge upheld her request, and when the Attorney General appealed the decision, the Superior Court issued a ruling in behalf of removing the tube. The judge stated that the decision to remove life support from an adult in a PVS who did not have an AD could be made by a close family member with the supporting testimony of two physicians. Unlike the Missouri decision, the court determined that the state's interest in preserving life did not outweigh Daniel's right of autonomy as expressed through a surrogate. (The fact that medical staff considered Daniel's condition to be a *permanent* vegetative state played an important role in the decision.)

Each of these cases is, of course, unique, yet they share some common features. Each patient was given a PVS diagnosis and received artificial nutrition and hydration by means of a feeding tube. None had executed a LW or AD and since they could not express their wishes, a healthcare surrogate(s) had to try to determine what they would have wanted. In each case disagreement over treatment options led to conflict and eventually to court.

ETHICAL ISSUES, QUESTIONS, AND PROBLEMS

When the above PVS cases were heard in court, they were examined in light of commonly held and accepted principles of bioethics, together with various state interests. First, there is the matter of *withdrawing* life support, in particular that which specifically helps the patient breathe and receive nourishment. It is usually more difficult to withdraw life-prolonging measures than to withhold them.

Recently the ethics committee of a SNF discussed the case of a patient who had been in a PVS for several years. When his condition seriously declined, the family requested the feeding tube be removed. The patient's attending physician acquiesced but could not escape the feeling that such an action would "starve him to death." Consultation with an ethicist helped him understand that the condition, not the withdrawing of the feeding tube, would lead to the patient's death.

In his book, *Caring to the End*, James Drane contends that the Missouri court's unwillingness to remove Nancy Kruzan's feeding tube (above) represented "an assault" on patients and their families by requiring life-prolonging treatment that is "burdensome and ineffective."[6] Likewise, we noted the Minnesota decision to keep Ms. Wanglie on life support even though her death was but days away.

Second, each of these cases touches upon the issue of *autonomy* or *self-determination* by means of a healthcare surrogate or surrogates. The basic right of self-determination allows a patient to have a say in his or her treatment. This issue stands behind *informed* consent and advance directive legislation. It also includes the principles of *substituted judgment* and *best interest*. For example, Daniel Fiori's mother came forward to express what she believed her son's wishes would be if he were able to communicate them to the court. Likewise Nancy Kruzan's family and friends testified to their belief that she would not have wanted to live indefinitely in a PVS. On the other hand the medical staff of the hospital that treated Helga Wanglie felt it was not in her best medical interest to continue life-prolonging measures.

On this subject Peter Singer writes, "So our decisions about how to treat such patients should depend . . . on the views of families and partners, who deserve consideration at a time of tragic loss."[7] This is true; however, each of the cases discussed above serves to remind us that self-determination is not *absolute*. It must be held in balance with other principles such as state interests in preserving life and that which is in the patient's best interest.

Third, each of the cases involves *medical futility*, which Cranston says is defined as the point at which a patient is "no longer serving a useful function."[8] He further contends that futility is a complex issue and cannot be easily determined. In another essay I discussed the case of a patient in a SNF who exhibited all of the classic PVS symptoms except for the fact that she was able to receive nourishment by mouth (as a total feed).[9] Her family requested that the facility prolong her life by any possible means. They were willing to come in every day to help the staff with her activities of daily living. Some of the staff questioned the family's judgment and wondered if further continuation of treatment was in the patient's best interest.

Fourth, *distributive justice* is relevant to each of the PVS cases examined above because of the enormous cost of treatment. Although it is difficult to put a 'dollar sign' on human life, we must nevertheless take the cost, availability, and distribution of resources into consideration. Is it 'reasonable' to keep PVS patients alive for whom there is no hope of recovery when others would benefit from the medical treatment those same dollars could provide?

Fifth, there is the matter of the *sanctity* or *preservation of life, quality of life*, and the '*totality*' principle. Sanctity of life is a time-honored principle of bioethics that has deep historical roots from Hippocrates to the Roman Catholic moral theologians. However, just as autonomy is not absolute, neither is the preservation of life. James Drane rightly reminds us of the historic distinction between a 'right to die' and 'a right to let die.'[10] The first permits an act of killing that for many ethical systems is not permissible. The second acknowledges that it is more merciful to permit a dying person to die without requiring the administration of overly burdensome treatments.

'Quality of life,' like futility, is difficult to determine. In the above PVS cases, some of the courts opted to preserve life, even where there was no hope for recovery. On the other hand, we noted that a Pennsylvania judge ruled that close relatives with the corroborating testimony of two physicians could decide such cases.

Alongside quality of life issues we should consider the concept of 'Totality' that says,

> The whole patient is considered to take priority, especially over lower biological levels of life. If the total person, including psychological and social functions cannot be benefited or sustained, then biological dimensions of life do not demand continued support. . . . The whole or total person is what is sacred and has rights.[11]

Although this would appear to be straightforward, all of the above court decisions demonstrate that families, surrogates, medical staff, and judges struggle with this and have a hard time agreeing with each other.

REFLECTIONS ON THE SCHIAVO CASE

There are many similarities between the PVS cases examined above and the Schiavo case. Perhaps one difference is that physicians are divided over whether Terri Schiavo truly was in a PVS condition. We have noted that diagnosing PVS can indeed be difficult.

In the final analysis, it would appear that those who make ethical decisions regarding PVS experience difficulty with *ambiguity*. Although we tend to view such decisions as between 'right and wrong' or 'good and evil' often in reality they are between competing 'goods.'[12] Under normal circumstances it is 'good' to preserve life, however, medical situations can arise in which it is 'good' to allow life to come to an end.

PVS by its very nature is ambiguous. Not only is it difficult to diagnose, but also even if the diagnosis appears to be certain there is the added difficulty of trying to determine whether or not to continue life-prolonging treatment. Recently, I read the story of a young woman who had been in a coma for months.[13] During that time her husband and other family members were faced with the prospect of removing her from life support. Each time when they were about to authorize the removal, they decided to wait a while longer. Her husband issued and rescinded two separate DNR orders. Eventually the woman awoke from the coma and appears to be on her way to a complete recovery. Of course this happy ending needs to be held in balance with a Daniel Fiori or Nancy Kruzan who wasted away in a PVS over many years.

The search for a decision in the matter of PVS amidst a multiplicity of situations and treatment options most often leaves one to decide between two or more 'good' possible directions. Since this is our reality we should learn to work harder at living and coping with ambiguity.

NOTES

This study was originally presented at the "Ethics and the Life Sciences" conference held at the University of Delaware on October 23, 2004. Shortly after that time, Terri Schiavo passed away when the court permitted her husband to remove life supports. The principles discussed here, however, remain valid for the current debate on PVS issues.

1. R. Cranston, "The Terri Schiavo Debacle: What Have We Learned?" available at: http://www.cbhd.org/resources/endoflife/cranston_2004-03-19_print.htm., p. 1.

2. J. Idziak, *Ethical Dilemmas in Long Term Care*, (Dubuque: Simon and Kolz, 2000) pp. 111–112.

3. "Court decisions in the highly publicized cases contained ethical arguments that themselves stimulated further ethical arguments. Later court decisions either approved or overturned earlier ones, and a whole corpus of legal bioethics literature came to be." (J. Drane, *More Humane Medicine*, [Edinboro, PA: EUP Press, 2003], p. 50).

4. All of these cases except Fiori are adapted from a May 31, 2001 article entitled, "3 Landmark Cases of the Persistent Vegetative State," available at: http://www.geocities.com.

5. In *Re: Daniel Joseph Fiori*. The Supreme Court of Pennsylvania Eastern District. 1995.

6. J. Drane, *Caring to the End*, (Erie, PA: Lake Area Health Education Center, 1997), p. 26.

7. P. Singer, "Rethinking Life and Death," in *Last Rights?Assisted Suicide and Euthanasia Debated*, ed. Michale Ullmann, (Grand Rapids: Eerdmans, 1998), p. 175.

8. R. Cranston, "Debacle," p. 2.

9. B. Mellon, "Faith-to-Faith at the Bedside: Theological and Ethical Issues in Ecumenical Clinical Chaplaincy," *Christian Bioethics* 9/1 (2003), p. 64.

10. J. Drane, *Caring*, p. 10.

11. J. Drane, *Caring*, p. 223.

12. C. Yusavits and B. Mellon, "Ethical Competencies for Pastors and Chaplains," a workshop presented at the Mennonite Health Assembly, San Francisco, CA. March 26, 2004.

13. L. O'Connor, "While I Was Sleeping," *Christianity Today*, Feb. 2004, pp. 45–47.

BIBLIOGRAPHY

Cranston, Robert. 2004. "The Terri Schiavo Debacle: What Have we learned?" [On-line]. Available: http://www.cbhd.org.

Drane, James. 1997. *Caring to the End*. Lake Area Health Education Center.
———. 2003. *More Humane Medicine*. Edinboro University Press.
Idziak, Janine. 2000. *Ethical Dilemmas on Long Term Care*. Simon and Kolz.
In Re: Daniel Joseph Fiori. 1995. The Supreme Court of Pennsylvania Eastern District.
Mellon, Brad F. 2003. "Faith-to-Faith at the Bedside: Theological and Ethical Issues in Ecumenical Clinical Chaplaincy." *Christian Bioethics* (9/1), pp. 57–67.
O'Connor, L. 2004. "While I Was Sleeping." *Christianity Today*, pp. 45–47.
Singer, Peter. 1998. "Rethinking Life and Death: A New Clinical Approach." In *Last Rights? Assisted Suicide and Euthanasia Debated*, ed. Michael Ullman. Eerdmans, 171–198.
"3 Landmark Cases of the Persistent Vegetative State." 2001. Available: http://www.essayshotline.com.
Yusavitz, Carl, and Brad F. Mellon. 2004. Ethical Competencies for Pastors and Chaplains. *Healing Ministry* (11/4), pp. 163–69.

ETHICS AND THE LIFE SCIENCES

PVS AND THE TERRI SCHIAVO CASE: A REPLY TO BRAD MELLON

GARY FULLER
CENTRAL MICHIGAN UNIVERSITY

ABSTRACT: Brad Mellon argues that persistent-vegetative-state cases, including the recent Terri Schiavo case, are ambiguous. By this he seems to mean that decisions about such cases are fraught with doubt and uncertainty and perhaps even that rational resolution of many such cases is impossible. Faced with such cases the most we can do is to live and cope with the ambiguity. I am more optimistic. With good will, and much clarification and discussion, rational agreement is possible in these cases, including the Schiavo case.

Brad Mellon argues that the Terri Schiavo case, and persistent-vegetative-state (PVS) cases in general, are ambiguous. By this he means at least that decisions about such cases are fraught with doubt and uncertainty. There is no simply moral principle or algorithm that can help us decide such cases. Instead, we are faced with the difficulties of defining, applying, and weighing many competing considerations. The central question in the Schiavo case is whether Terri Schiavo should be taken off medical life support, including artificial nutrition and hydration. The case raises questions about whether Terri is in a PVS, of what are her chances of recovery, of what is in her best interest, of what she would have chosen had she been mentally competent, and so on. These have to be clarified and weighed.

I certainly agree that PVS cases are difficult and require the handling of many competing considerations, but if Mellon is also suggesting that one person's handling of the considerations is as good as another's, or again that little progress can be made towards rational agreement in such cases, then I disagree. In many PVS cases, whether disputed or not, there are answers that are much better supported than their alternatives. In the Schiavo case, for example, the husband's claim that the right thing to do is to remove Terri's feeding tube and let her die is much better supported than the Schindler family's claim that the tube should not be removed.

The Florida courts were clearly right on this. There is little ambiguity here! Or so I shall argue.

Perhaps the best way to proceed in replying to Mellon is to raise various points and questions about his interesting paper. In what follows, I shall touch on the difference between PVS cases and other moral problems, the diagnosis and prognosis of PVS, the role of respect for patient autonomy, and other miscellaneous issues.

1. PVS CASES AND OTHER MORAL PROBLEMS

PVS cases are complex, but are they much different from many other familiar contemporary moral problems? They are not. To pick at random, take the issue of whether we should have capital punishment in Michigan (my home state). Deciding this issue involves clarifying and weighing considerations involving deterrence, retribution, cost, cruelty, and unfairness, to mention just a few. Or take what looks much simpler, the issue of abortion in the early stages of pregnancy. Deciding this turns out to involve clarifying, applying, and weighing considerations of human life, consciousness, potential, various rights of the mother, and so on. Does Mellon hold that these issues of capital punishment and abortion also involve "ambiguity"? What contemporary moral problems do not involve it?

2. THE DIAGNOSIS AND PROGNOSIS OF PVS

Mellon says that "[o]ne of the main problems in dealing with PVS is that it can defy analysis" (Mellon 2006, p. 291). The terms "vegetative state" and "persistent (or permanent) vegetative state" were introduced in the 1970s and their criteria have evolved over the years. The vegetative state (VS), which becomes PVS after a month or so, is now distinguished from brain death, which involves the loss of all brain function, and from locked-in syndrome and the minimally conscious state, which involve more or less cognitive function. The intuitive idea is that in VS the patient is alive and awake but unconscious (Jennett 2002, chap. 2; Pence 2004, pp. 44–46).

Diagnosis of VS, which may be mistaken if made too hastily, becomes much more reliable when an adequate period of behavioural observation is combined with such laboratory investigations as CAT and PET scans (Jennett 2002, pp. 20–29). Further, a diagnosis of VS, supplemented by facts about the causes of the condition, enables physicians to make highly reliable prognoses for recovery or lack of recovery. In one study (Multi-Society Task Force on PVS, cited in Jennett 2002, p. 60) none of fifty adults in VS for six months had recovered consciousness, much less independence, after a year (from diagnosis). Studies on later recoveries are fewer, but the consensus seems to be that such recoveries are very rare and "are almost always to sever disability. Most patients remain totally dependent, some reaching only the minimally conscious state or a little better" (Jennett 2002, p. 64).

There does remain the question of whether PVS patients have at least primitive experiences of pain, warmth, and so on. The consensus is that they do not—there is nothing that it is like to be them!—but there is room for discussion (especially among philosophers). The following two pieces of evidence against such patients having any experiences at all were endorsed by both the American Neurological Association and the American Medical Association. First, PVS patients have a low cerebral metabolic rate for glucose that is similar to that of deep surgical anaesthesia (Jennett 2002, p. 18). Second, experiences of pain require "the integrated functioning of the brain stem and cerebral cortex," and the extensive bilateral damage to the cortex in PVS patients rules this out (American Academy of Neurology, quoted in Pence 2004, p. 48; see also Jennett 2002, p. 18).

What does all this mean for Terri Schiavo? Behavioural tests have shown that she has been in a VS since her heart attack in 1990. Laboratory investigation confirms this. It is overwhelmingly plausible to agree with the Florida District Court that reviewed the Schiavo case that:

> By mid 1996, the CAT scans of her brain showed a severely abnormal structure.... [M]uch of her cerebral cortex is simply gone and has been replaced by cerebral spinal fluid. Medicine cannot cure this condition. Unless an act of God, a true miracle, were to recreate her brain, Theresa will always remain in an unconscious, reflexive state, totally dependent upon others to feed her and care for her most private needs. (Florida Second District Court 2001)

Further, if, contrary to the strong case against this view, Terri is having some minimal experiences, it is plausible to think that these will be mostly painful ones. The bodily responses to artificial feeding alone will surely produce unpleasant experiences, if they produce any at all. In the earlier case of Karen Ann Quinlin, the presence of the feeding tube caused vomiting and bodily agitation, including head writhing (Pence 2004, 32).

There does not seem to be much ambiguity, then, concerning the diagnosis and prognosis of PVS in the case of Terri Schiavo and, for that matter, in all four of the background cases discussed by Mellon.

3. RESPECT FOR THE PATIENT'S AUTONOMY

An unambiguous verdict in diagnosis and prognosis does not necessarily mean an unambiguous verdict about whether we can pull the plug, in the Schiavo case about whether we can remove the feeding tube. Other considerations have to be taken into account. Certainly a central consideration here is that of respect for the patient's autonomy. Again, however, people of good will can often reach rational agreement. Consider the following sequence of principles leading up to a principle relevant to the Schiavo case, each principle of which is plausible.

A. *If a fully competent and informed person, who is terminally ill and in unrelievable pain, requests to be removed from all life support, then it is permissible and right to withdraw the life support.*

Of course, arguments have been offered against (A). These have appealed to ideas such as that playing God is wrong, that there is always a chance of recovery, and that withdrawing support is always actively killing. It is not difficult, however, to criticize these arguments.

B. *If a fully competent and informed person requests, shortly before undergoing a dangerous medical procedure, that if a PVS results (and is confirmed by relevant tests) and persists for over a year, then all life support be withdrawn—then it is permissible and right to withdraw the life support.*

To persuade reasonable people to accept (B) it usually will be enough to go over all the information about the diagnosis and prognosis of PVS discussed above. If instead of prolonged PVS, the medical condition was quadriplegia, early AIDS, or early Alzheimer's, not to mention extreme psychological depression, then respecting the patient's autonomy would indeed be controversial. With prolonged PVS, however, I do not see much room for disagreement.

C. *If there is strong evidence—"clear and convincing evidence" in legal terminology—that had a person, such as Terri Schiavo, shortly beforehand known that she would go into a prolonged PVS, and been competent and fully informed about PVS, and had requested that life support be removed after a year—then it is permissible and right that life support be withdrawn.*

This principle seems close enough to (B) that there should be little difficulty in getting people to accept it. Indeed, even the rather conservative state of Missouri accepted (C) in the *Cruzan* case (Pence 2004, 41).

The difficulty here, of course, is in reaching consensus on what counts as strong evidence. Written documents, including advanced directives such as living wills; earlier oral statements with family, friends, or physicians about the dying process (Florida Second District Court); reasonable-person judgments, and appeal to what would be in the patient's best interests (now accepted by Florida, but not used in the Schiavo case [Florida Second District Court])—all these provide some evidence for what the patient would have chosen. Nevertheless, there is still some controversy, some ambiguity if you will, as to how to weigh these. With additional time and working through, which is going on both in the courts and in society generally, I see no reason why a rational consensus cannot be reached. I think, for example, that the combination of Terri's earlier oral statements, coherence with her other views, and reasonable-person judgments about her case suffice for holding that she would have chosen to have life support removed.

4. REMAINING CONTROVERSIAL ISSUES

No one denies that there remain many controversies about ending life in the areas of coma and PVS, assisted suicide, and euthanasia in general. Is giving a patient a lethal injection ever morally all right? If a prolonged-PVS patient has previously made it clear that he wants life support to be continued as long as possible, is it ever morally all right to pull the plug? Terri Schiavo was a mentally competent person before she went into a PVS; but what should we do in cases of infants who become vegetative but who have never been competent? Here again, I see no reason why rational discussion over time, should not result in consensus.

5. CONCLUSION

Cases of PVS, including the Terri Schiavo case, are complex and difficult. Initially, one might think that no rational resolution of these cases is possible and that we shall just have to live and cope with ambiguity here (Mellon 2006, 279). I deny this. With good will and much clarification and discussion, rational agreement is possible in a wide range of cases. In the Terri Schiavo case in particular, it is clear, I hope, that the reasons in favour of withdrawing life support are much better than those for keeping Terri alive.

BIBLIOGRAPHY

American Academy of Neurology. 1986. "Amicus Curiae Brief" in *Brophy v. New England Sinai Hospital, Inc.* Cited in Pence, 48.

Jennett, Bryan. 2002. *The Vegetative State*. Cambridge, England: Cambridge University Press.

Mellon, Brad. 2006. "Learning to Cope with Ambiguity: Reflections on the Terri Schiavo Case." *Ethics and the Life Sciences*. Charlottesville, VA: Philosophy Documentation Center, pp. 291–297.

Multi-Society Task Force on PVS. 1994. "Medical Aspects of the Persistent Vegetative State" (Part 2). *New England Journal of Medicine*, 330, pp. 1572–1579. Cited in Jennett 2002, pp. 59–70.

Pence, Gregory. 2004. *Classic Cases in Medical Ethics, 4th Edition*. Boston, MA: McGraw Hill.

Second District Court of Appeal of Florida. 2001. *Schindler and Schindler v. Michael Schiavo*. http://abstractappeal.com/schiavo/2dcaorder01-01.txt

ETHICS AND THE LIFE SCIENCES

TRULY HUMAN REPRODUCTION

ALEXANDER R. COHEN
UNIVERSITY OF VIRGINIA

ABSTRACT: For two million years, members of *Homo sapiens* (and the species from which it emerged) have shaped to their purpose almost everything they found in nature. Yet we are still reproducing by sex. This is a poor method of conceiving human beings, because it surrenders many of the future child's characteristics to luck. Both parents and children are better off the more parents control their children's genotypes. The emerging technologies that enable this do not reduce free will and will not eliminate human characteristics (such as certain forms of "mental illness") that are worth preserving; rather, they will match types of children to parents who can appreciate them. Technological reproduction, not reproduction by sex, should be regarded as truly human. It is the application of the power of thought to human reproduction, and, like the application of that power to external objects and to memes, it will enhance our capacities.

I.

Drummond: I mean, did people "begat" in those days about the same way they get themselves "begat" today?
Brady: The process is about the same. I don't think your scientists have improved it any.
—Lawrence and Lee 1957, p. 81

Over the two million years since our ancestors first smashed stones together and shaped them to the purposes of hominid life (Poirier 1993, p. 149), members of our species and of those from which it would emerge have applied thought to nature and ordered their world according to the judgments of their minds.[1] They domesticated fire, fostered the development of plants and animals they found valuable, and erected barns, factories, and skyscrapers. Applying thought to human bodies, they devised medicine and all its technology. And yet, for two million years, our ancestors in each generation begat (or at least conceived) progeny without technology, by a process remarkably similar to that used by australopithecines and, for that matter, by housecats.

Sex has its merits. But considered strictly as a method of producing babies,[2] sex has a lot of flaws. Consider: I propose to manufacture cars by a parallel method. Sometimes my factory will make cars, but more often it will just mangle the parts. Sometimes the cars will be sports cars, other times luxury sedans, but the customer will know which he is getting only after he has signed the contract. And buried in the contract will be a clause stating that if he gets a lemon, as some customers will, he must drive it anyway, no matter how much it costs to make it usable and how far below par his car remains even with repairs.

If I did this, the world would beat a path away from my door. This is not a suitable way to make and sell cars. Yet, when it comes to making babies, human beings have consistently used a method that is just as unpredictable: Sometimes, sex results in pregnancy, other times not; sometimes the fetus is male, other times female, but the parents find out which only after conception; and if a disabled child is born, the parents are expected to take care of him no matter what the impact on their lives—and no matter how little potential that child has.

I do not intend to suggest by this analogy that having a child is of no greater significance than buying a car. The significance of having a child is much greater. The attributes chosen at conception, whether by chance or by judgment, will influence the life of the new person from its beginning to its end and in every imaginable respect. And that person will be among the most significant figures in the lives of his parents.

Matters so vital as the genetic makeup of one's progeny should not be left to chance.

II.

Dolly'll never go away again.
—Herman 1964

In fact, few people voluntarily leave their children's genes completely to chance. Conception through sex does permit one choice, at least in the normal case in the modern Western world: the choice of a partner. Whether or not a person deciding to have sex intends that a child result, and whether or not he explicitly contemplates genetics in making the decision, sexual attraction rests on a number of traits that have, or at least may have, genetic origins. It is only when the prospective parents have sex that they surrender to luck.

Perhaps the simplest way to reduce the role of luck is selective abortion. Since 1955, it has been possible to determine the sex of a fetus (Rifkin 1998, p. 140).

More recently, physicians have begun testing for a variety of genetic diseases, using methods that can also determine whether a fetus has other genes the parents may not wish to transmit (Rifkin 1998, p. 133). With this information in hand, a woman can decide to abort her fetus, have sex again, and hope for a more desirable fetus—whatever that may mean to her.

If what is desired is a fetus of a particular sex, there is now an alternative to selective abortion: flow cytometry, which involves running a laser beam across specially dyed sperm in order to separate those that carry X chromosomes from those that carry Y chromosomes; artificial insemination can then be attempted with only X-bearing sperm if a female is desired and only Y-bearing sperm if a male is wanted. The technique is not perfectly reliable (Robertson 2001, p. 2), but if a gamete of the wrong sort (whichever that may be in a particular case) gets through and reaches the ovum, selective abortion is still available as a fallback.

In 1977, Patrick Steptoe took an egg from Lesley Brown and put it on a plastic dish. Her husband's sperm was added to the dish, and then, through a microscope, Robert Edwards watched the conception of Louise Brown. She was the first child conceived outside her mother's body. Luck controlled her genome as much as it controls that of any child conceived through sex, but her story is relevant here because, unlike any earlier human being, Louise Brown was conceived under a microscope. Science now had access to human conception—and the opportunity to examine and alter an embryo before implanting it in a woman's body (Silver 1997, pp. 67–68).

Now that embryos can be produced on a plate, genetic tests can be run on batches of embryos before one or two are introduced into somebody's womb. From among the embryos produced in the lab with genes chosen by luck, the best—according to the values of the person doing the selecting—can be chosen by judgment to develop into a person. How much control this affords depends in part on how many embryos are available (Silver 1997, pp. 203–209).

Twenty years after Louise Brown was born, Ian Wilmut and Keith Campbell introduced us all to a sheep that was glowing, crowing, and going strong (Kolata 1998; cf. Herman 1964). The sheep, of course, was Dolly—the first clone made from a mature mammalian cell (Silver 1997, p. 100). Shortly thereafter, rhesus monkeys—close cousins to *Homo sapiens*—were cloned (Kolata 1998, p. 186). That human reproductive cloning is achievable is almost no longer a question (Silver 1997, p. 93).

Cloning will make feasible a process that is now too inefficient to apply to human beings: genetic engineering, in which scientists add specific genes to an embryo or alter those that are already present. The simplest method of introducing a new gene works only half the time, and 5 percent of the time it causes genetic disease. In animals, the problem can be handled simply by applying the process to a large number of embryos and examining the animals that result; this would be unacceptable with humans. But with cloning, a genetic engineer could attempt to add the new material to a number of embryonic cells, determine which ones had accepted the material properly, and then, from the progeny of one cell that changed as desired, produce a child by cloning (Silver 1997, pp. 129–130).

III.

[A]s man is a being of self-made wealth, so he is a being of self-made soul.
—Rand 1957, p. 938

There are almost certainly limits to the control technology can give prospective parents over the characteristics of their children. If a trait is not and cannot be completely determined by genes, then it cannot be determined by genes no matter how, by luck or by judgment, those genes are selected.

Classic cases of identical twins developing similar behavior despite being separated show that genes have remarkable influence on human character, yet many identical twins emphasize the differences between themselves and their twins, which show that that influence is not complete determination. As Ian Wilmut (or his ghostwriter) put it, "Our genes provide the clay or the marble from which we are made, and this influences the kind of people we can become." But the environment, beginning with the uterine environment, has a substantial influence as well—and we should not forget the influence that each of us has on his own development. Non-genetic factors have a strong influence on the "personality" and physical characteristics even of rams, as shown by four rams that differ in size and behavior despite being cloned from the same source (Wilmut, Campbell, and Tudge 2000, pp. 276–280).

Cloning a person with an established record, whether oneself, a relative, or a celebrity, would therefore not guarantee the child a life like that of the clone parent. It has been suggested that cloned great men would probably have easier lives than their progenitors and might therefore be more productive (Kolata 1998, p. 72). But perhaps in at least some cases it was the great man's difficult early life that gave him the opportunity to find his path to achievement, and a coddled clone would merely have a lot of unrealized potential (Cf. Wilmut, Campbell, and Tudge 2000, p. 284). That said, he would at least have the genetic potential, which an individual conceived through sex might lack even with impressive parents.

Some opponents of choosing a prospective child's genes claim that an individual so conceived would, *ipso facto*, have less free will than anyone else. Perhaps it might be possible to create or select for a gene for servility, but it might also be possible to create or select for a gene for rugged individualism, and neither choice is implicit in any of these technologies. To suggest that the substitution of choice for chance in a person's conception inherently destroys free will is to imply that free will is nothing more than random chance—the randomness of which is all played out before the individual is born. Random chance is neither necessary for free will nor sufficient for, to borrow a phrase from Dennett, "free will worth wanting."

Free will is a mental phenomenon, and genes do not have minds. Neither does luck. Scientists and prospective parents choosing genes would think, but that exercise of *their* free will would have exactly the same influence on the free will of the future child as does the combination of the acts of free will that lead two people to become parents together through sex and the random chance that governs conception when they do.

For free will worth wanting to exist, random chance in the process of conception is not necessary. What is necessary is that we have the opportunity to think and thereby to choose our actions. This we have. And the extent to which our genes determine our choices is irrelevant to the present question because it does not differ according to the method by which we are conceived. Anything that must be determined by mind in technological reproduction is determined anyway in reproduction by sex—but by luck.

IV.

> A man cannot get a coat or a pair of boots to fit him, unless they are either made to his measure or he has a whole warehouseful to choose from: and is it easier to fit him with a life than with a coat, or are human beings more like one another in their whole physical and spiritual conformation than in the shape of their feet?
> —Mill 1859, p. 75

Mill was discussing the prospects for fitting a person with his own life, not with the life of another, but those closest to us have important places in our own lives. It would be a poor way of fitting a person with a life if his spouse and friends were chosen at random. Yet to the extent that a child's identity rests on his genes, reproduction by sex surrenders the choice of a child to random chance within the parameters set by the choice of a reproductive partner.

It is sometimes said that gender selection may be practiced for sexist reasons. Perhaps so; what follows? A person who strongly prefers a child of one sex or the other does well, for himself *and* the prospective child, to try to ensure that his child is of that sex. If he is stuck with a child of the wrong sex (whichever that may be in his case), the child is likely to provide him substantially less satisfaction than would a child of the sex he preferred; he may even see the child as an unwanted burden, especially if his desire for a child of a particular sex is so strong that he keeps trying until he gets what he wants.

Likewise, a child is better off with a parent who wants a child of his gender than with a parent who would rather have a child of the other gender. A prospective child who never comes to exist has neither rights nor interests and need not be considered; our concern should be for the parent and for the child who becomes actual. The child of the desired gender will have a better relationship with his parent than would a child of the undesired gender. He is likely to receive more support and more affection. He will have a parent who is enthusiastically pursuing his dreams of parenthood, rather than one who is dragging himself along out of duty. Certainly, it is not commendable to regard girls as less worthy of love and attention than boys (or vice-versa)—but it is even less commendable to take that attitude and then risk having a daughter who would be condemned to be raised by a parent who does not properly value her.

Similar considerations apply to other characteristics determined or influenced by a child's genetic material. Rifkin (1998, p. 140) expresses concern, for example, that parents may set too high a threshold for the quality of life of a child with Down syndrome, of which there are moderate cases, and that they might therefore abort fetuses that could develop into individuals with an acceptable quality of life. It may be that there are prospective parents who would find satisfaction in raising a child with moderate—or even severe—Down syndrome. It is certain, however, that there are prospective parents who would not. Even a child with moderate Down syndrome is likely to be more dependent for longer than a normal child; he will therefore impose higher costs in time, energy, and money. And if part of an intelligent prospective parent's motivation to have a child is the desire to guide the development of a mind as powerful as his own, he has little chance of achieving it with a Down child. He may take care of the child out of duty, but the relationship will satisfy neither party as well as would a relationship between a parent and a child of normal capacity. Parenthetically, let me remind you: A child, normal or otherwise, who does not come to exist has no rights and no interests, so any conclusions we draw about whether children of any particular sort *should* come into existence have no necessary consequences as to individuals who *do* exist, or come to exist. *None* of us had a right to be conceived; *none* of us had a right to be born—but that does not mean we do not have a right not to be killed now that we *have* been born.

What applies to the difference between the average and the sub-par applies equally to the difference between the exceptionally gifted and the average. A genius who hopes to have the opportunity to raise a genius may be quite frustrated with a child of merely normal intelligence. He may even come to regard the child as stupid and unworthy of intellectual attention and therefore fail to guide the child to that level of achievement which is within his reach. Again, neither parent nor child will be as happy[3] as he would be if the child had the capacity the parent wanted him to have.

And what applies to intelligence applies to other capacities as well: A child who is substantially inferior to his parent in any respect that the parent considers important is likely to bring his parent less happiness and to be less happy himself than would a child who has a capacity similar to or greater than that of his parent in that respect.

Attempts to shape a future child's character may seem more troubling than attempts to shape his abilities, but recall that gene selection does not imply violating free will. The child whose character is influenced by genes chosen by his parent does not make choice X at time t because his parent wishes at t that he choose X (unless the trait chosen was servility); he chooses X at t because at t he has (loosely speaking) the disposition to choose X, and he has that disposition because of the choice his parent made a long time before t. Yet the parent may be more likely to be pleased with his child's choices, or with his approach to making choices,[4] than he would be with the choices and approach of a child conceived through sex. And in that case, he has less reason to try to override the child's decision than otherwise—so he may develop *more* respect for his child's autonomy than he would with a child conceived through sex. Because he would therefore encourage the child's exercise of autonomy, the child is quite likely to turn out *more independent* than a child of the same parents conceived through sex.

One particularly interesting class of cases is "mental illness." Many parents are likely to seek to produce psychiatrically normal children, and there is some fear that the community will finally rid itself, to paraphrase Don McLean, of those too beautiful for it. But again, there is a cost in unhappiness to both parent and child when the child fails to meet the parent's expectations. That cost exists whether the parent's expectations are based on normal *or unusual* experiences, and some prospective parents who are regarded as mentally ill may see their ways—I should perhaps say *our* ways—of thinking and feeling as preferable to those of the majority and may want, and select for, children who will share them. These parents may be happier than they would be with "normal" children—and their children are likely to be much happier than similar children would be with normal parents, since parents who chose these traits are probably less likely to subject their children to the "therapeutic" traumas often inflicted on such children by "normal" parents.[5]

V.

> From this simplest necessity to the highest religious abstraction, from the wheel to the skyscraper, everything we are and everything we have comes from a single attribute of man—the function of his reasoning mind.
> —Rand 1943, p. 679

In addition to the potential of reproductive technologies to produce children who are better according to the subjective preferences of their parents, there is their potential to produce *objectively* better children—children who are more intelligent, who are physically healthier and stronger, who are more able in every genetically influenced respect. Part of this, of course, is that the results of individual choices will guide other individuals in making choices; we will learn from each other in this as in other aspects of our lives.

One biologist has predicted that by 2350, as genetic improvements build on genetic improvements, 10 percent of Americans will be members of a "genetic aristocracy," engineered for general health and for excellence in particular fields, beyond the possibility of competition from "Naturals," and on their way to forming a new and separate species (Silver 1997, pp. 4–7). But the assumption underlying that theory is that only the richest Americans will be able to afford genetic enhancements. That assumption flies in the face of the whole history of capitalism and technology. Almost all Americans, even graduate students, are, in a technological sense, richer than the czars: We buy cheaply things they could not have bought at any price. A few years ago, there was a lot of concern about a "Digital Divide" under which the poor would not have access to the Internet; all or almost all Americans are now able to get online, one way or another, though there is a "Digital Divide" in the Third World. As a general matter, technologies are adopted first by those who can afford, in light of their resources and their needs, to pay a high price for them—but these early adopters pave the way for those with less money or less urgent needs to obtain the same technologies at a lower price. In 1915, the cost of a phone call from New York to San Francisco was $20 (D'Souza 2001, pp. 72–104). It may be reasonable to think that a small fraction of our population will be different from the rest in 2350 because of reproductive technology—but the minority is more likely to be the Naturals than the genetically enriched.

Will mistakes be made? Yes. Some people are already talking about eliminating aggression (Silver 1997, p. 237), and parents who do that are likely to find their children pushed aside by the children of those with other ideas about enhancement—or even by Naturals. But assuming that we remain at least a moderately free society, with control over enhancements in the hands of individual parents and couples, the same mistake is not likely to be made in so many cases that the Republic will be unable to recover.

Prospective parents will make a wide variety of choices, and as people learn what choices have worked for others, they will slowly come to adopt them. Just as we have expanded our capacities with better tools and better ways of thinking, so we will expand our capacities with better genes. The improvement of our tools and our memes by rational thought is the hallmark of our species. We are beginning to apply to our genes the same power that has bent to our use every element of our surroundings, the same power by which we shape our souls, the power of the mind. Two million years after our ancestors began to shape rocks, we are at last beginning to practice truly *human* reproduction.

NOTES

1. This paper was originally written for a course taught at Mount Sinai School of Medicine by Rosamond Rhodes and Daniel A. Moros.

2. The discussion of sex in this paper is concerned exclusively with its role as a means of achieving conception. Nothing I say is intended to imply any views as to any other role of sex, including in the preservation of relationships between parents that may provide advantages to their children.

3. There is one potential element of happiness I specifically exclude from consideration throughout this paper: the happiness that some people gain by developing certain virtues precisely because their parents fail to value them properly. I omit it, not because I doubt that it is real, but because it would seem a perverse basis for prospective parents' choices or for commending parents. The benefits of parent-child conflict within a context of mutual love and respect are another matter, but because they may be considered in the process of making choices about one's future progeny, they do not disturb the argument.

4. A parent who genuinely values his child's independence will be pleased that his child has made an independent choice even when that choice goes against the parent's preference in the specific case: He would prefer that the child choose independently than that he slavishly follow the parent's preference.

5. For examples of such traumas, see Armstrong (1993).

BIBLIOGRAPHY

Armstrong, Louise. 1993. *And They Call It Help: The Psychiatric Policing of America's Children*. Reading, Mass.: Addison-Wesley.

D'Souza, Dinesh. 2001. *The Virtue of Prosperity: Finding Values in an Age of Techno-Affluence*. New York: Touchstone.

Herman, Jerry. 1964. *Hello, Dolly!* http://www.lyricsondemand.com/soundtracks/h/hellodollylyrics/hellodollylyrics.html.

Kolata, Gina. 1998. *Clone: The Road to Dolly and the Path Ahead*. New York: William Morrow & Company.

Lawrence, Jerome, and Robert E. Lee. 1960. *Inherit the Wind*. New York: Bantam.

Mill, John Stuart. 1859. *On Liberty*. Reprinted in *On Liberty and Other Essays*. New York: Oxford University Press, 1998. The page reference is to this Oxford World's Classics paperback edition.

Poirier, Frank E. 1993. *Understanding Human Evolution*, 3rd ed. Englewood Cliffs, N.J.: Prentice-Hall.

Rand, Ayn. 1943. *The Fountainhead*. Republished on *The Objectivism Research CD-ROM*. Indianapolis: Oliver Computing. The page reference is as given on this electronic edition.

———. 1957. *Atlas Shrugged*. Republished on *The Objectivism Research CD-ROM*. Indianapolis: Oliver Computing. The page reference is as given on this electronic edition.

Rifkin, Jeremy. 1998. *The Biotech Century: Harnessing the Gene and Remaking the World*. New York: Tarcher/Putnam.

Robertson, John A. 2001. Preconception gender selection. *Am J Bioethics* 1:2–9.

Silver, Lee M. 1997. *Remaking Eden: Cloning and Beyond in a Brave New World*. New York: William Morrow & Company.

Wilmut, Ian, Keith Campbell, and Colin Tudge. 2000. *The Second Creation: Dolly and the Age of Biological Control*. New York: Farrar Straus Giroux.

ETHICS AND THE LIFE SCIENCES

DEEP, CHEAP, AND IMPROVABLE: DYNAMIC DEMOCRATIC NORMS AND THE ETHICS OF BIOTECHNOLOGY

PETER DANIELSON, RANA AHMAD, ZOSIA BORNIK, HADI DOWLATABADI, AND EDWIN LEVY
UNIVERSITY OF BRITISH COLUMBIA

> ABSTRACT: A democratic ethics of biological technology must engage the public. This is not easy to do in a way that satisfies the demands of democratic ethics, or meets the pace of rapidly changing, complex technology. This paper describes a solution proposed by the University of British Columbia's Norms Evolving in Response to Dilemmas interdisciplinary research group. The solution, the NERD web survey, has three distinct advantages over other methods: it is *Deep*—the survey provides deep data, particularly when compared to alternatives such as polls and focus groups; *Cheap*—our survey is cost effective, which is important for a truly democratic tool; and *Improvable*—the NERD survey is a work in progress, improvable by design.

1. INTRODUCTION

There are three central problems for a democratic ethics of biological technology (biotech). First, biotechnology is complex. It forces us to examine not only our science, but also our motivations, our ethics and our society. Second, biotechnology is important to many people as it holds the promise of less disease and environmental degradation as well as more choices in food and resources. However, such promises also threaten some important individual and societal values (Daar et al. 2002). Third, biotech innovations contribute to a rapid rate of change as the uptake of applications of the science keep pace with the advancement of the science itself.

These three problems make a democratic ethics of biotechnology difficult to achieve. The technology is important to the public but difficult to comprehend, while our accepted ethical heuristics quickly show their age under the pressures of accelerated change. Our team, the University of British Columbia's Norms Evolving in Response to Dilemmas (NERD) interdisciplinary research group has developed a web-based survey instrument designed to address this set of problems in the following ways:

1. The NERD survey respects the importance and the complexity of the topic by providing a carefully constructed set of about twelve decision problems. These decision problems are based both on historical fact and established science and for each problem a respondent may consult five well-researched advisors. This structure allows us to provide a depth and variety of information.
2. Our survey is designed to stress test normative decisions with social and technological change and social pressure. This adds depth to the analysis.
3. The NERD approach is democratic. Internet-based technology allows us to include large numbers of people. This technology is inherently cheap—an important aspect of any democratic tool—allowing us to plan to survey thousands of people on three parallel issues of genomics in health, forestry, and human food agriculture. As with the previous point, this allows us an added depth of analysis through a rich data-set. This aspect also contributes to the survey's improvability, as relative to other methods, we are not constrained by cost.

Finally, we are part of a larger research project, *Democracy, Ethics, and Genomics: Consultation, Deliberation, and Modelling.* This project provides our normative focus on democratic ethics.[1] It also provides a comparative framework wherein we are committed to competitive testing of our survey instrument and models against our colleagues' focus groups and deliberative speculations as forms of public consultation applied to a shared set of issues.[2] Our tool engages the public by forcing individuals to come to terms with the complex facts of their norms, the subject matter, and the technological mediators we provide to allow the public to express their views.

The rest of this paper will introduce the NERD survey, explain its underlying motivations, and then analyze in greater detail the reasons why it is: deep, cheap, and improvable.

AN INVITATION AND CAUTION: To make what follows more rewarding, we invite you to participate in our survey at http://yourviews.ubc.ca?refer=020 NOW. However, if you choose to read this paper first, please exclude yourself from the survey by visiting this "sullied" url: http://yourviews.ubc.ca?refer=000

2. PRACTICAL AND THEORETICAL MOTIVATIONS

Our practical motivation is straightforward: to ground the ethics of biotechnology in knowledge of the norms people *actually* use. Using constructive methods we are building a platform for running controlled experiments on ethical evaluation in complex technological domains.

Our relation to mainstream philosophical ethics is less straightforward, however. On the one hand, we are influenced by the sociological strand in applied ethics and the game theoretic strand in theoretical ethics, both of which emphasize the role of social norms in a conception of ethics as convention. The mainstream, in contrast, gives more attention to abstract normativity rooted in individual rationality rather than social convention.

Philosophical ethicists are obviously less interested in surveying the content and dynamics of existing norms. However, the failure to appreciate the power of norms may also create problems. First, sampling one's colleagues' and students' "intuitions" in uncontrolled social conditions is likely to be highly selective and unreliable. Second, if norms have the impact we hypothesize they do, they change the way we do ethics. For example, the "precautionary principle," which appeals to many in both ethics and technology, is usually applied solely to technology, cautioning us against adverse technological "lock-in." Our approach broadens that caution, warning would-be democratic decision makers against "ethical lock-in." Thus, decision makers should not just ask "What if we get the technology wrong?" but also "What if we get our changing values wrong?" With this in mind, our surveys are designed to test the power of social norms as an empirical hypothesis (see Appendix 1, N1–3).

On the other hand, our approach is strongly committed to the philosophical ideals of rationality and generality. Rationality is served by our commitment to educating participants while also inquiring about their views. Colleagues in our larger research group, using alternative methods, criticize us for providing scientific, policy, and ethical information. We stress, in reply, that errors of omission (ignorance) are just as serious as errors of commission (bias) when consulting the public. We also claim that providing a variety of contesting advisors mitigates problems of bias. Again, while we begin with an assumption that information is important, we make this a testable hypothesis (See N4–9).

Generality is served by our strategy of constructing one survey instrument to explore many—initially three—diverse subjects: genomics and human heath, food, and forestry. This is contrary to the distinct approaches found in the traditions of both human bioethics and environmental ethics. Our surveys, however, will allow us to explore and unify research in both of these areas. We will be able to ask, and hopefully answer how the evaluation of genomic technology compares in these three diverse subjects (see N0).

3. THE PROJECT

Our project begins at the nexus of ethics, technology, and norm change. Briefly, it involves using an online survey tool[3] to better understand what information people use to make choices and what factors influence their decisions. Throughout the survey, all participants address the same questions and have links to the same fictional advisors offering information and explanations about the science, health, financial, moral, and other issues relevant to each question. Half of the participants also receive aggregated feedback information on how other participants answered each question and which advisors they consulted; the other half does not receive feedback. We conjecture that such information will affect the way respondents in the feedback group answer the questions.

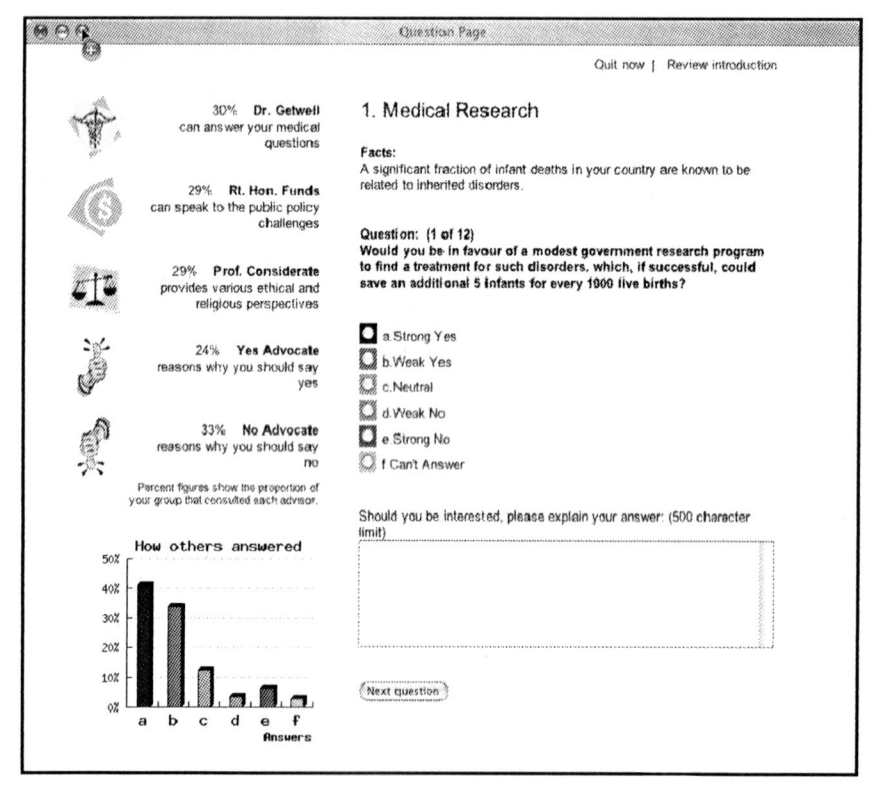

Fig. 1: Typical question dialogue with feedback

Figure 1 is a typical question page (for the feedback group). The question is designed to convey as much information as necessary in an accessible and easy to read manner. Figure 2 is an example of the information provided by the advisors. Through the advisors, we attempt to provide a balanced set of advice ranging from scientific/health information to philosophical perspectives. We include more "every day" advice—similar to the type of advice that might be provided by friends, family, or interest groups—through "Yes" and "No" advocates. Using the advisors is neither compulsory nor necessary for answering the questions. Participant's selection of and duration with each advisor are tracked and timed.[4]

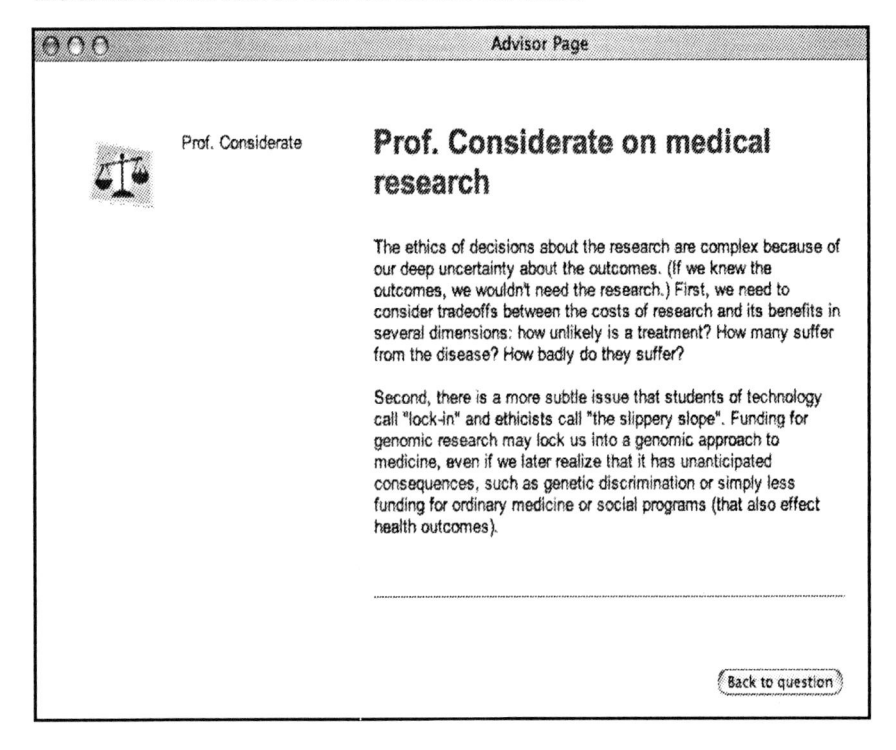

> **Prof. Considerate on medical research**
>
> The ethics of decisions about the research are complex because of our deep uncertainty about the outcomes. (If we knew the outcomes, we wouldn't need the research.) First, we need to consider tradeoffs between the costs of research and its benefits in several dimensions: how unlikely is a treatment? How many suffer from the disease? How badly do they suffer?
>
> Second, there is a more subtle issue that students of technology call "lock-in" and ethicists call "the slippery slope". Funding for genomic research may lock us into a genomic approach to medicine, even if we later realize that it has unanticipated consequences, such as genetic discrimination or simply less funding for ordinary medicine or social programs (that also effect health outcomes).

Fig. 2: Ethics (but please do not use the E-word) advisor

We have incorporated some unique features into the design of the survey to study norms under stress. For example, questions are presented in a "ratcheted" form: participants have an opportunity to reflect on all of their answers at the end of the survey, but cannot go back to a question they have previously answered. This provides data on how decisions change when their consequences become known. This paper was written when our first survey on human health and genomics was incomplete. It has since been completed and an analysis of its results can be found in (Ahmad et al 2005, Ahmad et al., forthcoming). Our primary interest is in building models of dynamic social norms using quantitative data, including both demographic data and information on: who decides what and in how much time; who consults which advisors and for how long; and who leaves when.

A key feature of our design is the bifurcation of the participants and the use of advisors and feedback. This allows us to discern whether those deprived of information about others' choices of options and advisors seek more factual and ethical (thick) information, or rather resort to norm-substitutes, the (thin) yes/no advocates (N8 and N9). The question is: "Does using informed advisors and the presence of feedback information make any difference to the way participants respond (N4 and N5)?" A distinct path for the feedback group signals the influence of (a proxy for) a social norm.

The first NERD survey ran from July 2004 to Feb 2005 with over 1000 participants. Figure 3 depicts the responses of the 316 participants who answered all twelve questions. Two aspects stand out. First, the answers fall predominantly on the pro-technology side, except for question 6 (about constraints on marriage), which was surprisingly negative. Second the answers of the feedback and no feedback groups are not significantly different; this is relevant to hypothesis N1 (see Appendix 1). However, further analysis reveals differences between the two groups in terms of extreme vs. moderate answers, and use of advisors (Ahmad et al. 2005).

A second key feature tracks and times advisor visits. Again the results of the first survey have been surprising. After the first question, participants visit advisors much less than we expected and advisor use falls off even as the scenarios become less familiar and more challenging. Obviously, this is disturbing for our view of democracy based on informed decision. We are planning further research to determine the reasons participants use advisors as little as we observed.

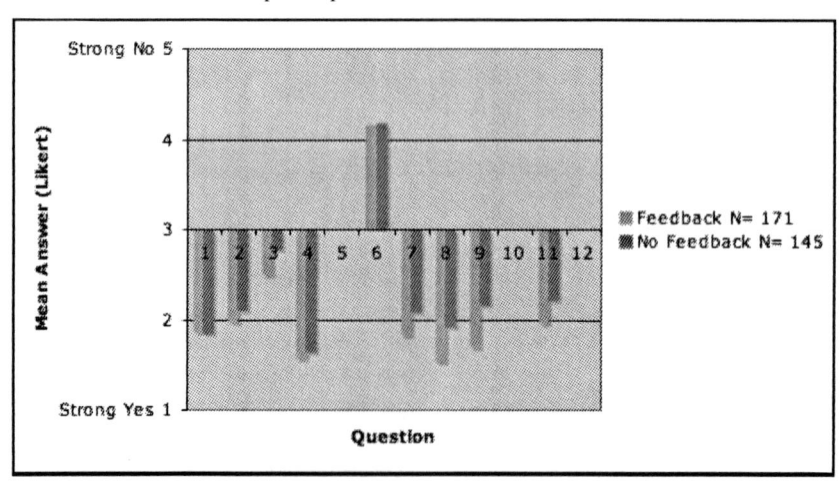

Fig. 3: Qualitative data output: answers show weak feedback effect

The quantitative data is supplemented by qualitative (i.e., textual) data. All participants have the opportunity to comment on each answer, and then revisit each answer at the end, and comment on each advisor. These opportunities were used far more than we expected.

DEEP, CHEAP, AND IMPROVABLE

• **Deep: History and Education**

Our first NERD survey is grounded in the ethically remarkable events that occurred in the middle of the last century when Cyprus was challenged by the genetic disorder Beta-Thalessemia (Bornik 2003; Bornik and Dowlatabadi 2004). By using a series of documented technological and social developments and projecting them into the future of genomic technology, we can calibrate our instrument's medical and social science fiction. This brings us close to the remarkable work of James Fishkin (Fishkin 1991; Fishkin 1992; Fishkin 1995; Fishkin and Laslett 2003), arguably the best current practitioner in democratic ethics, who aims to calibrate his on-line deliberative device against his face-to-face deliberations, which he considers canonical (Fishkin 2003).

We build on Fishkin's point that the raw data of much of politics—public opinion—is neither reliable nor ethically defensible.[5] Fishkin reports that conventional polls, in which a random sample of the public is asked to report on their opinions regarding a variety of issues, can result in mere "top of the head" responses (Converse, 1964). Such responses result in representative, but largely *uninformed* opinions and thus are of questionable value when attempting to identify *actual* opinions or norms (Fishkin, 2003, 132).

Our willingness to engage in educating the public aligns with Fishkin's motivation. Like Fishkin, we are critical of any normative role for superficial, uninformed public opinion. We extend this criticism to more "qualitative" methods, such as focus groups, which have been used by our colleagues in the Democracy, Ethics, and Genomics project. The information collected in these small, directed discussions may also be prone to "top of the head" initial responses. Additionally, focus groups necessarily involve the uncontrolled influence of social structure.

Our online survey project attempts to reduce instances of such "top of the head" responses and instead provide key information a respondent may not have access to on their own. It also attempts to collect data from an individual in the absence of potentially persuasive influences that might occur in focus groups. We do provide artificial "discussion groups" through the advisors, which, in effect, eliminates these uncontrollable elements. As such, we can measure the change in norms through informed decision-making and follow the path of dependence such norms take when under various feedback pressures.

• **Cheap Data: A Democratic Good**

It is unusual for philosophers or ethicists to think about data, much less its cost: our democratic goal and empirical aspirations do not afford us this luxury.[7] However, while cost alone is not decisive, it is unacceptable to have a consultation technology that cannot accommodate population sizes relevant to democracy at a reasonable cost. Additionally, a consultation technology needs to be cheap, easy and anonymous enough to allow diverse groups to access and afford it (e.g., mail and phone, or email and weblogs) in order to be truly democratic.

Although raw internet polling is the cheapest option,[7] we propose that we can provide better results on a cost/quality basis than either polling or focus groups. For example, Figure 4 compares the costs of NERD with a series of focus groups conducted by our colleagues in the *Democracy, Ethics and Genomics* project in the late winter of 2003. While the two methods are difficult to compare, our web-based tool is greatly (i.e., 100:1 in favor of NERD; 8 cents per individual data point) less expensive per person consulted, an important fact when examining instruments for democratizing decisions about technology. If anything, this quick comparison underestimates the differences, as NERD's technology is scalable; our software and server are now underutilized by 1 to 2 orders of magnitude.

Method	Total Cost	Number of participants	Number of data points	Cost per participant (C$)	Cost per data point (C$)
Focus groups	55,312[8]	40—70	N/A	1,382.80–790.17	N/A
NERD-I Web survey	6,600[9]	1,100	79,200	6.00	.083

Fig. 4: Costs Compared

While focus groups typically trade high costs for the promise of richer, more nuanced data, it is currently an open question whether our survey will collect more, and more meaningful, qualitative data through text input than our qualitatively focused colleagues. We are planning a parallel analysis of the two sets of textual data to answer this question.

• **Improvable: Limitations, Correctives**

Limitations to this project include:

1. Participation is voluntary creating a self-selection bias. Only those interested will take the survey, so it may not represent public opinion in the way a random sample does.
2. The "digital divide" limits participation to those who have some competency with computers and the internet. This is less of a problem in Canada than the US; however, it remains a limiting factor (Fishkin, 2003).
3. The advisors are limited in number and do not necessarily speak to the full range of philosophical and other (e.g., theological) approaches to ethics.
4. Our first survey is fictional. Beginning in Cyprus in the 1960s and ending somewhere in the near future, it might not "take": a participant may not (fully) accept the "shoes" of an alter-ego. A more general limitation of fiction—the bane of the electronic game violence literature—is that it evokes and legitimates fictive responses.

5. The NERD surveys are not fully interactive. A generation accustomed to more fully interactive games will likely expect their choices to lead to decision branching. The lack of interactivity may lead to frustration.
6. Democracy is, at best, a procedural value; people might still get important issues "wrong."

Possible correctives for these limitations include:

1. Seek a larger sample. This is the radical promise of internet-based social science: to swap sampling problems with massive amounts of data.
2. The digital divide, like the older numeracy and literacy "divides," involves both access and motivation. We see our comparative advantage in building attractive and easy-to-use instruments.
3. Use dynamic advisors. We invite you to participate in NERD by contacting us and contributing to the next generation of advisors.
4. Correlate the data via demographics. While our focus is Canadian, our survey is open to the world. Correlating via demographics might disclose whether participants closer to the conditions in Cyprus choose differently from Canadians.

Finally, the NERD survey is not a social decision device: it is a tool to extract data from experimental participants. Hopefully the models of social norms and moral agency we build with the data from NERD will allow us to advance democratic practice with future elaborations of NERD.

4. CONCLUSION

We are aware of the difficulties of the road ahead.[10] We welcome criticism but hope to be judged against relevant expectations. Playing with utopian simplifications and fanciful thought experiments can warp expectations. It should be recognized that democrats cannot do, or afford to do, Socratic dialogue, depth psychology, rational therapy, or clinical ethical counseling with millions of citizens.

Three large steps loom before us:

1. Moving from data to models (see Appendix for hypotheses to be tested).
2. Moving from social science to democratic social inquiry. Extending NERD to become a democratic decision device will mean opening up the advisors to the community of experts for input.
3. Opening the source of the NERD surveys to allow re-use and replication, consistent with quality control and fair return to the NERD research group.

We also hope our language and style communicate what is difficult to express directly: how much we have learned by involving a wide range of disciplines in a project of evident social worth and intellectual rigor, based on ideas that encapsulate the heritage of important work in modern philosophy.

APPENDIX 1:
NERD HYPOTHESES—LAST UPDATED JULY 27, 2004

N: Hypothesis is applicable to entire project
H: Hypothesis is specific to Phase One—Human Health and Genomics

1. Dynamic norms

N0: Individuals' values are problem-dependant. Technology offers participants different options for solving a given problem, and the set of traditional positions in normative theory will prove inadequate to explain our data set.

N0.1: There will be a break in the pattern of answers when "genetics" is used to change the solution options.

2. Group dynamics

N1: There will be a significant difference between participants' answers with group feedback and without.

N2: A homogeneous group of respondents has an identifiable pattern of answers.

N3: Subsequent to initiation by a homogeneous feedback group, we expect there to be a path-dependence in the pattern of answers.

3. Advice-seeking

N4: Individuals with extreme answers will have consulted the least number of advisors.

N5: There will be a correlation between the divergence of participants' answers with the script (world path) and the overall drop-out rate.

N6: There will be a correlation between participants who receive feedback and their frequency of advice-seeking.

N7: There will be a correlation between participants' advice-seeking and the popularity (%) of each advisor.

N8: There will be a correlation between positive answers and seeking advice from the Yes Advocate, and negative answers and seeking advice from the No Advocate.

N9: The overall frequency of advice-seeking will diminish over time (from Q1→Q12).

4. Demographics (a sampling of hypotheses)

N10: Regular voters will tend to drop out less.

N11: Participants who have familiarity with the issue will tend to spend more time with each advisor.

N12: Younger participants will spend less overall time seeking advice.

N13: Younger participants will seek more advice from Yes/No Advocates

H1: Parents/guardians will answer differently than others
H2: Older participants will seek more advice from Dr. Getwell.Endnotes

NOTES

This project is funded through grants from Genome Canada and the office of Genome British Columbia and through the National Science Foundation. We thank our colleagues in the W. Maurice Young Centre for Applied Ethics and the Institute for Resources, Environment and Sustainability, both at the University of British Columbia, who helped us at many stages.

Thanks to R. A. Brooks and A. M. Flynn, "Fast, Cheap and Out of Control: A Robot Invasion of the Solar System." *Journal of the British Interplanetary Society*, 1989, pp. 478–485. for the inspiration for the main title.

1. "Genomic research and the potential commercialization of its products raise hopes for future applications and concerns over appropriateness and regulation. If Canada is to determine policy and support research that is in the public interest, those mandated with overseeing the funding and regulation of genomic research must find ways of respecting the public and expert interests in discussions about ethical, environmental, economic, legal and social issues. These discussions must meet the democratic values of representation, transparency and accountability." The aim of the Democracy, Ethics and Genomics project, from our web site: http://gels.ethics.ubc.ca/.

2. A public consultation comparing the NERD survey to focus groups was held in Sept, 2005.

3. The survey has been made as accessible as possible through the use of an "ethics"-free url: http://www.yourviews.ubc.ca. To broaden participation we have peddled (discussion leads to bias) the project on local and national radio and press, the web and street venues served by public transit and libraries.

4. We are aware that the time that a visitor spends on an advisor page does not necessarily correspond to the time that the visitor spends reading and pondering about the advice. We will consider this gap in our analysis.

5. We suggest that much the same is true of the "intuitions" used in much ethics theorizing and teaching.

6. Worrying about costs is not intrinsically valuable. For years the lead author has done computer modeling ignoring costs of computation. Now storage, cpu cycles, and connectivity are so cheap that we likely ought to consider them free. So worry would have been wasted.

7. A local radio station, CFAX, runs a public issue poll every day. This is not unusual and evidently does not require a professional survey designer.

8. Costs include professional consultants for recruitment and execution, travel, hotel, honoraria, and equipment.

9. This overestimates costs in two ways: 1) by including the capital cost of our server, amortized over three years. With web hosting at $20/month, the lower estimate would be $5,040 and 2) by including the development of our web software, some of which will be spread over three surveys this year.

10. We have barely mentioned, for example, the burdens of securing participants identities while making the interface convenient or the burden of going against the conventions (the back button violates our unidirectional script constraint) carved into the dominant client side technologies.

BIBLIOGRAPHY

Ahmad, R., Z. Bornik, P. Danielson, H. Dowlatabadi, E. Levy, H. Longstaff and J.Wilkin. 2005. "Innovations in Web-based Public Consultation." *Proceedings of the First International Conference on e–Social Science*, 22–24 June 2005, Manchester, UK

———. Forthcoming. "A Web-based Instrument to Model Social Norms: NERD Design and Results" *Journal of Integrated Assessment* (forthcoming), 24 pages

Binmore, K. 1994. *Game Theory and the Social Contract: Playing Fair*. Cambridge, MA, MIT Press.

———. 1998. *Game Theory and the Social Contract: Just Playing*. Cambridge, MA, MIT Press.

———. In press. *Natural Justice*. Cambridge, Mass, MIT Press.

Bornik, Z. 2003. Decision Analysis and Genomics: Exploring Norms Shifts in a Complex Environment.

Bornik, Z., and H. Dowlatabadi. 2004. Explorations in regime change: b-Thalassaemia and the interplay of technological change and social norms.

Brooks, R., and A. M. Flynn. 1989. "Fast, Cheap and Out of Control: A Robot Invasion of the Solar System." *Journal of the British Interplanetary Society*: pp. 478–485.

Converse, P. 1964. "The Nature of Belief Systems in Mass Publics." *Ideology and Discontent*. D. E. Apter. London, Free Press of Glencoe: 342.

Daar, A. S., H. Thorsteinsdottir, et al. 2002. "Top Ten Biotechnologies for Improving Health in Developing Countries." *Nature Genetics* 32: pp. 229–232.

Fishkin, J. S. 1991. *Democracy and Deliberation: New Directions for Democratic Reform*. New Haven, Yale University Press.

———. 1992. *The Dialogue of Justice: Toward a Self-reflective Society*. New Haven, Yale University Press.

———. 1995. *The Voice of the People: Public Opinion and Democracy*. New Haven, CT, Yale University Press.

———. 2003. "Realizing Deliberative Democracy: Virtual and Face to Face Possibilities." Paper presented at the conference: *Démocratie délibérative: théorie et pratique*. Centre de recherche en éthique (CRÉUM), Université de Montréal.

Fishkin, J. S., and P. Laslett. 2003. *Debating deliberative democracy*. Malden, MA, Blackwell.

ETHICS AND THE LIFE SCIENCES

AN ETHICAL EVALUATION OF THE SUPREME COURT DECISION REGARDING ERISA INTERPRETATION

KRISTIN LEFEBVRE
WIDENER UNIVERSITY

ABSTRACT: Although the ethical and legal worlds are often at odds, a wealth of information is gained by evaluating legal decisions from an ethical perspective. Evaluating court decisions from an ethical viewpoint, increases our knowledge, and helps to beneficially influence future court precedent. Of particular importance to the relationship between the law, business, and ethics, is the ideal of beneficence and non-maleficence.
It is the court's role to protect the rights of individuals, especially with regards to their health care provision. These issues are especially present in conflicts that relate to the availability and access to health care and insurance coverage. Patient autonomy, physician malpractice and informed consent are all influenced by such current court precedent as addressed by the Employee Retirement Income and Security Act of 1974 (ERISA). This leads us to the central theme of this discussion on the ethical implications of the Supreme Court precedent on ERISA.

INTRODUCTION

Although the ethical and legal worlds are often at odds, there is a wealth of valuable perspective and information that can be gained by evaluating legal

© 2007 *Philosophy Documentation Center*   pp. 327–334

decisions from an ethical perspective. Not only can evaluating these decisions from an ethical viewpoint increase our knowledge, it could help to influence future court decisions by offering a distinct and unique view of court precedent. As M. Gregg Bloche tells us in his essay, "Medical Ethics and the Courts," "[C]ourts answer questions of distributive justice, disclosure and consent, and professional duty that are the staples of the medical ethics discussion" (Danis 2002). Unfortunately, even though famous court decisions (i.e., Roe v. Wade and Brown v. the Board of Education) are saturated with ethical issues; the courts have a history of maintaining a distinct separation from the field of ethics. "Medical ethics reasoning has made only a modest mark on the law of health-care provision in the managed care era, and judicial decisions routinely depart from medical ethics understanding" (Danis 2002).

Of particular importance to the relationship between the law, business, and ethics, is the ideal of beneficence and non-maleficence, for it is the court's role to protect the rights of individuals with regards to their health care provision. These issues are especially present in problems and conflicts that relate to patient autonomy, physician malpractice and informed consent. Which leads us to the central theme of this discussion, or the ethical implications of the Supreme Court precedent with regards to malpractice and the Employee Retirement and Income Security Act of 1974, otherwise known as ERISA.

HISTORY OF ERISA

From a legal perspective, the Employee Retirement Income and Security Act of 1974, otherwise known as ERISA, has had a greater impact on the regulation of Health Maintenance Organizations (HMOs), health insurance coverage, and the rights of individuals receiving health insurance through employer-based plans than ever intended at its conception. At its origin, the main focus of ERISA was to prevent corporations from underfunding pension plans and required them to create a system by which they would have to prove that they maintained the assets promised in pensions to their employees. Interestingly and almost by accident, in the process of protecting pension plans, ERISA also declared that regulation of employer-based health insurance plans would be preempted by the federal government, an area previously controlled by state governments according to the McCarron-Ferguson Act.

Because the Federal law under ERISA (section 514(a)) preempts all court cases that, as the statute says, "relate to" employer-based health insurance plans, the state insurance regulators lost control of the regulation of employer-based health insurance policies (Furrow et al. 2001). Thus, as HMOs and other managed care organizations emerged in the 1980s and became contracted with various employers throughout the country, they were not liable under state laws or in state courts regarding the decisions made by their utilization reviewers (often physicians).

So, the federal government preempted any claims of malpractice against the physicians performing utilization reviews at employer-based insurance companies.

And this is extremely important in relation to patient autonomy, because it is only in state court that one can sue for malpractice and receive compensation such as lost wages, pain and suffering, and punitive damages. A generalized example of this is as follows.

On the one hand, a patient goes to a physician complaining of a lump in the breast and the physician refuses to order a mammogram for that patient. That patient subsequently contracts breast cancer and passes away. The physician, in the above situation, can be sued in state civil court for compensation with regards to his malpractice or unwillingness to order the test. If found guilty, the compensation for the patient's family can include lost wages, pain and suffering, and punitive damages.

On the other hand, consider a patient with insurance provided through their employer and therefore preempted by ERISA. If the patient's physician orders a mammogram for the breast lump and the utilization reviewer (often a physician) denies the mammogram, and the patient gets breast cancer and dies, the patient's family, under ERISA, can only sue for the *cost* of the mammogram in federal court because of ERISA preemption. The patient's family cannot sue for punitive damages, lost wages or pain and suffering in federal court.

Unfortunately, in the last twenty-eight years there have been no federal provisions created to handle compensation in the case of malpractice of the utilization reviewer. Malpractice regulations with regards to insurance are enforced at the state and not the federal level. Because of federal preemption with regards to employer sponsored health insurance, the case cannot be tried in state court. Therefore, instead of receiving compensation as deserved or holding the physician reviewer liable for his actions, the family can be rewarded only with the cost or compensation of the test that should have originally been approved but wasn't. As you can see, many ethical issues regarding patient rights and autonomy surround such legislation.

UNDERSTANDING PREEMPTION

At the time of its inception, all regulatory statutes and common law complaints that "relate to" employee benefit plans were preempted by ERISA and thus heard and dealt with in Federal (section 502(a)), not state, courts of law (Furrow et al. 2001). As it states in *Health Law, Cases, Materials and Problems*, "[F]ederal jurisdiction is permitted when Congress has so completely preempted an area of law that any claim is brought under Federal law, and thus removable to Federal court" (ibid). In addition, section 502(a) of the ERISA statute tells us, "Thus state tort or contract, or even statutory claims that could have been brought as claims of benefits of breach of fiduciary duty under 502(a) are preempted by 502(a)" (ibid).

The importance of the legal questions regarding the interpretation of the ERISA act is that the Federal government did not create any statutes or regulations regarding how employer-based insurance companies would be held liable for their actions. Therefore, individual patient's rights and patient's autonomy are being violated with regards to malpractice claims against the "clinical" decisions made by their health insurance companies. This is a huge ethical dilemma which breaches not just the ethics behind legal decisions but also business ethics with relation to the provision of health insurance. Was the goal of the ERISA act of 1974 to interpret the individuals making the clinical decisions regarding acceptance or denial of care at health insurance companies or HMOs as fiduciaries and thus exempt from state malpractice laws?

Although the answer to the above questions regarding the interpretation of the ERISA act may seem obvious, this problem has not been an easy one to address from a legal perspective. Until recently, as stated above, most cases brought to court regarding decisions made by employer-based health insurance policies have been preempted by ERISA. An example of case as recent as 1993 illustrates these decisions. The legal problem that has existed is understanding whether or not ERISA preemption, or the interpretation of the act by common law judges and courts, was designed to prevent insurance based health insurance policies from being liable for their actions. Was the law really designed to prevent or block state common law actions against employer-based health insurance plans? Or does the interpretation of the act allow the clinical decisions regarding patient care to be covered under the state malpractice statutes?

CASES TO ILLUSTRATE THE ETHICAL DILEMMA

A case that illustrates the ethical dimensions of ERISA preemption is the 1993 decision of the United States Court of Appeals for the Eighth Circuit in the State of Missouri regarding the case of *Kuhl v. Lincoln National Health Plans of Kansas City*. The plaintiffs brought suit against the Lincoln National Health Plan in reference to Buddy Kuhl, covered through Lincoln National Health Plan via his employment with Belger Cartage Services (*Kuhl v. Lincoln National Health Plan* 1993).

On April 29, 1989, Buddy Kuhl suffered a heart attack and was placed under the care of Dr. Levi, a heart specialist at the Menorah Medical Center. Dr. Levi stated that an open-heart procedure was required to save Buddy Kuhl's life. Buddy Kuhl was at a high risk for sudden death and the surgery was recommended within the next few weeks. Because Kansas City Hospitals did not have the equipment to perform the necessary surgery, Buddy Kuhl was referred by his physician to Barnes Hospital in St. Louis, Missouri.

On June 23, 1989, Lincoln National refused to pre-certify payment because Barnes Hospital was outside of the Lincoln National Service area. The July 6, 1989 surgery was cancelled. Lincoln National then scheduled an appointment for Kuhl to see Dr. Brodine, a Lincoln National physician, to determine whether the procedure could be performed in Kansas City. Dr. Brodine agreed with Dr. Levi that the procedure needed to be performed at Barnes Hospital. The surgery team at Barnes Hospital in St. Louis would not be available again to perform the surgery until September 2, 1989.

Unfortunately, Kuhl's heart had deteriorated to the point that he was no longer appropriate for the option of surgery in St. Louis by September and was placed on the transplant list. Kuhl died while awaiting a heart transplant. On March 15th, Mary Kuhl, Marnie Kuhl, and Buddy Kuhl Jr. filed a petition asserting for claims against Lincoln National: medical malpractice, emotional distress, tortious interference with Buddy Kuhl's right to contract for medical care, and breach of contract. Lincoln National filed a Notice of Removal asserting that ERISA provides Federal question jurisdiction over each of the Kuhl's claims.

The district court held that the plantiffs' state law claims against Lincoln National were preempted by the ERISA act as they "relate to" the Belger Plan (employer-based insurance plan) and that ERISA did not authorize the recovery of monetary damages for Lincoln National's alleged misconduct. The appeals court agreed. The family was unable to sue for pain and suffering, punitive damages, and lost wages and the insurance company and utilization reviewer were not held liable.

Several other cases on record support this notion that all issues that "relate to" employer-based health plans are preempted by ERISA. These examples include *Shaw v. Delta Airlines*, in which the court, "[A]lmost always finding preemption when it found an ERISA plan to exist" (*Shaw v. Delta Airlines* 1982), and in *Metropolitan Life Insurance Company v. Taylor*, where the "complete preemption" is noted to the "well-pleaded compliant rule (*Metropolitan Life Insurance Company v. Taylor* 1987)." Under the well pleaded compliant rule, a cause of action arises under Federal law only if a Federal question is presented on the face of the plantiff's properly pleaded complaint. In the *Metropolitan v. Taylor* case as referenced above it is noted, "Congress may so completely pre-empt a particular area that any civil complaint raising this select group of claims is necessarily Federal in nature" (*Metropolitan Life Insurance Company v. Taylor* 1987).

THE BUSINESS OF HEALTHCARE

Although the precedent regarding ERISA preemption of state claims of malpractice with regards to utilization review is changing, the issues presented above pose several ethical questions. The first question involves the ethics of the judicial system. As stated earlier, it is the responsibility of the judicial system to assist in protecting the provision of healthcare and defend individual patient's rights with regards to that healthcare. With regards to the ERISA decision, it has taken twenty-five to twenty-eight years to begin to change the precedent with regards to this decision. If the judicial system was willing to more often consult with experts in ethics, so many apparent wrong doings may have been avoided.

Take, for example, the outline of "The Ethics Workup" of Daniel P. Sulmasy, O.F.M., M.D. of Georgetown University School of Medicine (Sulmasy 1999). If you evaluate the ERISA issue with a simple ethics work-up, it is clear that the legal precedent needed to change years ago. For example, we have already given "the facts" in the above paragraphs, but to reiterate, individuals were unable to sue their employer-based insurance utilization reviewers secondary to ERISA preemption.

"The issue" is that individuals lost their autonomy in addition to access to quality health care at the government's hands and it has taken twenty-five years for that to begin to change. We have "framed the issue" around the various cases involving ERISA preemption including the Buddy Kuhl case. But in framing the issue, we must also consider the "biomedical good of the patient," as Sulmasy refers to it, which has evidently been ignored for years with regards to ERISA preemption. When you "situate the issue" among other cases of ERISA preemption, you find many cases where families were wronged by insurance companies and received no remuneration. And finally in "deciding," the obvious decision would be for the law to change the precedent or create federal guidelines to regulate utilization review in employer-sponsored insurance plans. The criticizing of this issue is as follows.

FOR EVERY ACTION, A REACTION

On the surface, the issue regarding state control over tort reimbursement from employer sponsored insurance liability seems very black and white. When reflecting upon the previously mentioned cases, it seems obvious that people should be compensated for unjust decisions made by utilization reviews, and as is the job of the civil court, be made whole. But, as ERISA loses its ability to preempt lawsuits involving HMOs, what will happen to managed care as a result?

There is not one example in any of the above cases where common law judges felt that Congress meant to allow HMOs under employee based insurance plans to be immune from prosecution for negligent decisions based on the withholding of medical care. But, the time at which the ERISA statute was designed was before HMOs took such a significant role in health care. As it states in Medical Economics from April 10, 2000, "Including HMOs, PPOs and point of service plans, managed care now covers 90% of insured workers" (Terry 2000).

Since HMOs insure such a significant portion of the population, we must not ignore the possible impact this new interpretation of HMO liability will have on the business of insurance itself. In addition, what costs will insured workers now have to entail secondary to the increased costs that HMOs will attain from the prospect of litigation in state courts? As it states in *Health Affairs* from November/December 1999, "[H]ealth plans assert that litigation, whether actual or threatened, must result in premium increases. Employers add that such hikes, plus the fear of direct liability, will prompt them to trim benefit packages or terminate them all together" (Studdert 1999).

Although there are several methods by which health plans can protect themselves from liability, none are without the covered participant suffering some kind of recourse. The health plans can either self-insure, causing premiums rates to increase, or they can manage the threat by decreasing or liberalizing coverage. Liberalizing coverage would mean less dental or eye care, or less choice within the preferred network of providers. Regardless, just protection from liability is going to increase costs with regards to HMOs. As it states in *Health Affairs*, "[T]hey also anticipate greater use of attorneys and risk managers at every stage of business operations" (Studdert 1999).

CONCLUSION

As ethics is the study of conflicting goods, it is important to reflect on our critique regarding the ERISA act, as it seems that the common law courts are altering twenty years of wrongful decisions based on ERISA preemption. The broad interpretation of the ERISA statute could endanger the future stability of employer-based health insurance due to the rising costs of insurance providers defending themselves against healthcare liability claims. Still, from a utilitarian perspective, there is no doubt that states should maintain their right to regulate malpractice, especially in consideration of utilization review and rights to external review. However, continuing on with the existing ERISA statute is not the answer.

At this point, it is important to evaluate the provision of healthcare as seen as the business of healthcare from an ethical perspective. As Leonard Weber tells us, "To describe healthcare organization as a citizen is simply to recognize that it has a responsibility to promote the public good, particularly a responsibility to seek to improve the health status of the community" (Weber 2001). By reflecting on this quote we get to the heart of the ethical conflicts that exist with regards to ERISA. The question of ethical business practice within insurance companies, reflection on federal and state statutes and revisiting federal laws and regulations can lead us to a utilitarian ideal that can stand to benefit us all.

BIBLIOGRAPHY

Blue Shield Plans v. Travelers Ins. Co. 1995. 514 U.S. 645, 115 S.Ct.

Danis, M., C. Clancy, L. Churchil. 2002. *Ethical Dimensions of Health Policy*. Oxford University Press Inc.

Dukes v. U.S. Healthcare. 1995. 57 F 3d. 350; 1995 U.S.

Furrow, B. R., T. L. Greaney, S. H. Johnson, T. S. Jost, R. L. Schwartz. 2001. *Health Law: Cases, Materials and Problems*. St. Paul, MN: West Group.

Gaier, Matthew, T. A. Moore. 2003. "Medical Malpractice: Recent Second Circuit Decision on HMO Liability." *New York Law Journal*. March 229, p. 3.

Kuhl v. Lincoln National Health Plan of Kansas City. 1993. United States Court of Appeals for the Eighth Circuit, 999 F.2d 298.

Metropolitan Life Insurance Company v Taylor. 1982. 481 U.S. 58, 63, 95 L.Ed.; 1987.

Shaw v. Delta Airlines, 463 U.S. 85, 103 S.Ct.

Studdert, D., W. Sage, C. Gresenz, D. Hensler. 1999. "Expanded Managed Care Liability: What Impact On Employer Coverage?" *Health Affairs*. 18(6): pp. 7-24.

Sulmasy, Daniel. 1999. "The Ethics Workup" Georgetown University School of Medicine.

Terry, Ken. 2000. "Where Is Managed Care Headed?" *Medical Economics*. April 77(7), p. 244.

Weber, Leonard J. 2001. *Business Ethics in Healthcare: Beyond Compliance*. Indiana University Press.

ETHICS AND THE LIFE SCIENCES

PHARMACEUTICAL "GIFT-GIVING," MEDICAL EDUCATION, AND CONFLICT OF INTEREST

DALE MURRAY AND HEATHER CERTAIN
UNIVERSITY OF WISCONSIN-BARABOO AND
UNIVERSITY OF WISCONSIN-RICHLAND;
WILLIAM S. MIDDLETON MEMORIAL VETERANS HOSPITAL, GRECC AND
UNIVERSITY OF WISCONSIN CENTER FOR WOMEN'S HEALTH RESEARCH

ABSTRACT: In this essay, we argue that the acceptance of gifts by health professionals from the pharmaceutical industry is morally problematic. We conclude that whether physicians view the receipt of items from drug detailers as entitlements or gifts, this practice is unacceptable, as it constitutes a conflict of interest. In addition, we argue that these gifts are particularly problematic in academic hospitals. Physicians-in-training are inculcated with the belief that receiving gifts is morally acceptable. The cumulative effect of these worries should be sufficient to warrant the serious attention of medical associations worldwide.

I. INTRODUCTION

Recently, several physicians and bioethicists have published a rash of articles and commentaries concerning the moral appropriateness of pharmaceutical industry gift-giving to doctors. Previously, the literature had focused on large gifts from pharmaceutical detailers, with the AMA issuing a statement that gifts of only minimal value are appropriate (AMA 2000). However, Dana Katz, Arthur L. Caplan, and Jon F. Merz argue that even small gifts from pharmaceutical representatives to physicians can influence the latter's prescription decisions. Additionally, these decisions are not always in the best interests of patients and debase the fiduciary patient-doctor relationship. Thus, pernicious conflict of interest issues arise with small-scale gift-giving practices. Many of the commentaries on Katz et al.'s article agree in principle with its findings, putting forth only minor complaints and adding few emendations.

While Katz, Caplan, and Merz's work and the ensuing commentaries are valuable for bringing this once overlooked issue to light, the discussions are ultimately unsatisfying; they fail to examine important elements of this controversy. For the purposes of this essay, we will press two issues. Several key factors are identified as causal components for why physicians accept gifts even when presented with evidence that doing so may influence their prescription behavior. For example, Katz, Caplan, and Merz note that some physicians see the flattery and gifts bestowed upon them not really as gifts at all—but as entitlements and perks of their professional standing (Katz, Caplan, and Merz 2003, pp. 39–46). However, they include little discussion of either why physicians subscribe to this "entitlement" view, or why it is particularly problematic that they do.

A second subject mainly untouched in the recent literature relates to medical education. Little has been said about the effect pharmaceutical gift-giving has on *medical students and residents*—those new to the profession, who are in their "formative years." We shall take up these issues one at a time.

II. PROBLEMS OF GIFTS AS PROFESSIONAL ENTITLEMENTS

Katz, Caplan, and Merz note that many physicians meet drug detailers primarily to receive small gifts (Katz, Caplan, and Merz 2003, p. 40; Lexchin 1989). We can also conclude that such meetings are pervasive in medicine since they report that the "large majority of physicians meet with industry detailers several times a month" (Katz, Caplan, and Merz 2003, p. 40). In fact, interaction between drug representatives and health care professionals has been extremely high in several countries for at least the past fifteen years. Roughly 85 percent to 90 percent of doctors in Canada (Williams and Cockerill 1990), New Zealand (Benseman 1985), Britain (Greenwood 1989), and the United States (Bower and Burkett 1987), see drug detailers. Canadian (Williams and Cockerill 1990) and US (Lurie, Rich, and Simpson 1990) physicians are visited by drug detailers about once every two weeks on average. Physicians do not usually recognize the receipt of small gifts, such as pens, post-it notes, etc., as being morally problematic because the gifts are understood as mere tokens. Moreover, many physicians do not think that they raise ethical difficulties because they believe "reminder items" have marginal to no effect on their professional judgment (Katz, Caplan, and Merz 2003, p. 40). Additionally, J. Avorn, M. Chen, and R. Hartley found that 68 percent of physicians in their sample believed drug advertisements to have "minimal importance" in influencing their prescribing habits (Avorn, Chen, and Hartley 1982, p. 5). Doctors may surmise that they are "more reasonable and critical" than the average person who is more susceptible or more gullible to such bold-faced marketing ploys (Katz, Caplan, and Merz 2003). Therefore, most doctors do not see any problem with receiving free lunches, pens, and other drug marketing paraphernalia.

So, in this void, we can turn to a positive argument concerning why medical professionals might accept these items. For now, we will bracket the idea that such small items do indeed influence the prescription decisions of physicians. We are more interested in the notion stipulated earlier that medical professionals see themselves as being *entitled* to these items—that flattery and small gifts are acceptable due to one's standing as a professional.

Part of this expectation of entitlement is understandable if we recognize the traditional distinction between a profession and an occupation. Recognizing the important social contributions of certain sorts of work is supposed to confer onto professionals a status that is different from occupational workers. Physicians are acknowledged as professionals not only due to the importance of health care, but also because of the specialized knowledge necessary for their job. Those who wish to become doctors are among the most highly educated members of any society. The long years of study necessary for both the theoretical and practical training for the purpose of saving and maintaining life (and its quality) require enormous expenditures of time and money.

Of course, what also contributes to one's special standing as a medical professional are the long hours that physicians work, including late night calls that cannot be put off as in other lines of work. This says nothing about the special professional obligations that physicians must uphold. Special duties of physicians are deeply embedded in their codes of ethics, including confidential holding of their patients' information, special obligations to aid those in need, and duties to keep abreast of the latest medical procedures and pharmaceuticals, among others.

Given these factors, physicians as professionals are highly paid and socially respected, and respected professionals often receive entitlements. It is generally acceptable that attorneys who become partner in a firm receive box seats to the local baseball team, and that MBAs who attain a certain status receive keys to the proverbial executive bathroom. Therefore, it does not seem unfounded that physicians would also feel that they deserve entitlements, and that these perks may come in the form of pharmaceutical gifts.

But should these gifts be considered *entitlements* of a certain sort? Prima facie, this idea is specious if we simply consider the terms used. First of all, these small tokens are defended by drug representatives *as gifts*. Gifts, like charity, are supposed to be exempt from the *quid pro quo* of commercial exchanges in the first place. So, if providing lunches, pens, and drug samples to physicians are truly gifts, this cannot be due to any sort of privilege or entitlement. Entitlements are, by their nature, something that one is owed.

But this, it may be argued, is a matter of semantics. What physicians might *really* mean is that these are not gifts at all. Instead drug representatives and other members of society *should* provide items and services as tokens of appreciation. This is likely the sentiment that physicians are expressing: that because they are members of a respected profession, and one that provides a service to the public, they are entitled to certain benefits.

If we accept that physicians are entitled to certain perks afforded other professionals, we again need to address why receiving this "entitlement" from a pharmaceutical representative is problematic. When a lawyer receives box seats to the Knicks game from his employer, he is receiving a gift for a job well done. However, when a physician receives a gift from an industry detailer there is a proven conflict of interest that we can demonstrate in two ways.

On the one side, the professional obligation is to patients; physicians are supposed to play a professional role as patient advocate. A doctor who is influenced by drug detailers cannot be an advocate of the patient, but instead prescribes medication that may not be in the best interest of the patient. Perhaps there is a better drug on the market, or one that works as well as the one touted by a detailer, but at a lower cost.

The second way that conflict of interest develops is that the potential that the physician may be influenced by detailers erodes a trust vital to a patient-doctor relationship. As a patient, how can I have confidence that a physician is my advocate when she is susceptible to accepting gifts from the friendly detailer who has no interest in my care, but instead seeks to fill the corporate coffers?

Moreover, the pervasive acceptance of gifts may well compromise the integrity of the patient-physician relationship even when the patient's *own* physician refuses such gifts. To explain, we need to consider an analogous case from the field of political philosophy. If the goal of a liberal, democratic state is to be neutral in respect to the interests of its citizens, it cannot actually champion one conception of the good, or one way of life, over any other. However, this is not the only requirement of a just state if it is to be a neutral one. There is also a transparency requirement. It is not merely that justice *needs* to be done, it is also required that justice is *seen* as being done. Laws need to be made in a fashion that is open to the view of all citizens (even if few citizens care to view the legislative process). Citizens need to perceive that justice is being done, and part of that is the ability to monitor whether their lawmakers are "on the take." Analogous cases also come from the legal profession. Judges sometimes recuse themselves from cases in which there is even a possible hint of impropriety, such as the inability to make an unbiased judgment. As noted in *Stone v. Powell*, "The neutrality requirement helps to guarantee that life, liberty, or property will not be taken on the basis of an erroneous or distorted conception of the facts or the law" (*Stone v. Powell* 1976). Most importantly, note that it does not matter whether a judge believes herself to be impartial in regards to the outcome of the case. Again it is the possibility of impartiality that suggests her removal.

Likewise, physicians as professionals should be wary of appearing to select pharmaceuticals based on the biases of advertising. To deflect this appearance requires that physicians refuse to meet with detailers and reject their gifts—no matter how great or small. To retain one's standing as a professional, physicians must take neutrality and transparency seriously. To do so would appear to require limited contact with coercive factions such as drug detailers, along with the appearance of doing so. If it is common knowledge to my patients that doctors generally are susceptible to accepting the gifts of detailers, and that this may influence their prescribing practices, my professional status is weakened by the conduct (and perceived conduct) of my professional colleagues.

While physicians may believe themselves to be entitled to small gifts from pharmaceutical representatives, conflict of interest shows why the acceptance of gifts is morally wrong. It erodes the vital trust between physicians and patients, because physicians are influenced by detailers into prescribing their drugs, even if they are not in the best interests of their patients. As we have seen however, part of one's professionalism is giving the impression of performing in the clients best interests. Physicians should objectively choose drugs for patients purely on the basis of their medical needs. Physicians are not justified in accepting entitlements that work against this impression.

III. MEDICAL TRAINING

As we just discussed, the widespread belief held by physicians that small gifts from drug representatives are appropriate because they are *entitlements* is problematic. Here we will argue that this belief has important spillover effects on those who are not established doctors, but who are in the "apprenticeship pipeline." The ready acceptance of free lunches from pharmaceutical companies is part of the enculturation process of residents and medical students. Residents are also targeted as prospective future clients by drug companies. In both the United States and Canada, a significant amount of promotion is aimed at physicians-in-training (Education Council 1993). Lichstein, Turner and O'Brien surveyed 272 directors of internal medicine programs in the United States and found that residents met with sales representatives from pharmaceutical companies during working hours in 84 percent of programs (Lichstein, Turner, and O'Brien 1993). In addition, pharmaceutical companies sponsored conferences in 89 percent of programs (Hodges 1995, p. 554). We argue that pharmaceutical gift-giving to doctors is particularly problematic because of the possible corruption of professional character over time. Professional character is eroded when physicians display prescribing practices that are rife with conflict of interest, and this behavior influences future doctors.

The AMA already allows for pharmaceutical companies to play a supporting role in the education of medical students and residents, albeit a carefully constrained and structured one. The AMA Code of Ethics recommends some discretion in allowing pharmaceutical companies to fund student scholarships. In Section E-8.061 we find that:

> Scholarship or other special funds to permit medical students, residents, and fellows to attend carefully selected educational conferences may be permissible as long as the selection of students, residents, or fellows who will receive the funds is made by the academic or training institution. Carefully selected educational conferences are generally defined as the major

educational, scientific or policy-making meetings of national, regional, or specialty medical associations. (AMA 1991) Often, the role that pharmaceutical companies play in the education of medical students and residents goes beyond this minor financial support and comes in the form of small gifts. There have been a few studies that attempt to look at how residents feel about receiving gifts from pharmaceutical companies. Similar to their mentors, resident physicians argue that small gifts from industry detailers do not influence their prescribing practices. A study by Steinman, Shlipak, and McPhee indicated that 61 percent of residents felt that their prescribing behavior was not influenced by contact with pharmaceutical detailers (Steinman, Shlipak, and McPhee 2001, p. 554). Interestingly, this study also found that only 16 percent of residents felt that other doctors were uninfluenced by contact with pharmaceutical representatives (Steinman, Shlipak, and McPhee 2001, p. 554).

Residents also state that even if they were influenced by pharmaceutical representatives, it would not matter, as residents do not control what is on the hospital formulary (Steinman, Shlipak, and McPhee 2001, p. 554). In general, residents care for patients either as in-patients, or as patients in clinics that are associated with the hospital. Therefore, their prescribing practices are quite limited by the hospital formulary. However, studies have shown that the presence of pharmaceutical representatives in teaching hospitals has led to requests for changes in the hospital formulary (Chren and Landefeld 1994). While this is likely due to the attending physician's discussion with detailers (as residents usually do not request formulary additions), it is possible that residents influence the requests by discussing the drugs with the attending physician.

The influence of drug detailers on supervising physicians in teaching hospitals is also important, as these doctors control the training of residents. They direct medical education, guide the pharmaceutical research agenda, and sometimes formulate national policies regarding new drugs (Lurie, Rich, and Simpson 1990, p. 240). Supervising doctors who meet with industry detailers are more likely than other physicians to request that drugs produced by the detailers' companies be added to the formulary (Lurie, Rich, and Simpson 1990, p. 240). Certainly drug detailers hedge their bets and try to maintain good relations with all physicians. As Lurie, Rich, and Simpson put it, "good relations with residents can serve as the basis for future practitioner [accounts]—PR [pharmaceutical representative] interactions can provide access to their faculty" (Lurie, Rich, and Simpson 1990, p. 240).

In addition to feeling that they are not affected by drug lunches, residents and medical students may feel entitled to them just as some established doctors do. After all, residents are infamous for working long hours on little sleep and being relegated to the lowest spot on the totem pole. Many residents carry large debt burden and do not make significant sums of money in their first years out of school (relative to the possible income that first-year lawyers or MBAs could make). Medical students are actually paying to be in the hospital. Therefore, it seems that if anyone is entitled to a free lunch, it would be a medical student or a resident. In fact, a recent survey revealed that 80.3 percent of medical students feel entitled to gifts from pharmaceutical companies (Sierles, Brodkey, and Cleary 2005, p. 1038). It is also clear that free lunches draw in medical students and residents to meet with detailers. Residents admit that without gifts, their interactions with pharmaceutical representatives would be reduced (Wazana 2000, p. 375). Yet, just as full-fledged physicians should not use these as entitlements, the same conflict of interest exists for residents and even medical students. Given the above situation, residents may be even more subject to the influence of pharmaceutical representatives.

The presence of pharmaceutical companies in training hospitals is problematic not only because of the conflict of interest, but also because the training program is where a young doctor truly begins to cultivate his or her professionalism. Residents do this largely by observing the ethical choices of the teaching physicians. Codes of ethics for physicians are somewhat clear about the professional duties of doctors, and one might argue that despite what they observe around them, medical students and residents should be able to comprehend these obligations by themselves. Yet, students and residents cannot merely read about how to make ethically acceptable decisions any more than they can simply read about how to be a skilled doctor. Residents learn from watching certain procedures being done, and then trying them out for themselves.

Likewise, much of the moral training of residents comes from the way they see attending physicians handling difficult ethical choices that arise in the hospital and clinic. In fuzzy ethical situations, this is to whom they turn. Therefore, if residents see respected physicians taking small gifts from pharmaceutical representatives, and a presence of pharmaceutical companies is pervasive in their training program, it is not surprising that residents would start to accept gifts from detailers: they are vulnerable, they feel entitled to them, and it seems to be an accepted practice of their new profession. One study showed that even when residents found the presence of pharmaceutical representatives to be morally problematic, they still accepted gifts from them, which suggests that residents are influenced by both their peers and their supervisors (Steinman, Shlipak, and McPhee 2001: p. 555). Unfortunately, if the practice of accepting small gifts from pharmaceutical companies begins in residency, it is likely to persist. This may be exactly what the drug companies are hoping, and why they bring gifts to residents in the first place.

IV. CONCLUDING REMARKS

In this essay, we have revealed some reasons why the acceptance of gifts by health professionals is morally problematic. This paper has emphasized some particular problems with pharmaceutical industry gift-giving that have been largely ignored. We conclude that whether physicians view the receipt of items from drug detailers as entitlements or gifts—this practice is unacceptable. Since pharmaceutical companies clearly expect a return on the flattery they bestow upon physicians, they are not really giving gifts in the first place. Physicians are also not entitled to these items, since they compromise professional judgment and constitute conflict of interest.

Established physicians set the tone for ethical practice in the hospital for those who are learning from their mentors. Physicians-in-training are not merely receiving an education in the craft of healing, but also in practicing in an ethical manner. The ready acceptance of gifts, as practiced currently by many supervising physicians, promises to produce a new generation of future doctors who will do so, with all of its attenuating results. Thus again, the current acceptance of gifts is particularly pernicious because it perpetuates the cycle of compromise.

The cumulative effect of these worries should be sufficient to warrant the serious attention of medical associations worldwide. It would be convenient if we could easily delineate some threshold under which gifts would be acceptable. Yet empirical evidence has shown that all gifts great and small do influence physician prescribing practices, and philosophical investigation has revealed why this influence is morally problematic.

BIBLIOGRAPHY

American Medical Association. 1991. "Gifts to Physicians from Industry," *Journal of the American Medical Association* 265, p. 501.

American Medical Association Council on Ethical and Judicial Affairs. 2000. *Ethical Opinions/Guidelines.* E-8.061.

Avorn, J., M. Chen, and R. Hartley. 1982. "Scientific versus Commercial Sources of Influence on the Prescribing Behavior of Physicians," *American Journal of Medicine* 73, pp. 4–8.

Benseman, J. 1985. "The Great Paper Waste: The Use of Unsolicited Medical Literature by General Practitioners," *New Zealand Family Physician* 12, pp. 96–98.

Bower, A., and G. Burkett. 1987. "Family Physicians and Generic Drugs: A Study of Recognition, Information Sources, Prescribing Attitudes, and Practices," *Journal of Family Practice* 24, pp. 612–616.

Chren, M., and C. Landefeld. 1994. "Physicians' Behavior and Their Interactions with Drug Companies: A Controlled Study of Physicians Who Requested Additions to a Hospital Formulary," *Journal of the American Medical Association* 27, pp. 684–689.

Education Council, Residency Training Programme in Internal Medicine, Department of Medicine, McMaster University, Hamilton, Ont. 1993. Development of residency program guidelines for interaction with the pharmaceutical industry," *Canadian Medical Association Journal* 149, pp. 405–408.

Greenwood, J. 1989. *Pharmaceutical Representatives and the Prescribing of Drugs by Family Doctors.* Doctoral Thesis (Nottingham, U.K.: Nottingham University).

Hodges, B. 1995. "Interactions with the Pharmaceutical Industry: Experiences and Attitudes of Psychiatric Residents, Interns and Clerks," *Canadian Medical Association Journal* 153:5, pp. 553–559.

Katz, D., A. Caplan, and J Merz. 2003. "All Gifts Great and Small: Toward an Understanding of the Ethics of Pharmaceutical Industry Gift-giving," *American Journal of Bioethics* 3, pp. 39–46.

Lexchin, J. 1989. "Doctors and Detailers: Therapeutic Education or Pharmaceutical Promotion?" *International Journal of Health Services* 19:4, pp. 663–679.

Lichstein, P.R., R.C. Turner, and K. O'Brien. 1992. "Impact of Pharmaceutical Representatives on Internal Medicine Residency Programs," *Archives of Internal Medicine* 152, pp. 1009–1013.

Lurie, N, E. C. Rich, and D. E. Simpson. 1990. "Pharmaceutical Representatives in Academic Medical Centers: Interaction with Faculty and Housestaff," *Journal of General Internal Medicine* 5, pp. 240–243.

Sierles, F., A. Brodkey, L. Cleary. 2005. "Medical Students' Exposure to and Attitudes About Drug Company Interactions: A National Survey." *Journal of the American Medical Association* 294, p.1034–1042.

Steinman, M., M. Shlipak, and S. McPhee. 2001. "Of Principles and Pens: Attitudes and Practices of Medicine Housestaff toward Pharmaceutical Industry Promotion." *American Journal of Medicine* 110, p. 551–557.

Stone v Powell. 1976. 428 US 465, 483 n. 35, 96 S. Ct. 3037, 49 L. Ed. 2d 1067.

Wazana, A. 2000. "Physicians and the Pharmaceutical Industry: Is a Gift ever just a Gift?" *Journal of the American Medical Association* 283, p. 375.

Williams, A., and R. Cockerill. 1990. "Report on the 1989 Survey of the Prescribing Experiences and Attitudes toward Prescription Drugs of Ontario Physicians," *Prescriptions for Health: Report of the Pharmaceutical Inquiry of Ontario.* (Toronto: Pharmaceutical Inquiry of Ontario) pp. 1–102.

ETHICS AND THE LIFE SCIENCES

WHY SHOULDN'T INSURANCE COMPANIES KNOW YOUR GENETIC INFORMATION?

NEIL A. MANSON
THE UNIVERSITY OF MISSISSIPPI

ABSTRACT: In this paper I state and reject two of the most commonly given arguments for regulating access by insurance companies to the results of genetic tests. I then argue that since we cannot assume *a priori* that those genetically predisposed to disease will have worse health outcomes than those not so disposed, we cannot know *a priori* that genetic discrimination will emerge as a major problem in a free market health insurance system. Finally, I explore the possibility of a free-market solution to the problem of genetic discrimination: genetic insurance.

I. INTRODUCTION

For some diseases, genetic tests can demonstrate that test subjects or their offspring will be afflicted, or that there is an increased probability of contracting the disease. There are already many such tests in use, for conditions ranging from glaucoma and colon cancer to dwarfism and mental retardation, and with the completion of the Human Genome Project we can expect vastly more genetic tests will be developed. Genetic testing holds tremendous medical promise. It enables doctors to take preventive measures far earlier and to treat patients more effectively if they finally do get ill. With this great medical promise, however, comes the worry that health insurance will be denied or become more costly to people shown by genetic tests to be susceptible to certain diseases. The proper response to this problem, many think, is for the government to regulate insurance companies so as to prevent them from engaging in "genetic discrimination." Numerous state governments have already enacted such legislation, and so has the federal government within the Health Insurance Portability and Accountability Act (HIPAA). Concerned that these laws do not provide comprehensive protection against genetic discrimination, "The Genetic Information Nondiscrimination Act of 2005" was recently passed unanimously in the Senate (S. 306), though at the time of the writing of this paper (August 2005) the bill is still in committee in the House of Representatives.

Despite the opinion of Congress, I still regard it as an open question whether the possibility of genetic discrimination justifies governmental regulation of access to and use of genetic information by private health insurers. Ideally, the relationship between insured and insurer is a private one, so compelling arguments are necessary to justify intervention in the free market. My purpose in this paper is twofold. First, I aim to state and assess critically two of the most commonly given arguments—if not two of the best arguments—for the pro-regulation position. These are arguments often given by pro-regulation politicians, spokespersons, and advocacy groups. Second, I intend to explore the possibility of a free-market solution to the problem of genetic discrimination: genetic insurance. I will argue that it is misguided to assume *a priori* that those genetically predisposed to disease will have worse health outcomes than those not so disposed, and so we cannot know *a priori* that genetic discrimination will emerge as a major problem in a free market health insurance system. I also briefly present a positive case for genetic insurance. By doing all of these things I hope to make us less confident that genetic discrimination is a problem requiring governmental intervention in the insurance market.

II. THE SCOPE OF THIS PAPER

Let me also identify some issues I will not address. First, I will not address the larger question of whether there are any justifications whatsoever—justifications beyond the possibility of genetic discrimination—for governmental intervention with respect to the collection and use of genetic information by private health insurers. It could well be that, as the practice of genetic testing becomes more widespread, we will find a significant number of people refusing to submit to genetic tests for reasons having nothing to do with fear of genetic discrimination. For example, some might refuse because they think it would be too depressing to find out that they are genetically predisposed to some horrible disease. As we will see, a common pro-regulation claim is that the benefits, both personal and social, of genetic testing are so great that the government should take steps to encourage genetic testing (specifically, the step of banning genetic discrimination). If the benefits of genetic testing are so great, then perhaps those who refuse to take genetic tests will be seen as imposing a social cost large enough to justify the government's incentivizing or even requiring private insurers to collect genetic information on their customers. Such incentives or mandates would qualify as forms of governmental intervention

INSURANCE COMPANIES AND GENETIC INFORMATION 347

in the free market. Whether such interventions might be justified is not a question I will address here.

Second, I will not address pro-regulation arguments that question the general permissibility of standard underwriting practices. It seems to me that many pro-regulation arguments are not really specific to the issue of genetic testing. For example, there is the argument—often made by fans of the pro-regulation position—that it is wrong for insurers to drop a customer or charge that customer more on the basis of a genetic test, because genetic tests only indicate a propensity to disease rather than the actual presence of disease.[1] Yet if this makes it wrong for insurers to change rates or drop coverage for those for whom genetic tests show a predisposition for disease, then it also makes it wrong for insurers to change rates or drop coverage for smokers who have no symptoms of cancer, overweight people who are not yet diabetic, people with a family history of colon cancer, and so on. This is a pro-regulation argument that entangles us in the much larger question of whether it is permissible for insurers to underwrite on the basis of propensity for disease rather than actual disease. It is thus beyond this paper's scope.

III. SUICIDAL INSURERS

With those qualifications about scope in mind, let us look at the first oft-made pro-regulation argument.

The Suicidal Insurers Argument

(1) Almost all of us have a genetic predisposition to some disease or other, whether a test for that predisposition exists yet or not.

(2) Anyone with a genetic predisposition to some disease or other is subject to genetic discrimination.

So,

(3) [from 1 and 2] In the future, almost all us will be subject to genetic discrimination.

Therefore, governmental regulations are necessary to reform current insurance practices with respect to genetic testing.

Bundled together, premises (1) and (2) are a commonplace in pro-regulation writings. The July 1997 report from the Department of Health and Human Services, *Health Insurance in the Age of Genetics*, says

(A) "Each of us has between 5 and 30 misspellings or alterations in our DNA; thus, we could all be targets for discrimination based on our genes."[2]

On April 1, 2004 Dr. Francis Collins, Director of the National Human Genome Research Institute, said of an earlier version of the Genetic Nondiscrimination Act

(B) "Should it become law, S. 1053 will clearly protect all of us with disease-associated misspellings in our DNA—and that's ALL of us—from genetic discrimination in health insurance and employment."[3]

Collins has been making this point for a while now. In July 1997 he said:

> (C) "The targets of genetic discrimination today are those who carry gene alterations that, for better or worse, appear at the top of our list of scientific discoveries. But in time, we could all find ourselves in a similar situation. There are no perfect genetic specimens."[4]

In March 1994, Hillary Clinton stopped by Syracuse University as she toured the country promoting the Clinton administration's health-care plan; I was a student there at the time and braved the cold to hear her speak. She said:

> (D) "we want to eliminate unfair insurance practices that discriminate against people because they have pre-existing conditions—any kind of chronic illness, any sort of disability. Increasingly, as we learn more about our genetic makeup, we are learning that probably *all* of us have a gene for some kind of problem at some point in our lives, which means, therefore, that we have a pre-existing condition."[5]

And on March 6 of 1998, during a White House briefing given by Dr. Collins, she said:

> (E) the advancements in genetic testing "will make it even more imperative that we have universal health care because, otherwise, most of us will be uninsurable based on our genetic makeup."[6]

Except for (E), these statements are a bit ambiguous. Do they imply that, in the future, everyone all at once will be subject to genetic discrimination in health insurance? If so, then the authors of those statements are drawing sub-conclusion (3) of the Suicidal Insurers Argument. If the statements only imply that, in the future, it could be you who is genetically discriminated against, then the conclusion of the Suicidal Insurers Argument does not follow. I'd go farther and say I don't see that *any* sub-conclusion that might help in mounting a pro-regulation argument would follow from the weaker reading of (A)–(D). For any pro-regulation conclusion to follow, I would need to have some idea how likely I am to suffer genetic discrimination. Without that information—with only the information that there is a bare possibility I might suffer genetic discrimination—I cannot draw any pro-regulation conclusion at all.[7] Sub-conclusion (3) of the Suicidal Insurers Argument does tell me how likely I am to suffer genetic discrimination. It says that my likelihood of suffering genetic discrimination is very high—much greater than 50 percent. That is what Hillary Clinton said when she said "most of us will be *uninsurable*."

I think the problem with the Suicidal Insurers Argument is obvious. The health insurance business is lucrative, and the people who run it, even if all of them be callous and greedy, are nevertheless rational. If, in the future, genetic tests show that everyone has a genetic predisposition to disease, then any insurance company that denies or discontinues health insurance for those it knows to have a genetic predisposition to disease will lose most of its customers and will, if not go broke, at least be vastly poorer. Likewise, any insurance company that raises rates for almost all of its customers on the grounds that almost all of its customers are genetically predisposed to disease will be charging more for the exact same product. It is not as if people in the future will be more sick more often with genetic tests (quite the opposite, we hope); it is just that with genetic tests we will be better able to predict these illnesses. So if an insurer charges more but its customers are not getting more (in the way of additional medical services), that insurer will fall prey in a free market to insurance companies that offer the same coverage product at the old, pre-testing price. So it seems to me obviously false that, as Hillary Clinton said, "[without] universal health care . . . most of us will be uninsurable based on our genetic makeup," because a future in which most of us are uninsurable is a future in which insurers have put themselves out of business. The health insurance companies are not so stupid as to allow genetic tests to do this to them.

Another obvious objection to the Suicidal Insurers Argument is that, since not all medical problems are genetically predetermined, insurers will have a basis for providing some sort of coverage even of those with genetic predispositions to disease. Whether genetically unlucky or not, everyone risks broken bones, blown-out knees, pregnancies, slipped disks, and so forth. Furthermore, the genetically unlucky are rarely unlucky across the board, and so they can be insured against diseases for which they are not genetically predisposed. So insurers may still be willing to provide coverage for a wide array of conditions to the genetically unlucky, and likely at the same rate the genetically fortunate are charged (assuming the risk of broken bones and blown-out knees is basically independent of genetics). So, even if everyone suffers genetic discrimination, it still would not be true that everyone is uninsurable. It would only be true that not everyone is *fully* insurable.

Lastly, genetic testing might render insurable people who previously were uninsurable, simply by ruling out the presence of hereditary disease. For example, male children of men with Huntington's disease are currently ineligible for life insurance because about half of them go on to develop Huntington's, and no one knows which children will get Huntington's and which will not. Genetic testing could put half of these children back in the insurable camp while leaving the other half no worse off. So it is simply not true that in all cases genetic testing pushes people out of the insurable class. Sometimes it ropes them back in.[8]

The lesson here is that, if genetic testing becomes sufficiently widespread and if those tests show a sufficiently large percentage of the population has a genetic predisposition to disease, insurers are going to reach some accommodation other than policy cancellation with those who test positive for some predisposition or other, because doing so is the smart business move. Just because some insurers discriminate genetically some of the time now does not show that genetic discrimination will become prevalent as more tests are developed and more people take those tests.

IV. FEARFUL PATIENTS

Let us turn to our second oft-stated pro-regulation argument, then look at some examples of the argument in the advocacy literature.

The Fearful Patients Argument

(1) Maximizing the benefits of genetic testing requires that we maximize the number of people taking genetic tests.

(2) The fear of genetic discrimination makes some people avoid genetic tests. So, for the sake of maximizing the benefits of genetic testing, governmental regulation is necessary to assure potential test-takers that they will not suffer genetic discrimination.

The cancer advocacy group Facing Our Risk of Cancer Empowered [FORCE] says:

(F) "the fear of genetic discrimination is preventing people from benefiting from the information that genetic testing can provide. The only way to remove the fear of genetic discrimination is to pass comprehensive federal laws against it."[9]

Meanwhile, the National Workrights Institute says:

(G) "it is crucial that restrictions be placed on the accumulation of genetic information by employers and insurers. Without meaningful privacy safeguards and protections against discrimination, the benefits of genetic testing will ultimately be lost as individuals avoid tests in the fear of adverse consequences."[10]

Dr. Jennifer L. Howse, president of the March of Dimes, said of S. 306:

(H) "The March of Dimes recognizes the tremendous potential of genetic screening can only be realized if people feel secure that their genetic information will not be used to deny them health insurance or employment," and "(i)t would be a shame if parents were afraid to take advantage of the benefits of genetic testing and newborn screening because they feared retaliation from insurers."[11]

In Senate bill S. 306 (2005), section 2.5 ("Findings"), the U.S. Congress concludes that:

(I) "Federal legislation establishing a national and uniform basic standard is necessary to fully protect the public from discrimination and allay their concerns about the potential for discrimination, thereby allowing individuals to take advantage of genetic testing, technologies, research, and new therapies."

Dr. Collins, from the same April 1, 2004 comments mentioned earlier, says:

(J) "If this bill doesn't pass, my concern is that we won't be able to realize the full potential of advances in genetic science, because people will be afraid to participate in clinical trials or obtain genetic tests out of fear of discrimination."[12]

He adds later that his own lab found a gene that seems to increase the risk of type 2 diabetes about 30 percent, and said:

(K) "Someone testing positive for this variant could potentially incorporate preventive measures to avoid developing type 2 diabetes. Yet, if such a test is developed, some may be afraid to learn their own risks, for fear their insurance company might deny them insurance or raise their rates."[13]

Regarding this last claim, note that the incidence of type 2 diabetes in the U.S. is about 6 percent. A gene that increases the risk by about 30 percent thus increases the risk to about 8 percent. If an insurer wished to act rationally and was not prevented by regulation from doing so, how would that insurer react to the news that someone it covered tested positive for the gene Dr. Collins mentions? The smart thing to do, it seems to me, would not be to cancel the policy of the individual in question, but to raise the rates paid by that individual. And if, as a potential taker of a genetic test, I believe this of my insurer, and believe my insurer is free to act this way, then my fear that my insurer will cancel my coverage if I test positive for the gene would be irrational. At worst, I should fear a raising of my premium. And in that case I would have to ask myself whether the cost of this premium increase is worth the knowledge I would have gained that I am genetically predisposed to type 2 diabetes. That is certainly medically important information, because with it I can take early action to prevent the development of diabetes. Clearly that information has some value.

The case of the type 2 diabetes gene raises two questions for the Fearful Patients Argument. The first is whether governmental regulation is still justified if the fear in question is irrational—if the fear does not have any real basis. Probably we could all come up with a case of governmental regulation in which the ultimate justification for the regulation is that it quelled or quells an irrational public fear. But whether governments should be in the business of quelling irrational fears is, at the least, an open question. If the answer is a nice, utilitarian "yes," then to support the pro-regulation conclusion of the Fearful Patients Argument, its proponents may need to do nothing more than survey people to see what percentage fear genetic discrimination. If this really is a fairly widespread fear, then perhaps we should regulate away (although perhaps all we would need is a public education campaign to explain that the fear is unfounded). But if the answer is "no," then proponents of the Fearful Patients Argument will have to show not just that people *are* afraid of genetic discrimination, but that they are *rightfully* afraid.

The second question for the Fearful Patients Argument is whether it is irrational to avoid genetic tests even if the fear of genetic discrimination is rational. In statements (F)–(K) we are told about the great benefits to the individual of genetic testing. Might the benefits be great enough to outweigh the costs of having one's rates raised or one's policy canceled? We just considered this possibility in the type 2 diabetes case. Here is another example. If there is a gene which makes it more likely than not that I will develop colon cancer, and if I know that what is crucial to the effective treatment of colon cancer is early detection and removal, it seems to me I would be better off being tested for the presence of that gene, even if I risk a rise in rates or cancellation of my policy if I test positive. It might be more rational of me to take my chances finding a different insurer to cover me than to chance letting colon cancer spread undetected in me. Again, perhaps the information I stand to gain is worth more than the insurance I stand to lose.

My raising these questions should not be taken as asserting positively that the fear mentioned in premise (2) of the Fearful Patients Argument is irrational, or that the response mentioned in premise (2)—avoiding genetic testing—is irrational even given that the fears are rational. For some genetically based conditions, the fear may be eminently rational. What the questions show, I think, is that it is not obvious that a pro-regulation conclusion should be drawn from the premises of the Fearful Patients Argument. That argument would be a lot stronger if, instead of (2), it was based on a modified premise:

(2*) The fear of being denied health insurance in case of any positive tests is *reasonable* and gives someone *good reason* not to take any genetic tests.

Whether (2*) is true, however, will depend on empirical data—specifically on (a) just how frequent the different forms of genetic discrimination are, and (b) just how much disvalue should be attached to each of these different forms of discrimination relative to the value of the information a genetic test provides. In the pro-regulation literature, statistics regarding (a) are hard to find. The Council for Responsible Genetics (CRG) is one of the few advocacy organizations I have encountered to do more than just say "there are documented cases." In their program statement "Genetic Testing, Discrimination, and Privacy" they say there are "as many as five-hundred" well-documented cases of genetic discrimination.[14] That is a good start toward hard data, though the "as many as five-hundred" phrasing is a tip-off to me of either a lack of solid evidence or of significant disagreement within the CRG as to what to count as a case of genetic discrimination. In any case, the CRG provides neither a time span nor a base rate; that is, the CRG does not tell us how many years it took for these five-hundred cases to arise, nor does it tell us the total number of *potential* cases of genetic discrimination during that time span. The total number of potential cases of genetic discrimination will be some function of the total number of genetic tests taken in the relevant time frame, but that is information no one seems to present, or even estimate (and I've looked around, though not exhaustively).[15] As for (b)—just how much disvalue should be attached to each of these different forms of discrimination relative to the value of the information a genetic test provides—I have never seen the question addressed by any of the pro-regulation proponents. The politicians and advocacy groups who advance the Fearful Patients Argument just assume someone is better off avoiding the possibility of genetic discrimination than getting the information a genetic test provides. That is far from obvious, as the colon cancer example illustrates.

Instead of the ringing rejection I gave of the Suicidal Insurers Argument, then, I merely offer a simple "Beware!" in reply to the Fearful Patients Argument, for the reasons given above. Not every fear that prevents full achievement of some great good is one that governments should seek to address via regulation—especially since regulations themselves have costs. We need good evidence that the fear of genetic discrimination is such a fear.

V. GENETIC INSURANCE

At this point I expect some in the audience are chafing. "Isn't it just obvious," they are saying to themselves, "that, if insurers have full access to genetic information and no restrictions on what to do with that information, they will use the information to dump those customers who the insurers know will require expensive treatments? Generating equal risk pools is just what insurance companies *do*, so if they *know* some individuals are higher risks than others, they will remove those individuals from the old risk pool and either deprive them of insurance or lump them in a much more expensive risk pool. We do not need to know how many documented cases of genetic discrimination there have been till now. We can know *a priori* that genetic discrimination will emerge as a problem just from reflection on the nature of the insurance business."

Interestingly, this is basically the position of economist Alexander Tabarrok.

Genetic factors are unquestionably very important contributors to disease (sometimes they are the sole contributor). The probability of an individual developing a disease conditional on having a genetic defect is much larger than the prior probability of developing the disease. Competition will therefore separate individuals into genetic classes, the lucky will benefit from marginally lower insurance rates and the unlucky will face staggering bills or no insurance at all. The danger of genetic testing is the potential it has to create a 'genetic underclass' unable to afford health or life insurance, lacking a job, and facing dire medical problems.[16]

This statement is interesting, because Tabarrok proposes genetic insurance as the solution to this problem, which suggests that a pro-regulation conclusion does not follow even if we assume *a priori* that insurers will discriminate against those with positive genetic tests.

Since I will end up endorsing Tabarrok's solution to the problem of genetic discrimination, it may seem ungrateful of me to criticize his argument for thinking genetic discrimination will surely emerge as a problem. Yet it seems to me that Tabarrok is conditionalizing on the wrong information. In trying to figure out how insurers will behave in an era of genetic testing, the question we should expect rational insurers to ask is not this.

What is the probability that a patient will develop a disease conditional on the patient's having a genetic predisposition to that disease?

The question we should expect rational insurers to ask is, rather, this.

What is the probability that a patient will develop a disease conditional on *the total information we get when we learn that genetic tests have shown the patient is genetically predisposed to that disease*?

This total information will include lots of facts besides the fact that the patient is genetically predisposed to the disease. It will also include the facts that the patient knows he/she is so predisposed and that the patient's doctor knows this. Given this, it might not follow from the fact that a person is genetically predisposed to a certain disease that that person is more likely than the average person to develop the disease. Indeed, it might turn out in some cases that the positive tester is *less* likely to do so.

Take the case of the type 2 diabetes gene again. It could be that, on average, patients who learn that their risk of type 2 diabetes is 30 percent greater than average develop type 2 diabetes no more often than, or even less often than, the general population. This might result from a combination of two factors. First, the patients who get positive tests might modify their diets and lifestyles enough to compensate, or more than compensate, for their genetic predisposition. Second, it is entirely possible that some patients who get negative tests construe this as a license to continue with, or even ramp up, unhealthy diets and lifestyles. Consider the class of college students. Have all of them take a test for the alcoholism gene (we will presume for the sake of argument that there is such a gene and there is such a test). What would you predict the general behavior would be of college students who learn they have tested negative for the alcoholism gene? If I know my college students, they will take this as a license to drink away.

If this sort of dynamic emerges—and we cannot know *a priori* that it will not—then quite possibly those who test positive for the type 2 diabetes gene will have *better* health outcomes with respect to type 2 diabetes than those who test negative. Whether the class of those who test positive for the type 2 diabetes gene will do better or worse vis-à-vis type 2 diabetes than the class of those who test negative is an empirical question, it seems to me. This illustrates my complaint about Tabarrok's argument. We cannot tell how insurers will react to the widespread use of genetic tests until we know how patients and doctors will respond to the information genetic tests provide, thus we cannot know *a priori* that genetic discrimination will become common practice.

Having said that, it seems to me that even if insurers are prone to engage in genetic discrimination, there is a reasonably simple free-market solution: offer "genetic insurance." I will simply quote Tabarrok's proposal here.

> We are used to thinking of insuring against sickness but why not insure against a potential high probability of sickness? In other words why not sell genetic insurance? Consider the following proposal: before taking a genetic test, it should be made mandatory for every individual to purchase genetic insurance. For a small fee genetic insurance would insure against the possibility of a positive test result. If the test came back positive the customer would be paid a large sum of money, enough to cover the expected costs of his disease or equivalently enough to allow him to purchase health insurance at the new risk premium. If the test turns out negative the customer would lose his genetic insurance fee but would gain the results of the test and also lower health insurance premiums. Those who have positive test

results would be paid enough money to pay their health care costs and would also benefit from being able to plan in accord with the test results. Under this proposal average insurance rates will fall and everyone will be made better off.[17]

Tabarrok notes that, while this proposal does involve some minimal governmental regulation insofar as it mandates the purchase of insurance, the regulation is much less intrusive than anti-discrimination laws of the sort the Senate just approved. (I am not even sure why the proposal requires *mandating* purchase of the insurance, but we can set that aside.) Also, the cost of a system with genetic insurance plus regular insurance is guaranteed to be no greater than the cost of a system without genetic insurance, since we already insure against genetic disease. Genetic insurance

> takes advantage of new gene technologies to separate the genetic and non-genetic aspects of health insurance—it makes an implicit market explicit. Separating the genetic and non-genetic health insurance markets cannot raise the total price of health insurance because the same product is being sold.[18]

Thus at worst with genetic insurance no one is worse off in terms of premiums or access to insurance. Furthermore, genetic insurance involves only minimal governmental regulation, making it preferable to anti-discrimination laws.

VI. CONCLUSION

Genetic discrimination by insurers may emerge as a serious and widespread problem. But if it will, this fact cannot be known just by *a priori* reflection on the behavior of insurers. Furthermore, even if it will, good arguments are needed for thinking the solution is governmental regulation in the form of laws banning genetic discrimination. The Suicidal Insurers Argument and the Irrational Patients Argument are not good arguments. Meanwhile, the proposal of offering genetic insurance seems to be a perfectly workable free-market solution. Together, these considerations should, at the very least, reduce our confidence that the right response to the possibility of genetic discrimination is to pass laws banning it.[19]

NOTES

1. For example, the National Workrights Institute says "genetic information only indicates a predisposition or susceptibility to future illness; such information does not necessarily indicate when an individual will develop symptoms or how severe the symptoms will be. In fact, many people who test positive for genetic mutations associated with certain conditions will never develop those conditions at all. Many individuals identified as having a hereditary condition are, indeed, healthy. Genetic information does not necessarily diagnose disease." Online at www.workrights.org/issue_genetic/gd_congressional_letter.html.

2. Online at www.genome.gov/10000879. The "between 5 and 30" statistic may be on the low side, possibly by several orders of magnitude.

3. Online at www.genome.gov/11511396.
4. Online at www.genome.gov/10000882.
5. The *Syracuse Record*, vol. 24, no. 28, p. 7.
6. Online at www.netlink.de/gen/Zeitung/1998/980306.htm.
7. In other work of mine I bring out various pitfalls in reasoning about risk on the basis of bare possibilities without any regard to the probability of the occurrence of those possibilities. See "Formulating the Precautionary Principle," *Environmental Ethics* vol. 24, no. 3 (Fall 2002), pp. 263–274.
8. I thank Gregory Conko for pointing this out to me.
9. Online at www.facingourrisk.org/advocacy/nondiscrimination.php.
10. Online at www.workrights.org/issue_genetic/gd_congressional_letter.html.
11. Online at www.marchofdimes.com/aboutus/14817_15124.asp.
12. Online at www.genome.gov/11511396.
13. Ibid.
14. Online at www.gene-watch.org/programs/privacy.html.
15. Since completion of this article in 2005, I have come across the anthology *Genetics and Life Insurance: Medical Underwriting and Social Policy*, ed. Mark A. Rothstein (Cambridge, MA: The MIT Press, 2004). In their article "Perspectives of Consumers and Genetics Professionals," contributors Wendy R. Uhlmann and Sharon F. Terry provide an extensive literature review of studies seeking to document the incidence of genetic discrimination (see pp. 151–156). They fault many of the studies for flawed methodology, and they conclude their review by agreeing with the opinion of P. R. Reilly (author of *"Genetic Discrimination" in Genetic Testing and the Use of Information*, ed. C. Long [Washington, D.C.: American Enterprise Institute, 1999], p. 106) that "little evidence supports the widespread fear that people who undergo genetic tests to determine whether they are at increased risk for developing a serious disorder face a significant risk of genetic discrimination."
16. Alexander Tabarrok, "Genetic testing: an economic and contractarian analysis," *Journal of Health Economics* vol. 13 (1994), p. 77.
17. Ibid., pp. 87–88.
18. Ibid., p. 89.
19. I thank Gregory Conko for a careful review of this paper and numerous helpful comments. I also thank the audience at the 2005 Annual Meeting of the Association for Politics and the Life Sciences.